Handbook of Exponential and Related Distributions for Engineers and Scientists

Handbook of Exponential and Related Distributions for Engineers and Scientists

Nabendu Pal
University of Louisiana at Lafayette
Lafayette, LA

Chun Jin
Central Connecticut State University
New Britain, CT

Wooi K. Lim
William Paterson University
Wayne, NJ

CRC Press
Taylor & Francis Group
Boca Raton London New York

CRC Press is an imprint of the
Taylor & Francis Group, an **informa** business

A TAYLOR & FRANCIS BOOK

CRC Press
Taylor & Francis Group
6000 Broken Sound Parkway NW, Suite 300
Boca Raton, FL 33487-2742

First issued in paperback 2020

ISBN 13: 978-0-367-57796-4 (pbk)
ISBN 13: 978-1-58488-138-4 (hbk)

Library of Congress Cataloging-in-Publication Data

Pal, Nabendu.
　　Handbook of exponential and related distributions for engineers and scientists / Nabendu Pal, Chun Jin, Wooi K. Lim.
　　　　p. cm.
　　Includes bibliographical references and index.
　　ISBN 1-58488-138-0 (alk. paper)
　　1. Distribution (Probability theory) I. Jin, Chun. II. Lim, Wooi K. III. Title.

QA273.6.P35 2005
519.2'4--dc22　　　　　　　　　　　　　　　　　　　　　　　　　　　　　　　　2005051857

Visit the Taylor & Francis Web site at
http://www.taylorandfrancis.com

and the CRC Press Web site at
http://www.crcpress.com

To our parents and all our teachers for the wonderful things of life that we have today.

–Authors

Preface

The motivation to write this book came a few years back when a colleague from one of the engineering departments (University of Louisiana at Lafayette) approached us with a problem that required modeling a real life data set which looked highly positively skewed.

There are many good statistics books available on areas like survival analysis, reliability, and life testing, where one often encounters positively skewed data sets. There are a few recent books that devote entirely on one or a few specific positively skewed probability distributions. These books, which are quite thorough and rigorous, tend to be highly mathematical, and as a result may discourage many applied scientists who may lack a sound background in mathematical statistics. On the other hand, introductory level statistics books, which may appeal to a wider audience, place over-emphasis on the normal distribution (the *Bell Curve*) due to its popularity, but may not be suitable enough for modeling positively skewed data sets for obvious reasons. Hence we felt the need for a handbook which can explain and discuss the applications of some of the well known positively skewed probability distributions in a simple step by step manner.

We wanted to cover the *Exponential Distribution* due to its wide applications in modeling lifetime data and its famous *Lack of Memory* property. Immediately, we felt the need for generalizing this to a few other natural extensions, namely — the *Gamma*, the *Weibull*, and the *Extreme Value Distributions*, which are also popular in engineering fields. A quick glance at the CIS (Current Index in Statistics) shows more than 300 research papers on these few distributions, and naturally this put us in a bind to determine what to cover and what not to cover. As a result, we had to limit ourselves to the aforementioned distributions only, leaving out a few other common positively skewed distributions like the *Logistic* or the *Log − Normal*. Even for the above four distributions we wanted to cover as much material as possible, yet didn't want to sacrifice the conciseness and simplicity of the presentation which, we hope, would be one of the high points of this book. Since statistical computations with the help of numerous mathematical/statistical packages have become very convenient and less time consuming, we decided to trade some of the complex theoretical results with the tools based on simulations.

Statistical simulations give an applied researcher plenty of room to analyze and study the data without getting lost in the mish-mash of theory. Of course, no one (including us) can deny the importance of the theory, and a sound theoretical knowledge is very helpful in statistical analyses. But we had to walk a fine line between covering sufficient rigorous theoretical results (which may not evoke much appeal to a nonstatistician), and simulation techniques based on simple ideas by which anyone with basic statistical knowledge can model a data set and visualize its implications. One effect of this approach is that we ended up with a large number of graphs.

"In attempts to understand ... it is a good idea", — as Socrates once said — "to begin at the *beginning*". So we have started the book from scratch. If the reader needs a little brush-up of the elementary or the standard concepts of mathematics/statistics, he/she still finds those covered here briefly. The book has been divided in two parts. The first part has five chapters covering most of the basic concepts and general ideas needed (or used) in statistics — all the way up to inference. One chapter (Chapter 5) is devoted solely to statistical computations including the increasingly popular bootstrap method. The second part of the book, which also has another five chapters, covers the four positively skewed distributions mentioned above. For each distribution we have followed a clear-cut path (divided in four sections):

(i) Discuss the general properties and show the applicability with a few real data sets

(ii) Characterize the distribution (some theoretical results without much derivations or proofs)

(iii) Estimation of parameters and related inferences

(iv) Goodness of fit test (to see how well the specific distribution fits the given data)

The last chapter (Chapter 10) deals with system reliability, mainly for series and parallel systems, which may be of interest to the engineers. One can find a heavy usage of simulations (with a lot of graphs) since the theoretical results are hopelessly complicated. All the simulations and computations have been carried out by the package 'Mathematica' (with the addition package 'mathStatica.m'); and the graphs are drawn using GNUPLOT.

We sincerely hope that the book will be useful as a handbook (or as a reference book) for the applied researchers who are more interested in data analysis rather than mathematical developments of the methods. Apart from engineers and applied scientists, seniors and beginning graduate students of statistics can benefit greatly from this book. The references cited here, especially those with real life data sets, can be a treasure trove for any one who is interested in statistics.

We would like to thank many of our colleagues for their help. A special thanks to Dr. Robert Crouse (Central Connecticut State University) for his assistance with some of the computations and editorial help.

We greatly appreciate the encouragement that we have received from our spouses (Pallavi, Xiaoyan, and Guo-Zhen) and children (Nabapallav and Devpallav (Pal), Daniel (Jin), and Alice Wu and Gary Wu (Lim)) who understood the meaning of the lost family time. Without their continuous support this book could not have been possible.

Last but not least, we are thankful to Mr. Robert Stern and Ms. Theresa Delforn (of CRC Press) for their infinite patience. While we worked on this book at a very slow rate of convergence and missed the deadline repeatedly to submit the manuscript, they were generous enough to extend the deadline without any question every time we requested it, irrespective of the excuse we gave the previous time. Truly a memoryless property exhibited by the exponential distribution only.

Nabendu Pal, University of Louisiana at Lafayette

Chun Jin, Central Connecticut State University

Wooi K. Lim, William Paterson University

About the Authors

Nabendu Pal is Professor of Statistics in the Department of Mathematics, University of Louisiana at Lafayette (formerly University of Southwestern Louisiana), where he has been a faculty member since August 1989. Earlier (in Spring 1989), he was a visiting faculty in the Department of Mathematics and Statistics, University of Arkansas, Fayetteville. He received his master's in statistics (M. Stat) degree from the Indian Statistical Institute (ISI), Calcutta, in 1986, and Ph.D. in statistics from the University of Maryland Baltimore County in 1989. His areas of interest include Decision Theory, Biostatistics, Reliability, and Life Testing. He has served as the department's graduate coordinator from August 2001 to July 2005. He has also co-authored the book *Statistics: Concepts and Applications* (by Prentice-Hall of India, 2005).

Chun Jin is Professor of Statistics in the Department of Mathematical Sciences at Central Connecticut State University (CCSU). He received his MS degree in mathematics from the University of Southern Mississippi in 1988 and his Ph.D. in statistics from the University of Louisiana at Lafayette (formerly University of Southwestern Louisiana) in 1992. Before joining the faculty of CCSU in 1994, he taught in the Department of Mathematics at the University of New Orleans for two years. His main area of interest is Multivariate Analysis. He has also co-authored the book, *Ready, Set, Run! A Student Guide to SAS Software for Microsoft Windows* (by Mayfield Publishing Company, 1998).

Wooi K. Lim received his Ph.D. in statistics from the University of Louisiana at Lafayette (formerly University of Southwestern Louisiana) in December 1996. Since August 1999, he is an Assistant Professor in the Department of Mathematics, William Paterson University. Earlier (1995 to 1999), he taught in the Department of Mathematics, University of New Orleans. His areas of interest include Decision Theory, Statistical Simulations, and Numerical Computations.

Acronyms

AM : Arithmatic mean

BEL : Bayes expected loss

cdf : Cumulative distribution function (of a random variable)

cf : Characteristic function

CI : Confidence interval

CLT : Central Limit Theorem

$corr$: Correlation coefficient

df : Degrees of freedom

EV-I (II, III) : Extreme value distribution of Type-I (or Type-II, or Type-III)

GM : Geometric mean

HM : Harmonic mean

iid : Independent and identically distributed (for a set of random variables)

$ln\ c$: Natural logarithm of the value c

LRT : Likelihood ratio test

mgf : Moment generating function

MLE : Maximum likelihood estimator

MME : Method of moment(s) estimator

MMLE : Modified maximum likelihood estimator

MSE : Mean squared error

pdf : Probability density function (of a continuous random variable)

pmf : Probability mass function (of a discrete random variable)

PB-Bias (MSE) : Parametric bootstrap bias (mean squared error) *r.v.* : Random variable (or vector)

SB-Bias (MSE) : Semiparametric bootstrap bias (mean squared error)

SEL : Squared error loss function

UMVUE : Unique minimum variance unbiased estimator

Var : Variance

Notations

$A \cup B$: Union of events A and B.

$A \cap B$: Intersection of events A and B.

A^c : Complement of event A.

$A \subseteq B$: Event A is a subset of event B.

$B(n, \theta)$: Binomial distribution where n is the number of trials, and θ is the probability of success in a single trial.

$B(\alpha, \beta)$: Beta distribution with first parameter α and second parameter β.

$corr(X, Y)$: Correlation coefficient between the random variables X and Y.

$CU(\theta_1, \theta_2)$: Continuous uniform distribution over the interval (θ_1, θ_2).

$DU(\theta_1, \theta_2)$: Discrete uniform distribution over the countable set $\{\theta_1, \theta_1 + 1, \ldots, \theta_2\}$.

Example a.b.c. : Example number c of Section b of Chapter a. [Same for Theorem, Remark, Lemma, Corrollary].

$E(X)$: Expectation or expected value of the random variable X.

$E(X|Y)$: Conditional expectation of the random variable X for the fixed value of another random variable Y.

$EV - I(\beta, \mu)$: Type-I extreme value distribution with location parameter μ and scale parameter β.

$EV - II(\alpha, \beta, \mu)$: Type-II extreme value distribution with location parameter μ and scale parameter β and shape parameter α.

$EV - III(\alpha, \beta, \mu)$: Type-III extreme value distribution with location parameter μ and scale parameter β and shape parameter α.

F_{k_1, k_2} : (Snedecor's) F distribution with first (or numerator) df k_1 and second (or denominator) df k_2.

$\mathcal{F}$ (or $\mathcal{F}_X$) : Usually indicates a family of probability distributions (identified by *pmfs* or *pdfs* indexed by a parameter θ (possibly vector valued)).

$F_X(x)$ (or $F(x)$) : Usually denotes a *cdf* at x.

$\bar{F}(t)$: For a random variable X with *cdf* $F(\cdot)$, $\bar{F}(t) = 1 - F(t) = P(X > t)$, reliability at time t if X indicates the lifetime.

$\bar{F}_s(t)$: Reliability of a series system at time t.

$\bar{F}_p(t)$: Reliability of a parallel system at time t.

$G(\alpha, \beta, \mu)$: (Three parameter) Gamma distribution with location parameter μ, shape parameter α and scale parameter β.

$G(\alpha, \beta)$: (Two parameter) Gamma distribution with shape parameter α and scale parameter β. Also same as $G(\alpha, \beta, 0)$.

$G(\theta)$: Geometric distribution where θ is the probability of success in a single trial.

$LN(\mu, \sigma^2)$: Lognormal distribution with parameters μ and σ^2.

$L(\mu, \sigma)$: Logistic distribution with location parameter μ and scale parameter σ.

μ_X (or μ) : Same as $E(X)$.

$\mu'_{(k)}$: k^{th} raw moment, i.e., $E(X^k)$, expected value of the kth power of the random variable X.

$\mu_{(k)}$: k^{th} central moment, i.e., $E(X - E(X))^k$, expected value of the kth power of $(X - E(X))$.

$NB(r, \theta)$: Negative binomial distribution where r is the fixed number of successes to be obtained, and θ is the probability of success in a single trial.

$O(n)$: A sequence $\{c_n\}$ such that (c_n/n) is bounded as $n \longrightarrow \infty$.

$o(n)$: A sequence $\{c_n\}$ such that $(c_n/n) \longrightarrow 0$ as $n \longrightarrow \infty$.

$P(A|B)$: Conditional probability of the event A given the event B.

$P(\alpha, \beta)$: Pareto distribution with parameters α and β.

$Poi(\theta)$: Poisson distribution with (intensity) parameter θ.

$N(\mu, \sigma^2)$: Normal distribution with location parameter (also mean) μ and scale parameter (also variance) σ^2.

σ_X^2 (or σ^2) : Variance of the random variable X, i.e., $E(X - E(X))^2$.

σ_X (or σ) : $\sqrt{\sigma_X^2}$, standard deviation of the random variable X.

ρ_{12} (or ρ_{XY} or ρ) : Usually indicates the correlation coefficient between two random variables.

$((a_{ij}))$: A matrix where a_{ij} is the element of ith row and j^{th} column.

$diag(d_1, \ldots, d_n)$: A square matrix where (i, i)-element is d_i, $1 \leq i \leq n$; and all other elements are 0.

χ_k^2 : Chi-square distribution with k degrees of freedom (df). Same as $G(k/2, 1/2)$ distribution.

$t_k(\theta)$: (Student's) t-distribution with k degrees of freedom and location parameter θ.

t_k : Same as $t_k(0)$.

$W(\alpha, \beta)$: (Two parameter) Weibull distribution with shape parameter α and scale parameter β.

$W(\alpha, \beta, \mu)$: (Three parameter) Weibull distribution with location parameter μ, shape parameter α and scale parameter β.

$\Gamma(\alpha)$: $\int_0^\infty exp(-x)x^{\alpha-1}\, dx$, the gamma function evaluated at α, $\alpha > 0$.

$X_k \xrightarrow{d} X$: The sequence of random variables $\{X_k\}$ converges in distribution to the random variable X as $k \to \infty$.

$X_k \xrightarrow{p} X$: The sequence of random variables $\{X_k\}$ converges in probability to the random variable X as $k \to \infty$.

$X_k \xrightarrow{a.s.} X$: The sequence of random variables $\{X_k\}$ converges almost surely to the random variable X as $k \to \infty$.

$X \sim f(x)$: Indicates that the random variable X has *pmf* or *pdf* $f(x)$.

$\mathbb{R}$: Real line, the set of all real numbers.

$F_{k_1,k_2,\alpha}$: Right (or upper) tail α-probability cut-off point of the F_{k_1,k_2} distribution.

z_α : Right (or upper) tail α-probability cut-off point of the standard normal distribution.

$t_{k,\alpha}$: Right (or upper) tail α-probability cut-off point of the t_k distribution.

$\chi^2_{k,\alpha}$: Right (or upper) tail α-probability cut-off point of the χ^2_k distribution.

Θ : Usually indicates the unrestricted parameter space.

$\text{Exp}(\beta,\mu)$: (Two-parameter) exponential distribution with location parameter μ and scale parameter β.

$\text{Exp}(\beta)$: Same as $\text{Exp}(\beta,0)$.

$\Gamma_c(\alpha)$: $\int_0^c exp(-x)x^{\alpha-1}\,dx$.

$L(\delta(x),\theta)$: Loss function at θ for the decision $\delta(x)$ (with data $X = x$).

$R(\delta,\theta)$: Risk function of the decision δ at θ.

$r(\delta,\theta)$: Bayes risk of the decision δ with prior distribution π of θ on Θ.

$\widehat{\theta}_{\text{U}}$: Unbiased estimator of θ.

$\widehat{\theta}_{\text{ML}}$: MLE of θ.

$\widehat{\theta}_{\text{MML}}$: MMLE of θ.

$\widehat{\theta}_{\text{MM}}$: MME of θ.

Contents

Part-I
General Statistical Theory   1

Part-II
Exponential and Other Positively Skewed Distributions with Applications   149

List of Figures

List of Tables

Part-I
General Statistical
Theory

Chapter 1

Basic Concepts

1.1 The System of Real Numbers

A real number or a real value is one which can be experienced in real life. In fact, any value that can occur in our real life is indeed a real value. For example, it is possible to experience a temperature of (-22.4) degrees Fahrenheit. Hence the value (-22.4) is a real value. Take the example of a radio active element which emits, say, 623000000000 particles per unit of time. Hence the value 623000000000 is a real value.

To put it more rigorously, **real numbers** or **real values** are those which satisfy the following axioms (A.1 to A.5):

(A.1) : $x + y = y + x, \quad xy = yx.$

(A.2) : $x + (y + z) = (x + y) + z, \quad x(yz) = (xy)z.$

(A.3) : $x(y + z) = xy + xz.$

(A.4) : Given two real values x and y, there exists a real value z such that $x + z = y$. This z is denoted by $(y - x)$. For any real value x, $x - x$ is denoted by 0 which is independent of x. Also, for any real value x, there exists another real value $-x$ (called *negative x*) which means $0 - x$.

(A.5) : Given two real values x and y where x is not 0, there exists a real value z such that $xz = y$. This z is denoted by y/x. For any nonzero real value x, x/x is denoted by 1 which is independent of x.

Also, for any nonzero real value x, there exists another real value x^{-1} (called x-*inverse*) which means $1/x$.

From the above axioms all the usual algebraic rules can be derived. For instance, $-(-x) = x$, $\left(x^{-1}\right)^{-1} = x$, $-(x-y) = y - x$, etc.

The collection of all real numbers (or values) is called a real set and is denoted by $\mathbb{R}$. The following concepts and notations are widely used:

(1) The whole numbers $\ldots, -5, -4, -3, -2, -1, 0, 1, 2, 3, 4, 5, \ldots$ are called integers. $\mathbb{Z}$ denotes the collection of all integers; and $\mathbb{Z}^+(\mathbb{Z}^-)$ denotes the subcollection of all positive (negative) integers.

(2) A **rational number** is a real number which can be expressed as a quotient of two integers. $\mathbb{Q}$ denotes the collection of all rational numbers. A real number, which is not a rational number, is called an **irrational number**. The two popular irrational numbers are π and e (discussed later).

1.2   Some Useful Algebraic Results

If we have n real values, say, $x_1, x_2, \ldots, x_n$, their sum is denoted by $\sum_{i=1}^{n} x_i$ or by $\sum_{1}^{n} x_i$ or by $\sum_{i} x_i$ when the range of values of i is evident from the context. $\sum_{i=1}^{n} x_i$ is read as "the summation of x_i where i goes from 1 to n." The following simple results can be easily verified:

$$\text{(i)} \quad \sum_{i}(x_i \pm y_i) = \sum_{i} x_i \pm \sum_{i} y_i \qquad (1.2.1)$$

$$\text{(ii)} \quad \sum_{i} cx_i = c \sum_{i} x_i \qquad (1.2.2)$$

$$\text{(iii)} \quad \sum_{i}(c + x_i) = nc + \sum_{i} x_i \qquad (1.2.3)$$

where c is a real value.

Suppose we have $m \times n$ real values arranged in m rows and n columns as follows:

$$
\begin{array}{cccc}
x_{11} & x_{12} & \cdots & x_{1n} \\
x_{21} & x_{22} & \cdots & x_{2n} \\
\vdots & \vdots & \vdots & \vdots \\
x_{m1} & x_{m2} & \cdots & x_{mn}
\end{array}
\qquad (1.2.4)
$$

where x_{ij} is the real value occurring in both the ith row and the jth column. The sum of all the $m \times n$ real values can be obtained by first adding the values in each row and then adding each row totals:

$$(x_{11} + x_{12} + \ldots + x_{1n}) + (x_{21} + x_{22} + \ldots + x_{2n}) + \ldots$$

$$+ (x_{m1} + x_{m2} + \ldots + x_{mn})$$

$$= \sum_j x_{1j} + \sum_j x_{2j} + \ldots + \sum_j x_{mj}$$

$$= \sum_i \sum_j x_{ij} \qquad (1.2.5)$$

This grand total can also be obtained by first adding the values in each column and then adding the column totals:

$$(x_{11} + x_{21} + \ldots + x_{m1}) + (x_{12} + x_{22} + \ldots + x_{m2}) + \ldots$$

$$+ (x_{1n} + x_{2n} + \ldots + x_{mn})$$

$$= \sum_i x_{i1} + \sum_i x_{i2} + \ldots + \sum_i x_{in}$$

$$= \sum_j \sum_i x_{ij} \qquad (1.2.6)$$

The product of n real values $a_1, a_2, \ldots, a_n$ is denoted by $\prod_{i=1}^{n} a_i$ or by $\prod_1^n a_i$ or simply by $\prod_i a_i$ when the range of i is clear from the context. Like $\sum$, $\prod$ also obeys some simple rules as given below:

(i) $\quad \prod_i (a_i b_i) = \left(\prod_i a_i \right) \left(\prod_i b_i \right) \qquad (1.2.7)$

(ii) $\quad \prod_i (a_i / b_i) = \left(\prod_i a_i \right) \Big/ \left(\prod_i b_i \right)$ (for $b_i \neq 0$) $\quad (1.2.8)$

(iii) $\quad \prod_i (c a_i) = c^n \left(\prod_i a_i \right) \qquad (1.2.9)$

where c is a real value.

A **sequence** is a collection of quantities $c_1, c_2, \ldots$, called elements, which can be arranged in an order, so that when an integer k is given, the kth element of the sequence is completely specified. A sequence is denoted by $\{c_k\}$ with elements $c_1, c_2, \ldots$, etc.

A sequence $\{c_k\}$ is said to be bounded if there exists an arbitrary positive number M such that $|c_k| \leq M$ for all k (where $|c|$, called 'modulus c', denotes the **absolute value** of c, i.e., $|c| = c$ if $c \geq 0$; $= -c$ if $c < 0$).

For a sequence $\{c_k\}$ if there exists a value c_* such that for every choice of $\varepsilon > 0$, there exists an integer N such that $|c_k - c_*| < \varepsilon$ for all $k \geq N$, then the sequence $\{c_k\}$ is said to be **convergent** and has the limit c_*. It can be written as $\lim_{k\to\infty} c_k = c_*$. A sequence which is not convergent is called **divergent**.

A **series** is an expression of the form $c_1 + c_2 + \ldots + c_k + \ldots$, which may have a finite or an infinite number of terms. Its partial sums, $s_1 = c_1, s_2 = c_1 + c_2, \ldots, s_k = \sum_{i=1}^{k} c_i, \ldots$, constitute a sequence $\{s_k\}$. A series is said to be convergent if the sequence of these partial sums is convergent. If $\lim_{k\to\infty} s_k = s_*$ for some value s_*, then s_* is the sum of the convergent series and $s_* = \sum_{k=1}^{\infty} c_k$. If $\lim_{k\to\infty} s_k = \pm\infty$, then the series is said to be divergent to $\pm\infty$. A series is said to **oscillate** if the sequence of the partial sums does not tend to a definite limit, but oscillates between two values. A simple example of oscillating series is $\sum_i (-1)^i$, and it oscillates between 0 and -1.

Given two values a and b, and a positive integer k,

$$(a+b)^k = \binom{k}{0}a^k + \binom{k}{1}a^{k-1}b + \ldots + \binom{k}{r}a^{k-r}b^r + \ldots + \binom{k}{k}b^k \quad (1.2.10)$$

where $\binom{k}{r} = k(k-1)\ldots(k-r+1)/r!$ and $r! = r(r-1)\ldots 2\cdot 1$. The right-hand side (RHS) expansion of the above expression $(a+b)^k$ is called a 'Binomial series' and holds for any positive integer k.

The series

$$1 + \frac{1}{1!} + \frac{1}{2!} + \frac{1}{3!} + \ldots + \frac{1}{k!} + \ldots \quad (1.2.11)$$

is denoted by the letter 'e'. Also, for any positive value a,

$$a^x = 1 + \frac{x \, ln \, a}{1!} + \frac{x^2 (ln \, a)^2}{2!} + \ldots + \frac{x^k (ln \, a)^k}{k!} + \cdots \quad (1.2.12)$$

where x is any real value. As a particular case, by letting $a = e$, we get

$$e^x = 1 + \frac{x}{1!} + \frac{x^2}{2!} + \ldots + \frac{x^k}{k!} + \cdots \quad (1.2.13)$$

Two important results are

$$(1-a)^{-1} = 1 - a + a^2 - a^3 + a^4 - \ldots + (-a)^r + \ldots \quad (1.2.14)$$

$$\text{and} \quad (1+a)^{-1} = 1 + a + a^2 + a^3 + a^4 + \ldots + a^r + \ldots \quad (1.2.15)$$

which are valid for $|a| < 1$.

The logarithmic series which gives the expansion of $ln(1+a)$ is

$$ln(1+a) = a - \frac{a^2}{2!} + \frac{a^3}{3!} - \frac{a^4}{4!} + \ldots \quad (1.2.16)$$

Also,

$$ln(1-a) = -a - \frac{a^2}{2!} - \frac{a^3}{3!} - \frac{a^4}{4!} - \cdots \qquad (1.2.17)$$

Both the above logarithmic expansions hold for $|a| < 1$.

Some Algebraic Inequalities

1. If a and b are any two real values, then $|a+b| \leq |a| + |b|$.

2. Let $a_1, a_2, \ldots, a_n$ be a set of positive real values. The **arithmetic mean** (AM), **geometric mean** (GM), and **harmonic mean** (HM) of $a_1, a_2, \ldots, a_n$ are defined respectively as follows:

$$\text{AM} = \sum_i a_i \Big/ n, \ \text{GM} = \left(\prod_i a_i\right)^{1/n}, \ \text{and HM} = n \Big/ \sum_i (1/a_i)$$
$$(1.2.18)$$

It can be shown that $\text{HM} \leq \text{GM} \leq \text{AM}$.

3. If $a_1, a_2, \ldots, a_n$ and $b_1, b_2, \ldots, b_n$ are two sets of real values, then

$$\left(\sum_i a_i b_i\right)^2 \leq \left(\sum_i a_i^2\right)\left(\sum_i b_i^2\right)$$

$$\text{or} \ \left|\sum_i a_i b_i\right| \leq \sqrt{\left(\sum_i a_i^2\right)\left(\sum_i b_i^2\right)} \qquad (1.2.19)$$

This is known as **Cauchy-Schwartz** inequality. The equality holds when $a_i = c b_i$ for all i and a constant c.

Matrices

A matrix $A_{m \times n}$ (or simply A) of order $m \times n$ is a rectangular array of elements arranged in m rows and n columns as

$$A = \begin{pmatrix} a_{11} & a_{12} & \cdots & a_{1n} \\ a_{21} & a_{22} & \cdots & a_{2n} \\ \vdots & \vdots & \vdots & \vdots \\ a_{m1} & a_{m2} & \cdots & a_{mn} \end{pmatrix} = \left(\left(a_{ij}\right)\right) \qquad (1.2.20)$$

where a_{ij} represents the element in the ith row and jth column.[1] Also, a matrix of order $m \times n$ is a set of $m \times n$ elements arranged in a convenient

---

[1] A is called a row vector when $m = 1$; and a column vector when $n = 1$.

way. For instance, closing stock prices of m publicly held companies at the end of n different quarters may be represented by a matrix $A = \left(\left(a_{ij} \right) \right)$ of order $m \times n$ where a_{ij} is the closing stock price of ith company at the end of jth quarter.

Two matrices $A = \left(\left(a_{ij} \right) \right)$ and $B = \left(\left(b_{ij} \right) \right)$ are called equal (written as $A = B$) if and only if they are of the same order and $a_{ij} = b_{ij}$ for all i and j.

A matrix with the same number of rows as of columns is called a **square matrix**. The elements a_{ii} of a square matrix are called the main diagonal elements. A square matrix is **symmetric** if $a_{ij} = a_{ji}$ for all i and j. In the following, some **standard matrix operations (properties)** are defined (discussed):

1. The product of a matrix $A = \left(\left(a_{ij} \right) \right)$ and a real value c is defined as:
$$cA = c\left(\left(a_{ij} \right) \right) = \left(\left(ca_{ij} \right) \right) \qquad (1.2.21)$$
Hence $cA = Ac$, and $(-1)A$ is written as $(-A)$.

2. The sum of $A_{m \times n}$ and $B_{k \times l}$ is defined only when $m = k$ and $n = l$, and $A + B = \left(\left(a_{ij} + b_{ij} \right) \right)$. Similarly, $A - B = \left(\left(a_{ij} - b_{ij} \right) \right)$.

3. The product of $A_{m \times n}$ and $B_{k \times l}$ is defined only when $n = k$, and $A_{m \times n} B_{n \times l} = C_{m \times l} = \left(\left(c_{ij} \right) \right)$ where $c_{ij} = \sum_{r=1}^{n} a_{ir} b_{rj}$. Note that $A_{m \times n} B_{k \times l}$ is undefined unless $n = k$. Also, for square matrices of same order, AB need not be same as BA.

4. The **transpose** A' of a matrix A is obtained by interchanging the rows and columns of A. Therefore,
$$\left(A' \right)' = A, \quad \left(A \pm B \right)' = A' \pm B', \quad \left(AB \right)' = B'A' \qquad (1.2.22)$$
Thus a square matrix A is symmetric if $A = A'$.

5. A **diagonal matrix** is a square matrix where all off-diagonal elements are zero, and hence it is symmetric too. A square matrix is called an identity matrix if it is diagonal and all diagonal elements are equal to 1. An **identity matrix** of order n (or $n \times n$) is denoted by I_n (or simply I). It can be verified that
$$I_m A_{m \times n} = A_{m \times n} = A_{m \times n} I_n \qquad (1.2.23)$$

6. A square matrix A is called **idempotent** if $A^2 = A \cdot A = A$. Note that if A is idempotent, then for any positive integer k,

$$A^k = \underbrace{A \cdot A \ldots A}_{k \text{ times}} = A \qquad (1.2.24)$$

7. A matrix is called a **null matrix** (denoted by $O_{m \times n}$ or simply O) when all elements are equal to zero. It can be seen that for any matrix $A_{m \times n}$,

$$A_{m \times n} O_{n \times l} = O_{m \times l}, \quad O_{k \times m} A_{m \times n} = O_{k \times n}$$

$$\text{and } A_{m \times n} \pm O_{m \times n} = A_{m \times n} \qquad (1.2.25)$$

8. A square matrix Q is called an **orthogonal matrix** if $QQ' = I = Q'Q$.

9. For every square matrix $A_{n \times n}$, the determinant of A (denoted by $|A|$ or $\det(A)$) is defined as $|A| = \sum_\pi \left(\pm a_{1i_1} a_{2i_2} \ldots a_{ni_n} \right)$, where π is the collection of $n!$ different permutations $\{i_1, i_2, \ldots, i_n\}$ of $\{1, 2, \ldots, n\}$ and the sign of each term is plus $(+)$ if the permutation consists of even number of cycles, whereas it is minus $(-)$ if the number of cycles[2] is odd.

   The **minor** of an element a_{ij} in A is defined to be the determinant of a matrix of order $(n-1)$ obtained from A by deleting the ith row and the jth column. The **cofactor** of a_{ij}, denoted by A_{ij} is $A_{ij} = (-1)^{i+j} \left(\text{the minor of } a_{ij} \right)$.

   The determinant of a matrix $A_{n \times n}$ can also be obtained as

$$|A| = \sum_{j=1}^{n} a_{ij} A_{ij} \quad \text{for any } i = 1, 2, \ldots, n \qquad (1.2.26)$$

   In the simplest case, when A is of order 2×2, $|A| = a_{11}a_{22} - a_{12}a_{21}$. Also, it can be verified that $|A| = |A'|$ for any square matrix $A_{n \times n}$.

10. A square matrix $A_{n \times n}$ is called **nonsingular** if $|A| \neq 0$. If $A_{n \times n}$ is nonsingular then said to have a rank n. $A_{n \times n}$ is called a **singular** matrix if $|A| = 0$. For a nonsingular matrix A, there exists another

---

[2]A cycle of length k means that, given integers $\{i_1, i_2, \ldots, i_k\}$, we have a permutation $\{i_2, i_3, \ldots, i_k, i_1\}$.

matrix (of order $n \times n$), denoted by A^{-1} such that $AA^{-1} = I_n = A^{-1}A$. Such matrix A^{-1} is called the **inverse** of A, and it can be obtained as

$$A^{-1} = \left(a^{ij}\right) \quad \text{where } a^{ij} = \frac{1}{|A|}A_{ji} \qquad (1.2.27)$$

It can be shown that:

(a) $\left|A^{-1}\right| = 1/|A|$

(b) If A is symmetric, then so is A^{-1}

(c) For an orthogonal matrix Q, $Q^{-1} = Q'$

11. For a symmetric matrix A of order $n \times n$, and variables $x_1, x_2, \ldots, x_n$, the expression

$$Q_A(x) = x'Ax = \sum_{i=1}^{n}\sum_{j=1}^{n} a_{ij}x_i x_j \qquad (1.2.28)$$

where $x = (x_1, x_2, \ldots, x_n)'$, is called a **quadratic form** in x. If $Q_A(x) \geq 0$ for all $x \in \mathbb{R}^n$, then A is called positive semidefinite; and positive definite if $Q_A(x) > 0$ for all $x \neq 0$. Likewise negative semidefinite and negative definite. For two symmetric matrices A and B of same order, if $Q_A(x) = Q_B(x)$ for all $x \in \mathbb{R}^n$ (the n-dimensional real space), then $A = B$.

Calculus

If x and y are two variables such that the value of y depends on the value of x, then y is said to be a **function** of x. The variable x, whose value y depends on, is called the independent variable; and y is called the dependent variable. To denote the dependence of y on x, we usually write $y = f(x)$, and $f(x_0) = y_0$ indicates the value of y when $x = x_0$.

Suppose the variable y is a function of the variable x. The collection of all possible values of x, for which the function is defined, is called the **domain** of the function, and the collection of all possible values of y is called the **range** of the function. A function $y = f(x)$ is called **bounded** if there exists a real value $M > 0$ such that $|y| < M$ for all x in the domain.

If there is a finite value y_0 to which $f(x)$ tends as x tends to the value x_0, then y_0 is called the limit of $f(x)$ as x approaches to x_0. This is denoted as:

$$\lim_{x \to x_0} f(x) = y_0$$

More rigorously, it means, given any $\varepsilon > 0$, there exists a $\delta > 0$ such that $|f(x) - y_0| < \varepsilon$ whenever $|x - x_0| < \delta$.

For two functions $f_1(x)$ and $f_2(x)$ of x, **common limit operations** are:

1. $\lim_{x \to x_0} \{f_1(x) + f_2(x)\} = \lim_{x \to x_0} f_1(x) + \lim_{x \to x_0} f_2(x)$

2. $\lim_{x \to x_0} \{f_1(x) \cdot f_2(x)\} = \lim_{x \to x_0} f_1(x) \cdot \lim_{x \to x_0} f_2(x)$

3. $\lim_{x \to x_0} \{f_1(x)/f_2(x)\} = \left\{ \lim_{x \to x_0} f_1(x) \right\} / \left\{ \lim_{x \to x_0} f_2(x) \right\}$
    provided $\lim_{x \to x_0} f_2(x) \neq 0$

Some **common limit results** are:

1. $\lim_{x \to \infty} (1 + 1/x)^x = e$

2. $\lim_{x \to 0} (1 + x)^{1/x} = e$

3. $\lim_{x \to a} \left(x^k - a^k \right) / (x - a) = ka^{k-1}$

4. $\lim_{x \to 0} \dfrac{\sin x}{x} = 1$

A function $f(x)$ of x is said to be **continuous** at a point $x = x_0$ if $\lim_{x \to x_0} f(x) = f(x_0)$. Otherwise, $f(x)$ is said to be discontinuous at $x = x_0$. If $f(x)$ is continuous at $x = x_0$, then this point x_o is called a continuity point of $f(x)$.

Given a function $y = f(x)$, the limiting value

$$\lim_{h \to 0} \left(f(x_0 + h) - f(x_0) \right)/h$$

when it exists, is called the **derivative** of $f(x)$ with respect to x at $x = x_0$. In this case, we say that $f(x)$ is **differentiable** at $x = x_0$. If the limit does not exist, then $f(x)$ is not differentiable at $x = x_0$. The derivative of $f(x)$ at $x = x_0$ is denoted as $f'(x_0)$.

If $y = f(x_1, x_2, \ldots, x_k)$ is a function of several variables $x_1, x_2, \ldots, x_k$, then the limit,

$$\lim_{h \to 0} \left(f(x_1, \ldots, x_{i-1}, x_i + h, x_{i+1}, \ldots, x_k) - f(x_1, x_2, \ldots, x_k) \right) \Big/ h$$

if it exists when $x_1, \ldots, x_{i-1}, x_{i+1}, \ldots, x_k$ are treated as constants, is called the (first) partial derivative of f with respect to x_i and is denoted by $f^{(i)}(x_1, x_2, \ldots, x_k)$ or $\partial f / \partial x_i$.

Let $y = f(x)$ be a bounded function on the interval $[a, b]$. Let the interval $[a, b]$ be partitioned into n subintervals with a width $h = (b-a)/n$, i.e.,

$$[a, b] = [a, a + h] \cup [a + h, a + 2h] \cup \ldots \cup [a + (n - 1)h, b]$$

such that $h \to 0$ as $n \to \infty$. Let n arbitrary points, $x_1, x_2, \ldots, x_n$, be chosen from the above subintervals, respectively. If for any selection of points $\{x_1, x_2 \ldots, x_n\}$, the sum $h \sum_{i=1}^{n} f(x_i)$ tends to a fixed limit, then that limit is called the definite integral of $f(x)$ over the interval $[a, b]$ and is denoted by $\int_a^b f(x)\,dx$. Geometrically, the value $\int_a^b f(x)\,dx$ represents the area bounded by the curve $y = f(x)$, the x-axis, and two ordinates at a and b.

The procedure of finding the integral of a function is called **integration**, and it is the inverse of differentiation. Therefore, if $y = f(x)$ is a function with derivative $f'(x)$, then the integral of $f'(x)$ with respect to x between a and b is $f(b) - f(a)$. Hence, for integrating a function $f(x)$, we look for an expression $F(x)$ such that $F'(x) = f(x)$. The function $F(x)$ is called the indefinite integral of $f(x)$ and is denoted by $\int f(x)\,dx$.

Since $\frac{d}{dx}\left(F(x) + c\right) = F'(x)$ for any constant c, the integral of $F'(x)$ is $\left(F(x) + c\right)$. Hence $F(x) + c$ is also an indefinite integral and therefore the constant c must appear in any indefinite integral, i.e., $\int f(x)\,dx = F(x) + c$.

Order 'big oh' and 'small oh':

A function $f(x)$ is of order:

(a) $O(x)$ (called, 'big oh' of x) if $(f(x)/x)$ is finite as $x \to \infty$

(b) $o(x)$ (called, 'small oh' of x) if $(f(x)/x) \to 0$ as $x \to \infty$

1.3   Set Theory

A **set** is a collection of well-defined distinct objects. Usually, a set is denoted by a capital letter (with or without a suitable subscript), like A, B, or C, etc., or $A_1, A_2, \ldots$, etc.

For example , let A be the collection of all positive integers less than 6, i.e., A contains the integers 1, 2, 3, 4, and 5. We can write this set A as $A = \{1, 2, 3, 4, 5\}$. Each member of the set A is called a **set element**. Note that this collection is well defined and all set elements are distinct.

Another example could be the collection of stocks listed on NYSE (New York Stock Exchange) at the end of December 31, 2000. We can denote this collection as $B = \{$IBM, IP, XRX, NU, T, $\ldots\}$. Again, this collection is well defined and all elements are distinct.

If we consider the collection of students enrolled at a particular college, then it is not well defined because the student body changes from semester to semester, and hence this collection, in true sense, cannot be a set. However, if we specify the collection of students enrolled in Spring 2005 at a particular college, then it makes sense; we can identify all qualifying students in the collection, and this collection is a set.

A few more of examples of a set are listed below:

(a) The collection of U.S. cities with reported homicides not less than 1 per thousand people in a particular year

(b) The collection of 'Dot-Com' start-ups world wide in 2004

(c) The collection of rational numbers between 0 and 1

A set is called an **empty set** or a **null set** if it does not have any element in it. Such a set is denoted by ϕ. For example, let A denote the set of integers between $1/3$ and $2/3$. It can be seen that $A = \phi$ because there is no integer between $1/3$ and $2/3$.

Given two sets, say A and B, B is said to be a **subset** of A if each element of B is also an element of A (i.e., B is a sub-collection of A). This is denoted as '$B \subseteq A$'. For instances, (i) let $A = \{1, 2, 3, 4, 5\}$ and $B = \{2, 4\}$, here $B \subseteq A$. (ii) Let A_1 be all dollar bills issued by the U.S. Federal Reserve in 2000, and let A_2 be all dollar bills issued by the U.S. Federal Reserve in 2000 with even serial numbers. Here, $A_2 \subseteq A_1$.

For any two sets A and B:

- $A \cup B$ (called 'A **union** B') is the collection of all elements which are members of A or B or both

- $A \cap B$ (called 'A **intersection** B') is the collection of those elements which are common to both A and B

Example 1.3.1: (i) Let $A = \{1, 2, 3, 4, 5\}$ and $B = \{4, 5, 6, 7\}$, then $A \cup B = \{1, 2, 3, 4, 5, 6, 7\}$ and $A \cap B = \{4, 5\}$. (ii) Let $A_1 = \{1, 2, 3, 4\}$ and $A_2 = \{2, 3\}$, then $A_1 \cup A_2 = \{1, 2, 3, 4\}$ and $A_1 \cap A_2 = \{2, 3\}$.

For any set A, A^c (called 'A-**complement** ') is the collection of all elements which are **not** members of A. The following set theoretic rules are easy to follow:

(a) For any set A, $A \cup \phi = A$ and $A \cap \phi = \phi$

(b) For any two sets A and B,

$$(A \cup B)^c = A^c \cap B^c \text{ and } (A \cap B)^c = A^c \cup B^c$$

(c) For any three sets A, B, and C,

$$A \cap (B \cup C) = (A \cap B) \cup (A \cap C) \text{ and } A \cup (B \cap C) = (A \cup B) \cap (A \cup C)$$

(d) For any set A, $(A^c)^c = A$

1.4   Introduction to Probability Theory

In statistics, we are concerned with collection, presentation, and interpretation of 'chance outcomes' which occur in a scientific study. Chance outcomes are those which cannot be predicted with certainty. For example, (i) we may record the number of burglaries that take place per month in a neighborhood to justify more resources for the law enforcement agency; or (ii) we may be interested in investigating voters' mood for an upcoming election where two candidates, say Candidate-D and Candidate-R, are in the fray. Note that the number of burglaries in a particular month can be any nonnegative value and cannot be predicted beforehand. Similarly, whether a particular voter votes for D or R is unknown unless or until he/she casts his/her ballot. But once the particular month is elapsed, the number of burglaries are known; and once the election is over, a voter's choice is known. [Here we are dealing with **quantitative data** representing counts or measurements, or perhaps with **qualitative data** which can be categorized according to some criterion.]

From now on we shall refer to any recording of information, quantitative or qualitative, as an observation. Thus the quantitative information of 3, 5, 7, 1, representing the number of burglaries per month in a neighborhood for four consecutive months, constitutes a set of observations or a quantitative data set. In a similar way, the qualitative information of R, R, D, R, D, D, R, D, D, R, representing the preferences of ten randomly selected registered voters in a town, constitutes a set of observations or a qualitative data set.

In statistics, we often use the word **experiment** to describe a process or a mechanism that generates a set of observations or a data set. Some simple examples of statistical experiments are: (i) Tossing a coin only once. In this experiment, there are only two possible outcomes, Head (H) or Tail (T). (ii) Rolling a die only once and recording its face value. Here the possible outcomes are 1, 2, 3, 4, 5, or 6. (iii) Selecting a student at random from the local community college and recording his or her ethnic background. The possible outcomes of this experiment are — African American, Caucasian, Asian, Native Indian, etc. Note that the outcome of an experiment can not be predicted with certainty and depends on the chance or probability of each outcome.

Before we discuss 'probability' or 'chance' of outcomes, let us introduce some commonly used notations and terminology.

Definition 1.4.1 *(a) A statistical experiment is a mechanism whose outcome is uncertain.*
(b) The set of all possible outcomes of a statistical experiment is called the sample space, and it is denoted by S.

Each outcome in a **sample space** S is called an element or a member of the sample space, or simply a **sample point**. A sample space may have a finite number of sample points or can have infinitely many sample points. For example, if a geologist measures the magnitude of an earthquake at the foothills of the Himalayan Mountain range, then he/she might record any possible magnitude between 0 and 10 on the Richter scale. Hence, the sample space of the geologist's experiment contains infinitely many sample points. On the other hand, if an experiment consists of drawing a card at random from a standard deck of 52 cards, then the sample space is finite and has only 52 possible sample points.

If the sample space has a finite number of sample points, then we can list all elements separated by commas and enclosed in brackets. Thus the sample space S of possible outcomes when a die is rolled, is given as

$$S = \{1, 2, 3, 4, 5, 6\}$$

Suppose a coin is tossed three times consecutively. The sample space S can be written as

$$S = \{\text{HHH, HHT, HTH, HTT, THH, THT, TTH, TTT}\}$$

where H and T denote 'head' and 'tail' respectively.

It is convenient to describe a sample space with a large or infinite number of sample points by a statement or rule. For instance, if an experiment consists of drawing a card at random from a standard deck of 52 playing cards, then the sample space can be written as

$$S = \{\text{all 52 playing cards in a standard deck}\}$$

Similarly, consider the experiment where a geologist records the magnitude of an earthquake, the sample space may be described as

$$S = \big\{x \big| \, 0 < x < 10\big\}$$

where the vertical bar '$|$' means 'such that'; or in short, $S = (0, 10)$ (the open interval from 0 to 10).

In any given experiment, we may be interested in the occurence of certain types of outcomes rather than a single outcome. For instance, when a die is rolled, we might be interested in those outcomes which are divisible by 3. This will happen if the outcome is an element of the subset, say, $A = \{3, 6\}$, of the sample space $S = \{1, 2, 3, 4, 5, 6\}$. Such a subset is called an **event**. [3]

Definition 1.4.2 *An event is a subset of the sample space.*

---

[3]Events are usually denoted by capital letters (with or without subscripts).

Let us consider an experiment of observing the lifespan of a 60-watt (120 volts) electric bulb. The sample space $\mathcal{S} = \{t \mid t \geq 0\}$, where t is the life-time in hours. The manufacturer claims that the average life of such bulbs is 750 hours. Hence, we are interested in the event that a bulb fails before completing 750 hours. Therefore, the event of interest can be described by the subset $B = \{t \mid 0 \leq t < 750\}$.

It is possible that an event can be the entire sample space that includes all sample points of $\mathcal{S}$. The empty set ϕ is called the empty event and contains no sample points at all.

Consider an experiment in which the drinking habit of airline pilots are being recorded. The sample space of possible outcomes might be classified as a collection of nondrinkers, light drinkers, moderate drinkers, and heavy drinkers. Let all the nondrinkers correspond to a subset of $\mathcal{S}$ which is called the complement of the subset of all drinkers.

Definition 1.4.3 *The complement of an event A, with respect to the sample space $\mathcal{S}$, is the event of all elements of $\mathcal{S}$ that are not in A. The complement of A is denoted by A^c.*

Basically A^c indicates 'nonoccurance' of A. For example, let R be the event in which a red card is drawn from a standard deck of 52 playing cards. Then R^c is the event that a black card is drawn from the deck.

Since events are subsets of the sample space, we can invoke the set theoretic operations discussed earlier (in Section 1.3) for events also.

Meaning of Probability

The word **probability** may be used in two different contexts. First, it may be used regarding some proposition. Consider the following statement: "It is very probable that the digital divide between the Hispanics and nonHispanics in the United States will decrease in the next twenty years." Probability here means the degree of belief in the above proposition of the individual making the statement. This is usually called **Subjective Probability**.

Alternatively, the word **probability** may be used regarding an event given rise by an experiment that can conceivably be repeated infinitely many times under similar conditions. The probability of an event here

refers to the proportion of cases in which the event occurs in such repetitions of the experiment. This type of probability is called **Objective
Probability**.

Objective probability, which reflects the real world, is the sense that
we should be concerned with most of the time in this handbook. Subjective probability will be discussed later when we deal with Bayesian
analysis.

Classical Definition of Probability

To develop the concepts of probability, we first consider only those experiments for which the sample space $\mathcal{S}$ has a finite number of sample points.
This is done for the sake of simplicity and will not be applicable to the
situations where $\mathcal{S}$ consists of infinitely many elements (either countable
or noncountable).

Secondly, it is assumed that the experiment is such that all the possible outcomes in $\mathcal{S}$ are equally likely, i.e., when all relevant evidence
is taken into account, no single outcome can be expected to occur in
preference to the others.

The probability of any event A, denoted by $P(A)$, is defined as

$$P(A) = \frac{\#\text{ elements in } A}{\#\text{ elements in } \mathcal{S}} \qquad (1.4.1)$$

The above ratio is called the **classical definition** of probability. Thus
probability can be treated as a function P defined on the class of all
events which assigns a unique value between 0 and 1 to an event A. Note
that:

(a) $0 \leq P(A) \leq 1$

(b) If $A = \phi$, then $P(A) = 0$

(c) If $A = \mathcal{S}$, then $P(A) = 1$

(d) If $A \subseteq B \subseteq \mathcal{S}$, then $P(A) \leq P(B)$

(e) $P(A) = 1 - P(A^c)$

(f) For any two events A and B, $P(A \cup B) = P(A) + P(B) - P(A \cap B)$[4]

---

[4]This is usually called the *additive rule*.

Example 1.4.1: An unbiased coin is tossed twice. What is the probability that at least one head occurs?

The sample space for this experiment is $S = \{$HH, HT, TH, TT$\}$. Since this is a fair coin, each of these outcomes would be equally likely to occur. If A represents the event that at least one head is observed, then $A = \{$HH, HT, TH$\}$, and $P(A) = 3/4$.

Limitations of Classical Definition

The classical definition of probability has some obvious drawbacks. Surely, direct application of this definition is not possible if the total number of elements in the sample space is infinite. Even if the number of elements in S is finite, they must have equally likely outcomes in order to apply the classical definition of probability.

In Example 1.4.1, it was crucial to know that the coin was balanced or unbiased. Because of the unbiasedness of the coin, all possible outcomes were equally likely and hence we could compute the probability of the event A of at least one head occuring. But what about an experiment where outcomes are not equally likely? Can we still find the probability of an event? Well, we can, provided we know what kind of bias is involved in the experiment, and this is explained through the following example.

Example 1.4.2: A die is loaded in such a way that an even number is three times as likely to occur as an odd number. If the die is rolled only once, then find the probability of the event that the face value is less than or equal to 3.

The sample space of the above example is $S = \{1, 2, 3, 4, 5, 6\}$. We assign a probability, say 'p', to each odd number and a probability of $3p$ to each even number. The total probability (of the whole sample space) must add up to 1, i.e.,

$$p + p + p + 3p + 3p + 3p = 12p = 1$$

or $p = 1/12$. Let $B = \{1, 2, 3\}$ be the event of interest, then

$$P(B) = \frac{1}{12} + \frac{3}{12} + \frac{1}{12} = \frac{5}{12}$$

Probability as Long Term Relative Frequency

In Example 1.4.2, we knew the type of bias involved in the experiment. We could use the known bias to assign a probability to every sample point in the sample space. If we have a reason to believe that a certain sample point is more likely to occur when the experiment is conducted, the probability assigned should be close to 1. On the other hand, a probability closer to 0 is assigned to a sample point that is less likely to occur. Still, if we do not know the nature of the experiment (whether or not the outcomes are equally likely), then it is impossible to assign probabilities to the elements of S. Therefore, we conceptualize the notion of probability as a relative frequency **in the long run**. By doing this, we avoid the difficulties of applying the classical definition of probability when the sample space is not finite.

Suppose we perform a sequence of n repetitions (or replications) of an experiment. Let f_n be the number of times an event A is occurring among these n repetitions. The value f_n is called the absolute frequency (or simply, frequency) of A in n repetitions, and the ratio, f_n/n, is called the relative frequency of A.

In many cases, it is observed that the successive values of the sequence $\{f_n/n\}$ differ from one another by small amounts when n gets large. In such cases, there is a tendency in $\{f_n/n\}$ to stabilize around a fixed value. This limiting value of $\{f_n/n\}$ as $n \to \infty$ is regarded as the probability of A in the experiment.

As we adopt the above notion of probability, it must be kept in mind that, practically speaking, it is impossible to determine the exact probability of an event (without any further information or assumption as we had in the last two examples). We cannot even prove the existence of a limit to the relative frequency in any given case. On the other hand, we can treat any observed relative frequency as an estimate of the supposed limit. In this sense, we can speak of the probability that it would rain in Washington, D.C., on 4th of July, or of the stock market acting positively in an election year. If we observe the weather on 4th of July at Washington, D.C. for many years, then the proportion of times it rained would be an approximation of the actual probability under consideration. Similarly, one can approximate the actual probability of the stock market acting positively in an election year.

With probability being interpreted in this way, the following relations are taken as axioms:

(**A1**) for any event A, $P(A) \geq 0$

(**A2**) $P(\mathcal{S}) = 1$

(**A3**) if $A_1, A_2, \ldots$ are pairwise disjoint events, or $A_i \cap A_j = \phi$ for $i \neq j$,

$$\text{then } P\left(\bigcup_{i=1}^{\infty} A_i\right) = \sum_{i=1}^{\infty} P(A_i)$$

Thus, any function P, defined on the events of a sample space $\mathcal{S}$, that obeys axioms (A1)-(A3) is called a **probability function**.

Conditional Probability

The probability of an event B occuring when it is known that some other event A has occured is called a conditional probability and is denoted by $P(B|A)$. The notation $P(B|A)$ is usually read as **the probability of B, given A**.

Definition 1.4.4 *The conditional probability of B, given A, denoted by $P(B|A)$ is defined as*

$$P(B|A) = \frac{P(A \cap B)}{P(A)} \quad provided \; P(A) > 0 \tag{1.4.2}$$

As an illustration, suppose that our sample space consists of all students enrolled at a local community college in Fall 2004. We categorize the students according to their smoking habit (at least one cigarette per day or not) and gender in the following table.

	Smoker		
Gender	Yes	No	Total
Male	312	918	1230
Female	117	773	890
Total	429	1691	2120

Table 1.4.1: Student Smoking Habit vs. Gender.

One of these students is selected at random for a campaign to publicize the ill effects of smoking on health. We are concerned with the following two events:

$$F : \text{ the selected student is a female}$$

$$N : \text{ the selected student is a nonsmoker}$$

Note that the sample space $\mathcal{S}$ consists of all 2120 students. We want to find the conditional probability $P(F|N)$, i.e., the student we selected at random is a female given that the student is a nonsmoker.

Using the above definition,

$$P(F|N) = \frac{P(F \cap N)}{P(N)} = \frac{(\# \text{ elements in } F \cap N)/(\# \text{ elements in } \mathcal{S})}{(\# \text{ elements in } N)/(\# \text{ elements in } \mathcal{S})}$$

$$= \frac{(\# \text{ elements in } F \cap N)}{(\# \text{ elements in } N)} = \frac{773}{1691} = 0.4571$$

Definition 1.4.5 *Two events A and B are independent if and only if*

$$P(B|A) = P(B) \quad or \quad P(A|B) = P(A)$$

Otherwise, A and B are called dependent events.

Theorem 1.4.1 *If in an experiment the events A and B can both occur, then*

$$P(A \cap B) = P(A) \cdot P(B|A) = P(B) \cdot P(A|B)$$

Theorem 1.4.2 *Two events A and B are independent if and only if*

$$P(A \cap B) = P(A) \cdot P(B)$$

The following theorem, called the multiplication law, is concerned with the probability of the joint occurrence of several events.

Theorem 1.4.3 *Let $A_1, A_2, \ldots, A_m$ be events of an experiment, then*

$$P(A_1 \cap A_2 \ldots \cap A_m) = P(A_1)P(A_2|A_1)P(A_3|A_1 \cap A_2) \cdots$$
$$P(A_m|A_1 \cap A_2 \ldots \cap A_{m-1})$$

Independence of several events can be defined as follows.

Definition 1.4.6 *A collection of events $A_1, A_2, \ldots$ are called independent provided for any finite subcollection $A_{i_1}, A_{i_2}, \ldots, A_{i_k}$, $k \geq 2$,*

$$P(A_{i_1} \cap A_{i_2} \ldots \cap A_{i_k}) = P(A_{i_1})P(A_{i_2}) \cdots P(A_{i_k}), \quad for \ 1 \leq i_1 < \ldots < i_k$$

Bayes' Principle

Note that for a given event A, we can write

$$A = (A \cap B) \cup (A \cap B^c)$$

where B is any other suitable event. This is illustrated through the Figure 1.4.1.

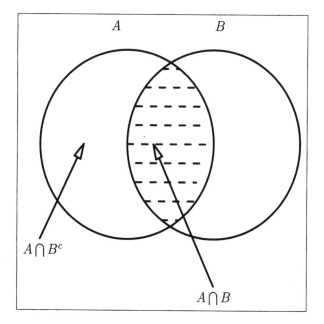

Figure 1.4.1: Venn Diagram for the Events $(A \cap B)$ and $(A \cap B^c)$.

Since $A \cap B$ and $A \cap B^c$ are mutually exclusive, using the addition and multiplication laws of probability, one can obtain the following equation:[5]

$$
\begin{aligned}
P(A) &= P(A \cap B) + P(A \cap B^c) \\
&= P(B)P(A|B) + P(B^c)P(A|B^c)
\end{aligned}
$$

The above probability equation is useful in many real life problems. The following example is one of such applications.

---

[5]This equation is called the total probability formula.

Example 1.4.3: A medical test given to an individual is known to be 90% reliable when the person is HIV positive and 99% reliable when the person is HIV negative. In other words, 10% of the HIV positive individuals are diagnosed as negative, and 1% of the HIV negative individuals are diagnosed as positive by the medical test. If a person is selected from a population where 2% of the individuals are known to be HIV positive and the test indicates he or she is positive, what is the probability that he or she is actually HIV negative?

Consider the experiment of drawing an individual at random from a population consisting of 2% HIV positive people. The sample space S is the collection of all individuals in the population. Let B be the event of HIV positive individuals. It is known that $P(B) = 0.02$. Let A be the event that a person is diagnosed as HIV positive. From the reliability information about the test we have

$$P(A|B) = 0.90 \quad \text{and} \quad P(A^c|B^c) = 0.99$$

Now we need to find $P(B^c|A)$.

From the conditional probability formula,

$$
\begin{aligned}
P(B^c|A) &= \frac{P(B^c \cap A)}{P(A)} \\
&= \frac{P(B^c \cap A)}{P(B \cap A) + P(B^c \cap A)} \\
&= \frac{P(B^c)P(A|B^c)}{P(B)P(A|B) + P(B^c)P(A|B^c)} \\
&= \frac{(1 - 0.02)(1 - 0.99)}{(0.02)(0.90) + (1 - 0.02)(1 - 0.99)} \\
&= 0.8448
\end{aligned}
$$

The above concept is generalized in the Bayes' theorem as given below.

Theorem 1.4.4 (Bayes' Theorem) *Let the collection of events* $\{B_1, B_2, \ldots, B_m\}$ *constitute a partition of the sample space* S *(i.e., $B_i \cap B_j = \phi$ for $i \neq j$; and $S = B_1 \cup B_2 \ldots \cup B_m$). Then given any event A,*

$$P(B_i|A) = \frac{P(B_i)P(A|B_i)}{\sum_{j=1}^{m} P(B_j)P(A|B_j)}, \quad \text{for any } B_i, \ 1 \leq i \leq m \qquad (1.4.3)$$

Remark 1.4.1 *In many situations, $B_1, B_2, \ldots, B_m$ may be looked upon as the possible causes of the effect A. Bayes' theorem then gives the probability of the cause B_i when the effect A has been observed, or the probability of the 'hypothesis' B_i in the light of the 'data' A. In Bayesian analysis, Theorem 1.4.4 provides the 'posterior probability' of B_i in terms of the 'prior probabilities' $P(B_i)$, $i = 1, 2, \ldots, m$, and the conditional probabilities of A.*

1.5   Random Variables

A sample space $\mathcal{S}$ could be difficult or inconvenient to describe if the sample points are not numbers. Often we are not interested in the details associated with each sample point but only in some numerical description of the outcome. We shall now discuss how we can use a rule by which an element s of $\mathcal{S}$ may be associated with a real value x.

Consider the example of tossing a coin three times in a row. The sample space $\mathcal{S}$ is given as

$$\mathcal{S} = \{\text{HHH, HHT, HTH, HTT, THH, THT, TTH, TTT}\}$$

If one is concerned only with the number of heads observed, then a numerical value of 0, 1, 2, or 3 will be assigned to each sample point. Notice that the numerical values 0, 1, 2, and 3 are random quantities determined by the outcome of the experiment. They can be thought of as values assumed by some **random variable** X, which in this case represents the number of heads when a coin is tossed three times.

Generally speaking, given an experiment with a sample space $\mathcal{S}$, a function X that assigns to each element s in $\mathcal{S}$ one and only one real number $X(s) = x$, is called a random variable. The range (or space) of X is the set of real numbers $\{x \mid x = X(s), s \in \mathcal{S}\} = R_X$ (say).

Definition 1.5.1 *A random variable (r.v.) is a function that associates a real number with each element in the sample space.*

We shall use capital letters with or without subscripts, like X, Y, Z or $X_1, X_2, \ldots$ etc. to denote random variables.

Example 1.5.1: Let an experiment be rolling a die only once and observing the face value of the die. The sample space is known as

$\mathcal{S} = \{1, 2, 3, 4, 5, 6\}$. For each $s \in \mathcal{S}$, define a random variable Y as $Y(s) = s$. The range (or space) of the random variable Y is then $R_Y = \{1, 2, 3, 4, 5, 6\}$.

Definition 1.5.2 *If the range of a random variable contains a finite number of values or an infinite sequence of values indexed by natural numbers (i.e., countable infinite), then the random variable is called a discrete random variable.*

Definition 1.5.3 *If the range of a random variable contains infinitely many values equal to the number of points on a line segment or a collection of line segments (i.e., uncountably many values), then the random variable is called a continuous random variable.*

For notational convenience, given a random variable X, we denote the event $\{s \mid s \in \mathcal{S}$ and $X(s) = a\}$ by $\{X = a\}$. That is, the event $\{X = a\}$ is the set of elements in the sample space that is mapped onto the real value, a, by the function X. Similarly, the event $\{s \mid s \in \mathcal{S}$ and $a < X(s) < b\}$ will be denoted by $\{a < X < b\}$. Now if we want to find the probabilities associated with the events described in terms of X, such as $\{X = a\}$ or $\{a < X < b\}$, we use the probabilities of those corresponding events in the original sample space $\mathcal{S}$. In other words, the probabilities for a random variable X is defined as

$$P(\{X = a\}) = P(\{s \mid s \in \mathcal{S} \text{ and } X(s) = a\}) \quad \text{or}$$

$$P(\{a < X < b\}) = P(\{s \mid s \in \mathcal{S} \text{ and } a < X(s) < b\})$$

and these are called **induced probabilities**.

For a discrete random variable X, the induced probability $P(X = x)$, denoted by $f(x)$, is called the probability mass function (*pmf*). A *pmf* $f(x)$ of X with range R_X satisfies the following properties:

(a) $f(x) > 0, \quad \forall\, x \in R_X$

(b) $\displaystyle\sum_{x \in R_X} f(x) = 1$

(c) $P(X \in A) = \displaystyle\sum_{x \in A} f(x), \quad$ where $A \subseteq R_X$

Usually the *pmf* $f(x) = 0$ when $x \notin R_X$. Since the probability $f(x) = P(X = x) > 0$ for $x \in R_X$ where R_X contains all values of X with positive probabilities. R_X is also referred to as the **support** of X.

Example 1.5.2: In an introductory statistics class of 40 students, there are 6 math majors, 11 nursing majors, 16 engineering majors, 3 criminal justice majors, and 4 undecided. One student is selected at random and the random variable X is assigned a score 0 for a student with undecided major, 1 for math, 2 for nursing, 3 for engineering, and 4 for criminal justice.

The range of X is thus $R_X = \{0, 1, 2, 3, 4\}$. The *pmf* of X is given by

$$
\begin{aligned}
f(0) = P(X = 0) &= \frac{4}{40} \\[4pt]
f(1) = P(X = 1) &= \frac{6}{40} \\[4pt]
f(2) = P(X = 2) &= \frac{11}{40} \qquad\qquad (1.5.1) \\[4pt]
f(3) = P(X = 3) &= \frac{16}{40} \\[4pt]
f(4) = P(X = 4) &= \frac{3}{40}
\end{aligned}
$$

It is understood that $f(x) = 0$ when $x \notin R_X$.

Definition 1.5.4 *The set of ordered pairs* $(x, f(x))$ *is called the probability distribution of the discrete random variable* X.

It is often helpful to look at a probability distribution in terms of a graph that depicts the *pmf* of X. Two types of graphs can be used to give a visual representation of the *pmf*, namely, a bar graph and a probability histogram.

A bar graph of the *pmf* $f(x)$ of a discrete random variable X is a graph of $f(x)$ having a vertical line segment joining the points $(x, 0)$ and $(x, f(x))$ for each x in R_X.

Instead of plotting the points $(x, f(x))$, we can also draw a rectangle of height $f(x)$ and a base of length 1 centered at x for each $x \in R_X$. Such a graphical representation is called a **probability histogram**. Even if the bases are not of unit width, we can still adjust the heights of rectangles to obtain areas that would be equal to $f(x)$ at x. This concept of using areas to represent probabilities is necessary for our consideration of the probability distribution of a continuous random variable.

Figure 1.5.1 and Figure 1.5.2 display a bar graph and a probability histogram, respectively, for the *pmf* $f(x)$ (1.5.1).

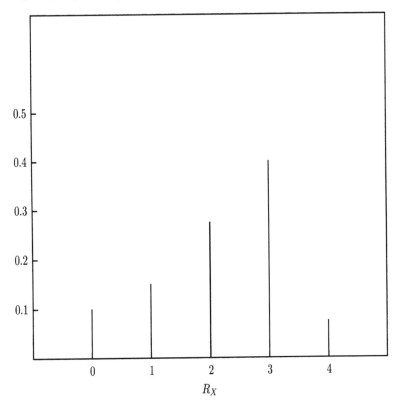

Figure 1.5.1: Bar Graph of the *pmf* (1.5.1).

In many problems, instead of considering events such as $\{X = x\}$ and the corresponding induced probabilities $f(x) = P(X = x)$, we may wish to consider events of the form $\{X \le x\}$ and the corresponding probabilities $P(X \le x)$. Writing $F(x) = P(X \le x)$ for every real number x, we define $F(x)$ to be the cumulative distribution function (*cdf*) of the random variable X.

Definition 1.5.5 *The cumulative distribution function (cdf) $F(x)$ of a discrete random variable X with the probability mass function $f(x)$ is given by*

$$F(x) = P(X \le x) = \sum_{t \le x} f(t) \quad \text{for } -\infty < x < \infty \qquad (1.5.2)$$

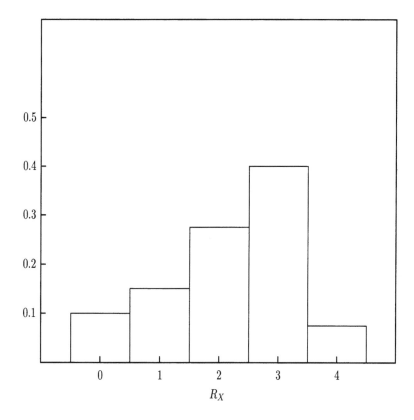

Figure 1.5.2: Probability Histogram of the *pmf* (1.5.1).

In Example 1.5.2, the random variable X has the range $R_X = \{0, 1, 2, 3, 4\}$ with the *pmf* $f(x)$ in (1.5.1). Therefore,

$$
\begin{aligned}
F(0) &= f(0) = \frac{4}{40} \\
F(1) &= f(0) + f(1) = \frac{10}{40} \\
F(2) &= f(0) + f(1) + f(2) = \frac{21}{40} \\
F(3) &= f(0) + f(1) + f(2) + f(3) = \frac{37}{40} \\
F(4) &= f(0) + f(1) + f(2) + f(3) + f(4) = 1
\end{aligned}
$$

Hence

$$F(x) = \begin{cases} 0 & \text{if } x < 0 \\ 4/40 & \text{if } 0 \le x < 1 \\ 10/40 & \text{if } 1 \le x < 2 \\ 21/40 & \text{if } 2 \le x < 3 \\ 37/40 & \text{if } 3 \le x < 4 \\ 1 & \text{if } x \ge 4 \end{cases} \tag{1.5.3}$$

The graph of the *cdf* in (1.5.3) is obtained by plotting points $(x, F(x))$ in Figure (1.5.3) which looks like a step function.

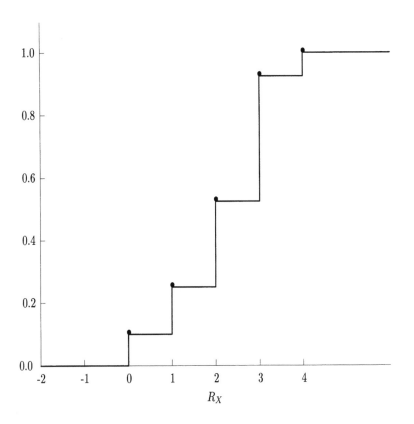

Figure 1.5.3: Graph of the discrete *cdf* (1.5.3).

For a continuous random variable X, the induced probability $P(X = x)$, $x \in R_X$, is zero. This may seem a bit puzzling, but it becomes more plausible when we consider the following example.

Consider a random variable X representing the percentage change in the NASDAQ stock index at the end of a business day. Between any two fixed values, say 1.5 and 2.9%, there are infinitely many other possible values. What is the probability of the NASDAQ index going up at the end of a business day by exactly 2%? Note that the value 2% is only one of those infinitely many possible values, and hence a probability of zero is assigned to this event. This is not the case if we try to find the probability of the NASDAQ stock index going up by an amount that is strictly between 1.75 and 2.25% of the value on the previous day; i.e., we are talking about the probability of the event $\{1.75 < X < 2.25\}$. Also, by the above argument,

$$
\begin{aligned}
P(1.75 < X < 2.25) &= P(1.75 < X < 2.25) + P(X = 2.25) \\
&= P(1.75 < X \leq 2.25)
\end{aligned}
$$

because it does not matter whether we include an end point of the interval or not. Probability of an interval $(a, b]$ can be expressed as

$$
P(a < X \leq b) = P(X \leq b) - P(X \leq a)
$$

and similar to the discrete case here too for a continuous random variable X, we have the *cdf* of X at 'a' and 'b' respectively as $F(a) = P(X \leq a)$ and $F(b) = P(X \leq b)$. Hence,

$$
\begin{aligned}
P(a < X \leq b) &= P(X \leq b) - P(X \leq a) \\
&= F(b) - F(a)
\end{aligned}
$$

In the case of a continuous random variable X, quite often it is possible to express the above probability $F(b) - F(a)$ in terms of the integration of another function $f(x)$ as

$$
P(a < X \leq b) = F(b) - F(a) = \int_a^b f(x)\, dx \qquad (1.5.4)
$$

Such a function $f(x)$, if it exists, is called the *probability density function (pdf) of* X.

Definition 1.5.6 *A function $f(x)$ is called a probability density function (pdf) for a continuous random variable X defined over* $\mathbb{R}$, *the set of real numbers, if*

(a) $f(x) \geq 0$ for all $x \in \mathbb{R}$

(b) $\int_{-\infty}^{\infty} f(x)\, dx = 1$

(c) $P(a < X \leq b) = \int_a^b f(x)\, dx$ for all $-\infty < a < b < \infty$

Note that for a continuous random variable, we cannot define the range exactly the way it was done for a discrete random variable since, as we discussed earlier, a continuous random variable can assume a single value with probability zero. However, the concept of *support* can be extended to the continuous case as the set of all values for which the *pdf* is strictly positive, i.e., for a continuous random variable X with *pdf* $f(x)$,

$$R_X = \text{support of } X = \{x \mid f(x) > 0\}$$

Example 1.5.3:   A continuous random variable X is defined as $2 \leq X \leq 5$, with the *pdf* $f(x) = 2(1+x)/27$. Find that the probability of the event $(X \leq 3)$.

It is known that the *pdf* $f(x)$ is zero where $x \notin [2,5]$. Therefore, using Definition 1.5.6, we have

$$P(X \leq 3) = \int_{-\infty}^{3} f(x)\, dx = \int_{2}^{3} \frac{2}{27}(1+x)\, dx = \frac{7}{27}$$

Similar to the discrete case, we define the *cdf* of a continuous random variable as follows.

Definition 1.5.7 *The cumulative distribution function (cdf) $F(x)$ of a continuous random variable X with the pdf $f(x)$ is given by*

$$F(x) = P(X \leq x) = \int_{-\infty}^{x} f(t)\, dt \quad \text{for } -\infty < x < \infty$$

An immediate consequence of the above definition is that $f(x) = F'(x) = dF(x)/dx$ provided the derivative exists. It is noticed that for any random variable X, the *cdf* always exists, but the existence of the *pdf* is not guaranteed.

Any cumulative distribution function $F(x)$ possesses the following four properties:

(a) $0 \leq F(x) \leq 1$ because $F(x)$ is a probability

(b) $F(x)$ is a nondecreasing function of x

(c) $\lim\limits_{x \to -\infty} F(x) = 0$ and $\lim\limits_{x \to +\infty} F(x) = 1$

(d) $F(x)$ is right-continuous, i.e., $\lim\limits_{x \to x_0^+} F(x) = F(x_0)$ for any x_0

["$x \to x_0^+$" means that x approaches to x_0 only from the right side of x_0.]

Definition 1.5.8 *Let X be a random variable with pdf or pmf $f(x)$. A value (a) M is called the* **median** *of X provided $P(X \le M) = 0.5 = P(X > M)$; (b) M_0 is called a* **mode** *of X provided $f(M_0) \ge f(x)$, for all x in the range (or support) of X.*

Expectations of Random Variables

An important concept in summarizing basic characteristics of the probability distribution of a random variable is that of mathematical expectations. We shall introduce this concept through an example.

Example 1.5.4: Suppose we device a game where a participant pays a certain fee to toss a fair coin three times. If the person gets no heads, he/she receives nothing; if exactly one head occurs, he/she receives \$1; if exactly two heads occur, he/she receives \$2; and if all the three heads occur, he/she receives \$3. Therefore, the events of interest are:

$$
\begin{aligned}
A_1 &= \{0 \text{ heads}\} = \{\text{TTT}\} \\
A_2 &= \{1 \text{head}\} = \{\text{HTT, THT, TTH}\} \\
A_3 &= \{2 \text{ heads}\} = \{\text{HHT, HTH, THH}\} \\
A_4 &= \{3 \text{ heads}\} = \{\text{HHH}\}
\end{aligned}
$$

So, let the random variable X be the \$ amount the participant receives. X has range $R_X = \{0, 1, 2, 3\}$. The *pmf* of X can be obtained as

$$
P(X = 0) = \frac{1}{8}; \; P(X = 1) = \frac{3}{8}; \; P(X = 2) = \frac{3}{8}; \text{ and } P(X = 3) = \frac{1}{8}
$$

If the game is played a large number of times, then about $1/8$ of the times need a payment of \$0, about $3/8$ of the times need a payment of \$1, about $3/8$ of the times need a payment of \$2, and $1/8$ of the times need a payment of \$3. Thus, the approximate average payment (in \$) is

$$
(0)\left(\frac{1}{8}\right) + (1)\left(\frac{3}{8}\right) + (2)\left(\frac{3}{8}\right) + (3)\left(\frac{1}{8}\right) = 1.5
$$

Note that a participant never receives exactly $1.5; the rewards are either $0, $1, $2, or $3. However, the weighted average $1.5 is what we expect to pay per participant if the game is played for a large number of times. This weighted average is called the mathematical expectation of the participant's reward. Thus, if we decide to charge a fee of $2 to each participant per play, then we can make, on average, a profit of 50 cents per play.

Definition 1.5.9 *If X is a random variable (either discrete or continuous) with pmf or pdf $f(x)$ over a suitable range (or support) R_X, then the expectation (or the expected value) of X, denoted by $E(X)$ or μ_X or simply μ (when the subscript is clear from the context), is defined as*

$$\mu_X = E(X) = \begin{cases} \displaystyle\sum_{x \in R_X} x f(x) & \text{if } X \text{ is discrete} \\ \displaystyle\int_{R_X} x f(x)\, dx & \text{if } X \text{ is continuous} \end{cases} \tag{1.5.5}$$

More generally, if $Y = g(X)$ is a real valued function of X with *pmf* or *pdf* $f(x)$, then

$$\mu_Y = E(Y) = E\big[g(X)\big] = \begin{cases} \displaystyle\sum_{x \in R_X} g(x) f(x) & \text{if } X \text{ is discrete} \\ \displaystyle\int_{R_X} g(x) f(x)\, dx & \text{if } X \text{ is continuous} \end{cases}$$
$$\tag{1.5.6}$$

The expectation of a random variable is also called its mean (or mean value).

Definition 1.5.10 *Let X be a random variable with pmf or pdf $f(x)$ and mean μ_X. The variance of X, denoted by σ_X^2 or simply σ^2, is defined as*

$$\sigma_X^2 = E\big[(X - \mu_X)^2\big] = \begin{cases} \displaystyle\sum_{x \in R_X} (x - \mu_X)^2 f(x) & \text{if } X \text{ is discrete} \\ \displaystyle\int_{R_X} (x - \mu_X)^2 f(x)\, dx & \text{if } X \text{ is continuous} \end{cases}$$
$$\tag{1.5.7}$$

The positive square root of the variance, σ_X, is called the standard deviation of X.

Unless mentioned otherwise, we shall use μ and σ^2 to denote the mean and variance of a random variable X. An alternative expression of the variance of X is given as

$$\sigma^2 = E(X^2) - \mu^2 \tag{1.5.8}$$

The first term in the above expression, $E(X^2)$, is called the second raw moment of the random variable X. Thus, for any integer k, the kth **raw moment** of X, denoted by $\mu'_{(k)}$, is

$$\mu'_{(k)} = E(X^k) \tag{1.5.9}$$

provided the summation or integration involved is meaningful. Similarly, kth **central moment** of X, denoted by $\mu_{(k)}$, is

$$\mu_{(k)} = E[(X - \mu)^k] \tag{1.5.10}$$

Hence, we have $\mu = \mu'_{(1)}$ and $\sigma^2 = \mu_{(2)}$.

Some simple yet useful properties of expectation are shown in the following result.

Theorem 1.5.1 *Given a random variable X with either pmf or pdf $f(x)$, the expectation operation 'E' satisfies the following properties whenever it exists.*

(i) If c is a constant, then $E(c) = c$.

(ii) If c is a constant and $u(X)$ is a real valued function of X, then

$$E[cu(X)] = cE[u(X)]$$

(iii) If c_1 and c_2 are constants, and $u_1(X)$ and $u_2(X)$ are real valued functions of X, then

$$E[c_1 u_1(X) + c_2 u_2(X)] = c_1 E[u_1(X)] + c_2 E[u_2(X)]$$

1.6   Joint Probability Distributions

So far, we have restricted ourselves to one-dimensional sample spaces, and recorded outcomes of an experiment in terms of numerical values assumed by a single random variable. But in many real life situations, we may find it useful to record simultaneous outcomes of several random variables. For instance, one can measure the height (H) and weight (W) of a patient selected at random in a clinical trial, thus giving rise to a two-dimensional range, say R_{HW}, consisting of all possible outcomes of $(H, W) = (h, w)$.

If X_1 and X_2 are two discrete random variables, the probability distribution for their simultaneous occurrence can be represented by a function $f(x_1, x_2)$ for any pair of values (x_1, x_2) within the range (say, $R_{X_1 X_2}$ or simply R_{12}) of the random variables X_1 and X_2. It is customary to refer this function as the joint probability distribution of X_1 and X_2. Therefore, in the discrete case,

$$f(x_1, x_2) = P(X_1 = x_1, X_2 = x_2)$$

Suppose X_1 and X_2 represent, respectively, the number of children and the number of pets in a household. Then $f(3, 2)$ is the probability that a randomly selected household has 3 children and 2 pets.

Definition 1.6.1 *The function $f(x_1, x_2)$ is a joint probability distribution of the discrete random variables X_1 and X_2 with range R_{12} if*

(i) $f(x_1, x_2) > 0$ for all $(x_1, x_2) \in R_{12}$;

(ii) $\displaystyle\sum_{(x_1, x_2) \in R_{12}} f(x_1, x_2) = 1$;   and

(iii) $P\big((X_1, X_2) \in A\big) = \displaystyle\sum_A f(x_1, x_2)$ for any subregion $A \subseteq R_{12}$.

Example 1.6.1: Consider a small town where the joint probability distribution of $X_1 = \#$ *children per household* and $X_2 = \#$ *pets per household* is given through the Table 1.6.1. If a household is selected at random, what is the probability that the total number of kids and pets in the household is less than or equal to 2?

	$f(x_1, x_2)$	X_2 0	1	2	3
	0	6/30	5/30	1/30	1/30
X_1	1	6/30	3/30	1/30	0
	2	4/30	1/30	0	0
	3	1/30	1/30	0	0

Table 1.6.1: Joint Probability Disribution of (X_1, X_2).

Note that in terms of X_1 and X_2, we are interested in finding the probability $P(0 \leq X_1 + X_2 \leq 2)$. The subregion $A = \{(x_1, x_2) \mid 0 \leq$

$x_1 + x_2 \leq 2\}$ includes the pairs of $(0,0), (0,1), (0,2), (1,0), (1,1)$, and $(2,0)$. Therefore,

$$
\begin{aligned}
P(0 \leq X_1 + X_2 \leq 2) &= f(0,0) + f(0,1) + f(0,2) \\
&\quad + f(1,0) + f(1,1) + f(2,0) \\
&= \frac{25}{30}
\end{aligned}
$$

When X_1 and X_2 are continuous random variables, the joint probability density function $f(x_1, x_2)$ is a surface lying above the x_1-x_2 plane, and $P((X_1, X_2) \in A)$, where A is any subregion in the x_1-x_2 plane, is equal to the volume of the right cylinder bounded by the base A and the surface $f(x_1, x_2)$.

Definition 1.6.2 *The function $f(x_1, x_2)$ is a joint probability distribution of the continuous random variables X_1 and X_2 with range R_{12} if*

(i) $f(x_1, x_2) > 0$ for all $(x_1, x_2) \in R_{12}$;

(ii) $\int_{R_{12}} f(x_1, x_2)\, dx_1 dx_2 = 1$;   and

(iii) $P((X_1, X_2) \in A) = \int_A f(x_1, x_2)\, dx_1 dx_2$ for any subregion $A \subseteq R_{12}$.

Example 1.6.2: Two continuous random variables , X_1 and X_2, have the joint density given by

$$
f(x_1, x_2) = \begin{cases} k(x_1^2 + x_2^2), & 0 < x_1 < 2,\ 1 < x_2 < 4 \\ 0, & \text{otherwise} \end{cases}
$$

Find $P(X_1 > X_2)$.

First, we need to find the value of k. Using (ii) of Definition 1.6.2, we have

$$
\int_0^2 \int_1^4 k(x_1^2 + x_2^2)\, dx_2 dx_1 = 1
$$

After a straight forward integration, we obtain that $k = 1/50$. Hence, using (iii) of Definition 1.6.2 (also see Figure 1.6.1),

$$
P(X_1 > X_2) = \int_0^2 \int_1^{x_1} \frac{1}{50}(x_1^2 + x_2^2)\, dx_2 dx_1
$$

A further simplification gives that $P(X_1 > X_2) = 7/150$.

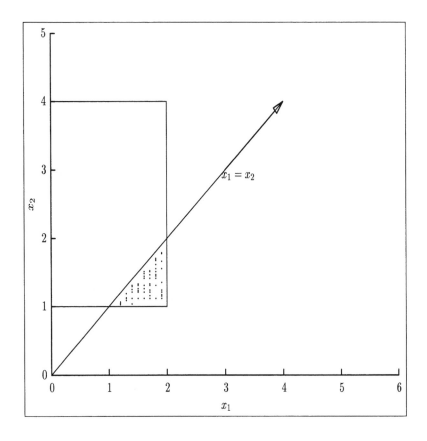

Figure 1.6.1: The Event $(X_1 > X_2)$ is the shaded area.

Given the joint probability distribution $f(x_1, x_2)$ of the discrete random variables X_1 and X_2, the probability distribution $f_1(x_1)$ of X_1 alone is obtained by summing $f(x_1, x_2)$ up over the values of X_2. Similarly, the probability distribution $f_2(x_2)$ of X_2 alone is obtained by summing $f(x_1, x_2)$ up over the values of X_1. The functions $f_1(x_1)$ and $f_2(x_2)$ are called the marginal distributions of X_1 and X_2, respectively. For continuous random variables, summations are replaced by integrals.

The marginal distributions $f_1(x_1)$ and $f_2(x_2)$ are indeed the probability distributions of the individual random variables X_1 and X_2, respectively.

Definition 1.6.3 *The marginal distributions of individual random variables X_1 and X_2 are given by*

$$f_1(x_1) = \sum_{x_2 \in R_{12}^{x_1}} f(x_1, x_2) \quad and \quad f_2(x_2) = \sum_{x_1 \in R_{12}^{x_2}} f(x_1, x_2)$$

for the discrete case, and by

$$f_1(x_1) = \int_{x_2 \in R_{12}^{x_1}} f(x_1, x_2)\, dx_2 \quad and \quad f_2(x_2) = \int_{x_1 \in R_{12}^{x_2}} f(x_1, x_2)\, dx_1$$

for the continuous case; where $R_{12}^{x_1}$ is the set of possible values that X_2 can assume when $X_1 = x_1$, and $R_{12}^{x_2}$ is the set of possible values that X_1 can assume when $X_2 = x_2$.

The concept of conditional probability can now be invoked in the context of joint probability distributions.

Define the two events A_1 and A_2 as $A_1 = \{X_1 = x_1\}$ and $A_2 = \{X_2 = x_2\}$ where X_1 and X_2 are two discrete random variables. Using Definition 1.4.4 we can obtain

$$
\begin{aligned}
P(X_2 = x_2 \mid X_1 = x_1) &= \frac{P(X_1 = x_1, X_2 = x_2)}{P(X_1 = x_1)} \\
&= \frac{f(x_1, x_2)}{f_1(x_1)} \\
&= f_{2|1}(x_2 \mid x_1) \quad \text{(say)}
\end{aligned}
$$

provided $f_1(x_1) > 0$.

It can be shown that $f_{2|1}(x_2 \mid x_1)$, which is strictly a function of x_2 when x_1 is fixed, is also a probability distribution, and this is true even if the random variables are continuous.

Definition 1.6.4 *Let X_1 and X_2 be two random variables (discrete or continuous). The conditional distribution of the random variable X_2, given that $X_1 = x_1$, is given by*

$$f_{2|1}(x_2 | x_1) = \frac{f(x_1, x_2)}{f_1(x_1)}, \quad provided\ f_1(x_1) > 0$$

Similarly, the conditional distribution of the random variable X_1, given that $X_2 = x_2$, is given by

$$f_{1|2}(x_1 | x_2) = \frac{f(x_1, x_2)}{f_2(x_2)}, \quad provided\ f_2(x_2) > 0$$

Example 1.6.3: Consider the Table 1.6.1 representing the joint probability distribution of X_1 and X_2, the number of children and the number of pets per household, respectively. The marginal distributions are

$$f_1(0) = \frac{13}{30}, \quad f_1(1) = \frac{10}{30}, \quad f_1(2) = \frac{5}{30}, \quad f_1(3) = \frac{2}{30}$$

and

$$f_2(0) = \frac{17}{30}, \quad f_2(1) = \frac{10}{30}, \quad f_2(2) = \frac{2}{30}, \quad f_2(3) = \frac{1}{30}$$

Given that a household has two children, what is the probability that the household has no pets? In terms of X_1 and X_2, it is seen that

$$P(X_2 = 0 | X_1 = 2) = \frac{f(2,0)}{f_1(2)} = \frac{4}{5}$$

As it can be seen in the above example,

$$f(2,0) = \frac{4}{30} \neq f_1(2) \cdot f_2(0) = \frac{5}{30} \times \frac{17}{30}$$

But there are cases where $f(x_1, x_2) = f_1(x_1) \cdot f_2(x_2)$ for all (x_1, x_2) in the range R_{12}.

Definition 1.6.5 *Given two random variables X_1 and X_2 (discrete or continuous) with joint probability distribution $f(x_1, x_2)$ and two marginal distributions $f_1(x_1)$ and $f_2(x_2)$, X_1 and X_2 are said to be independent if and only if $f(x_1, x_2) = f_1(x_1) \cdot f_2(x_2)$ for all (x_1, x_2) within the range R_{12} of (X_1, X_2).*

Example 1.6.4: Two continuous random variables have the joint density given by

$$f(x_1, x_2) = \begin{cases} 4x_1 x_2, & 0 < x_1 < 1, \ 0 < x_2 < 1 \\ 0, & \text{otherwise} \end{cases}$$

Are X_1 and X_2 independent?

The marginal densities of X_1 and X_2 are respectively

$$f_1(x_1) = \int_0^1 4x_1 x_2 \, dx_2 = 2x_1$$

$$f_2(x_2) = \int_0^1 4x_1 x_2 \, dx_1 = 2x_2$$

Hence, $f(x_1, x_2) = f_1(x_1) \cdot f_2(x_2)$ for all (x_1, x_2) such that $0 < x_1 < 1$ and $0 < x_2 < 1$. Therefore, X_1 and X_2 are independent.

It is also worth mentioning that two random variables are independent if and only if

$$f_{1|2}(x_1|x_2) = f_1(x_1) \quad \text{or} \quad f_{2|1}(x_2|x_1) = f_2(x_2)$$

for all (x_1, x_2) in the range R_{12}.

Definition 1.6.6 *Let X_1 and X_2 be random variables with joint probability distribution $f(x_1, x_2)$. The covariance of X_1 and X_2, denoted by σ_{12}, is defined as*

$$Cov(X_1, X_2) = \sigma_{12} = E\left[(X_1 - E(X_1))(X_2 - E(X_2))\right]$$

If we denote $E(X_1)$ and $E(X_2)$ by μ_1 and μ_2, respectively, then σ_{12} can be rewritten as

$$\sigma_{12} \quad = \quad \begin{cases} \displaystyle\sum_{(x_1,x_2)\in R_{12}} (x_1 - \mu_1)(x_2 - \mu_2)f(x_1, x_2) & \text{discrete case} \\[2mm] \displaystyle\int_{R_{12}} (x_1 - \mu_1)(x_2 - \mu_2)f(x_1, x_2)\,dx_1 dx_2 & \text{continuous case} \end{cases}$$

$$= \quad E(X_1 X_2) - \mu_1 \mu_2$$

It can be shown that for constants a_1, a_2, b_1, and b_2,

$$Cov(a_1 X_1 + b_1, a_2 X_2 + b_2) = a_1 a_2\, Cov(X_1, X_2)$$

Therefore, covariance of two random variables can be affected by the choice of scales for recording X_1 and X_2. Covariance can also be viewed as a measure of association between the random variables X_1 and X_2. But since the covariance depends on the units being used for X_1 and X_2, a unit free measure of the association between X_1 and X_2 is found through *correlation coefficient* .

Definition 1.6.7 *The correlation coefficient between two random variables X_1 and X_2 is defined as*

$$\rho_{12} = \frac{Cov(X_1, X_2)}{\sqrt{Var(X_1) \cdot Var(X_2)}}.$$

Properties of the correlation coefficient:

(a) Obviously, the correlation coefficient ρ_{12} is a pure number, that is, ρ_{12} is independent of units of measurement of both X_1 and X_2.

(b) Let $X_1^* = (X_1 - a_1)/b_1$ and $X_2^* = (X_2 - a_2)/b_2$ where a_1, b_1, a_2, and b_2 are four arbitrarily chosen constants. Let ρ_{12}^* be the correlation coefficient between X_1^* and X_2^*. It can be shown that $\rho_{12}^* = \rho_{12} =$ correlation coefficient between X_1 and X_2.

(c) For any two random variables, X_1 and X_2, it is true that $-1 \leq \rho_{12} \leq +1$. The correlation coefficient ρ_{12} takes the lowest value -1 when $X_2 = \mu_2 - \frac{\sigma_2}{\sigma_1}(X_1 - \mu_1)$, and the highest value $+1$ when $X_2 = \mu_2 + \frac{\sigma_2}{\sigma_1}(X_1 - \mu_1)$. Here, μ_i and σ_i are mean and standard deviation of X_i, respectively, $i = 1, 2$.

(d) Positive correlation (i.e., $0 < \rho_{12} \leq 1$) implies a positive association between X_1 and X_2. In other words, large values of X_1 tend to be associated with large values of X_2. On the other hand, negative correlation (i.e., $-1 \leq \rho_{12} < 0$) implies a negative association between X_1 and X_2, i.e., large values of X_1 tend to be associated with small values of X_2 and vice-versa. If X_1 and X_2 are independent, then $\rho_{12} = 0$, but the converse is not always true.

All the preceding concepts concerning two random variables can be generalized to the case of n random variables.

Let $f(x_1, x_2, \dots, x_n)$ be the joint probability distribution (*pmf* or *pdf*) of the random variables $X_1, X_2, \dots, X_n$. The marginal distribution of X_i is given by

$$f_i(x_i) = \sum_{x_1, \dots, x_{i-1}, x_{i+1}, \dots, x_n} f(x_1, x_2, \dots, x_n)$$

for the discrete case, and by

$$f_i(x_i) = \int \dots \int f(x_1, x_2, \dots, x_n)\, dx_1 \dots dx_{i-1} dx_{i+1} \dots dx_n$$

for the continuous case where the summation or integration is taken over all possible values of $(x_1, x_2, \dots, x_{i-1}, x_{i+1}, \dots, x_n)$ where the ith component is fixed at x_i. Similarly, we can obtain the joint marginal distribution of $(X_1, X_2, \dots, X_k)$, $1 \leq k < n$, as

$$f_{1,2,\dots,k}(x_1, x_2, \dots, x_k)$$

$$= \begin{cases} \displaystyle\sum_{x_{k+1}, \dots, x_n} f(x_1, x_2, \dots, x_n), & \text{discrete case} \\[2mm] \displaystyle\int \dots \int f(x_1, x_2, \dots, x_n)\, dx_{k+1} \dots dx_n, & \text{continuous case} \end{cases}$$

As a result, the joint conditional distribution of $(X_{k+1}, \ldots, X_n)$ given $(X_1 = x_1, \ldots, X_k = x_k)$ is

$$f_{k+1, \ldots, n|1, \ldots, k}(x_{k+1}, \ldots, x_n \mid x_1, \ldots, x_k) = \frac{f(x_1, x_2, \ldots, x_n)}{f_{1,2, \ldots, k}(x_1, x_2, \ldots, x_k)}$$

(1.6.1)

provided $f_{1,2, \ldots, k}(x_1, x_2, \ldots, x_k) > 0$. The random variables, $X_1, X_2, \ldots X_n$ with joint distribution $f(x_1, x_2, \ldots, x_n)$ and marginal distributions $f_1(x_1), f_2(x_2), \ldots, f_n(x_n)$, respectively, are said to be mutually independent if and only if

$$f(x_{i_1}, x_{i_2}, \ldots, x_{i_k}) = f_1(x_{i_1}) f_2(x_{i_2}) \ldots f_k(x_{i_k})$$

for all $1 < k \le n$ and $(x_{i_1}, x_{i_2}, \ldots, x_{i_k})$ within the underlying range.

Note that with the above setup of n random variables $X_1, X_2, \ldots, X_n$ with joint probability distribution $f(x_1, x_2, \ldots, x_n)$, the mean of X_i is

$$\mu_i = \mu_{X_i} = E(X_i) = \begin{cases} \sum x_i f_i(x_i), & \text{discrete case} \\ \int x_i f_i(x_i)\, dx_i, & \text{continuous case} \end{cases}$$

where the summation or integration takes over the appropriate range (unconditional) of the random variable X_i. The vector $\boldsymbol{\mu} = (\mu_1, \mu_2, \ldots, \mu_n)'$ is called the mean vector of the random vector $\boldsymbol{X} = (X_1, X_2, \ldots, X_n)'$. The matrix $\Sigma_{n \times n}$ defined as

$$\Sigma = E\left[(\boldsymbol{X} - \boldsymbol{\mu})(\boldsymbol{X} - \boldsymbol{\mu})'\right]$$

$$= \begin{pmatrix} Var(X_1) & Cov(X_1, X_2) & \cdots & Cov(X_1, X_n) \\ Cov(X_2, X_1) & Var(X_2) & \cdots & Cov(X_2, X_n) \\ \vdots & \vdots & \ddots & \vdots \\ Cov(X_n, X_1) & Cov(X_n, X_2) & \cdots & Var(X_n) \end{pmatrix}$$

is called the dispersion matrix of the random vector $\boldsymbol{X}$. Using the notations $\sigma_{ij} = Cov(X_i, X_j)$, $i \ne j$, and $\sigma_{ii} = Var(X_i) = \sigma_i^2$ (say), $1 \le i \le n$, the matrix Σ can be denoted as $\Sigma = \left(\!\left(\sigma_{ij}\right)\!\right)$. It is easy to decompose Σ as[6]

$$\Sigma = diag(\sigma_1, \ldots, \sigma_n) \cdot \boldsymbol{\rho} \cdot diag(\sigma_1, \ldots, \sigma_n)$$

---

[6] $diag(\sigma_1, \ldots, \sigma_n)$ is a $n \times n$ diagonal matrix with diagonal elements of $\sigma_1, \ldots, \sigma_n$.

where the matrix ρ is given as

$$\rho = \begin{pmatrix} 1 & \rho_{12} & \cdots & \rho_{1n} \\ \rho_{21} & 1 & \cdots & \rho_{2n} \\ \vdots & \vdots & \ddots & \vdots \\ \rho_{n1} & \rho_{n2} & \cdots & 1 \end{pmatrix}$$

and is called the correlation matrix. For $i \neq j$, ρ_{ij} is the correlation coefficient between X_i and X_j.

1.7   Moment Generating Function

Earlier, we discussed the kth moment (raw) $\mu'_{(k)}$ of a random variable X given as $\mu'_{(k)} = E(X^k)$, provided it exists. A unified approach to find all moments (if they exist) of X can be obtained through the moment generating function (*mgf*) of X.

The moment generating function of X, denoted by $\phi_X(t)$, is defined as

$$\phi_X(t) = E(e^{tX}) \tag{1.7.1}$$

provided the above expectation exists for all $t \in (-\varepsilon, \varepsilon)$, $\varepsilon > 0$, a small neighborhood of 0.

The *mgf* $\phi_X(t)$ of X acts as a tool to derive all moments of X. It can be shown that

$$\phi_X^{(k)}(0) = \phi_X^{(k)}(t)\Big|_{t=0} = \left(\frac{d^k}{dt^k}\phi_X(t)\right)\Big|_{t=0} = \mu'_{(k)} \tag{1.7.2}$$

As a consequence, we have

$$\mu_X = E(X) = \mu'_{(1)} = \frac{d}{dt}\phi_X(t)\Big|_{t=0} = \phi_X^{(1)}(t)\Big|_{t=0}$$

$$E(X^2) = \mu'_{(2)} = \frac{d^2}{dt^2}\phi_X(t)\Big|_{t=0} = \phi_X^{(2)}(t)\Big|_{t=0}, \; \ldots, \; \text{etc.}$$

It should be pointed out that the existence of the *mgf* of a random variable X is a stronger condition than the existence of its moments. If the *mgf* $\phi_X(t)$ exists, then all moments $\mu'_{(k)}$ of X exist and can be obtained by (1.7.2). However, all or a few moments of X may exist, yet $\phi_X(t)$ may not exist. A few such examples will be discussed later in Chapter 2.

Apart from its ability to generate moments, a major usefulness of the moment generating function is that it characterizes a probability distribution uniquely. Note that if the *mgf* exists, it characterizes an infinite collection of moments. Therefore, a natural question is whether characterizing the infinite set of moments uniquely determines a probability distribution. The answer to this question is 'no'. Characterizing the set of moments is not sufficient to determine a probability distribution uniquely since there can be two distinct probability distributions having the same set of moments. This can be illustrated through the following example.

Example 1.7.1: Consider two continuous random variables, X_1 and X_2, having pdfs $f_1(x_1)$ and $f_2(x_2)$, respectively, given by

$$f_1(x_1) \quad = \quad \frac{1}{\sqrt{2\pi x_1}} \exp\left(-(lnx_1)^2/2\right), \quad 0 < x_1 < \infty$$

$$\text{and} \quad f_2(x_2) \quad = \quad \left(1 + \sin(2\pi lnx_2)\right)f_1(x_2), \quad 0 < x_2 < \infty$$

It can be shown that for any positive integer k, $E(X_1^k) = E(X_2^k) = \exp(k^2/2)$. Thus, even though X_1 and X_2 have distinct *pdfs*, they have the same moments.

If the *mgf* exists then it actually says more than the sequence of moments it can generate.

The above problem of nonuniqueness of the distributions does not arise if the probability distributions have bounded support. In that case, the infinite sequence of moments does uniquely determine the probability distribution. Also, if the *mgf* $\phi_X(t)$ of a random variable X exists in a neighborhood of 0, then the probability distribution of X is uniquely determined regardless of the support. The next theorem summarizes the above discussion.

Theorem 1.7.1 *Let* X_1 *and* X_2 *be two random variables with probability distributions,* $f_1(x_1)$ *and* $f_2(x_2)$, *respectively. Let the corresponding mgfs be* $\phi_1(t)$ *and* $\phi_2(t)$, *respectively (assuming that they both exist).*

(i) If f_1 *and* f_2 *have bounded support, then* $f_1(x) = f_2(x)$ *for all* x *if, and only if,* $E(X_1^k) = E(X_2^k)$ *for all integers* $k = 0, 1, 2, \ldots$.

(ii) If $\phi_1(t) = \phi_2(t)$ *for all* t *in some neighborhood of 0, then* $f_1(x) = f_2(x)$ *for all* x.

The *mgf* of a random vector $\boldsymbol{X} = (X_1, X_2, \ldots, X_n)'$ is defined as

$$\phi_{\boldsymbol{X}}(\boldsymbol{t}) = E\left(e^{\boldsymbol{t}'\boldsymbol{X}}\right) = E\left(\exp\left(\sum_{i=1}^{n} t_i X_i\right)\right) \tag{1.7.3}$$

provided the above expectation exists for all $\boldsymbol{t} = (t_1, t_2, \ldots, t_n)'$ in a neighborhood of $\boldsymbol{0} = (0, \ldots, 0)'$. In the special case, when the random variables $X_1, X_2, \ldots, X_n$ are mutually independent,

$$\phi_{\boldsymbol{X}}(\boldsymbol{t}) = \prod_{i=1}^{n} E\left(e^{t_i X_i}\right) = \prod_{i=1}^{n} \phi_i(t_i)$$

where $\phi_i(t_i)$ is the *mgf* of X_i, $1 \le i \le n$.

Some Moment Inequalities

The following moment inequalities are widely used in mathematical statistics, and they can help to explain the characteristics of many statistical experiments.

(a) **Cauchy-Schwarz Inequality**: For any two random variables X and Y,

$$E|XY| \le \sqrt{E(X^2)E(Y^2)} \tag{1.7.4}$$

(b) **Jensen's Inequality**: For any random variable X and a convex function $h(x)$, [7]

$$E[h(X)] \ge h(E(X)) \tag{1.7.5}$$

(c) **Holder's Inequality**: For any two random variables X and Y, and two positive real values a and b such that $a + b = 1$,

$$E|XY| \le \left(E(|X|^{1/a})\right)^a \left(E(|Y|^{1/b})\right)^b \tag{1.7.6}$$

It is seen that (1.7.4) is a special case of (1.7.6) with $a = b = 1/2$.

(d) **Liapounov's Inequality**: If one takes $Y \equiv 1$ in (1.7.6), then the Holder's Inequality reduces to:

$$E|X| \le \left(E(|X|^{1/a})\right)^a, \quad \text{for } 0 < a \le 1$$

---

[7]A function $h(x)$ is called convex if $h\left(\lambda x_1 + (1 - \lambda)x_2\right) \le \lambda h(x_1) + (1 - \lambda)h(x_2)$ for all x_1 and x_2, and $\lambda \in (0, 1)$. Also, if $h''(x) \ge 0$, then $h(x)$ is convex. On the other hand, $h(x)$ is said to be concave if $\left(-h(x)\right)$ is convex, or $h''(x) \le 0$.

(e) **Minkowski's Inequality**: For any two random variables X and Y, and any real value a, $0 < a \le 1$,

$$\left(E\left(|X+Y|^{1/a}\right)\right)^a \le \left(E\left(|X|^{1/a}\right)\right)^a + \left(E\left(|Y|^{1/a}\right)\right)^a \qquad (1.7.7)$$

1.8 Order Statistics

If $X_1, X_2, \ldots, X_n$ are random variables, and $X_{(1)} \le X_{(2)} \le \ldots \le X_{(n)}$ are the same variables arranged in ascending order of magnitude such as

$$
\begin{aligned}
X_{(1)} &= \text{the smallest observation of } \{X_1, X_2, \ldots, X_n\} \\
X_{(2)} &= \text{the second smallest observation of } \{X_1, X_2, \ldots, X_n\} \\
&\vdots \\
X_{(n-1)} &= \text{the second largest observation of } \{X_1, X_2, \ldots, X_n\} \\
X_{(n)} &= \text{the largest observation of } \{X_1, X_2, \ldots, X_n\}
\end{aligned}
$$

then $X_{(1)} \le X_{(2)} \le \ldots \le X_{(n)}$ are called the order statistics of the random variables $X_1, X_2, \ldots, X_n$.

Ordered values of the random observations have many real life applications. For instance, to study the strength of steel cable used in heavy-duty cranes, a random sample of ten specimens has been obtained from a manufacturer. Each specimen is then tested in the lab where the stress level is measured at which a cable breaks (and this is called the 'breaking point' of a cable). For the sample of ten cable specimens, we thus obtain $X_1, X_2, \ldots, X_{10}$ where X_i represents the ith breaking point. For field applications of this particular type of cable, one may be interested in the probability distribution of either $X_{(1)}$ or $X_{(10)}$, or $\left(X_{(5)} + X_{(6)}\right)/2$, the sample median of these ten 'breaking point' values.

If the random variables $X_1, X_2, \ldots, X_n$ are *iid* with *cdf* $F(x)$ and *pdf* $f(x)$, then the *pdf* of $X_{(1)}$ is

$$f_{(1)}(x) = nf(x)(1 - F(x))^{n-1} \qquad (1.8.1)$$

and that of $X_{(n)}$ is

$$f_{(n)}(x) = nf(x)(F(x))^{n-1} \qquad (1.8.2)$$

More generally, the *pdf* of $X_{(j)}$, $1 \leq j \leq n$, is

$$f_{(j)}(x) = \frac{n!}{(j-1)!(n-j)!} f(x)(F(x))^{j-1}(1 - F(x))^{n-j} \qquad (1.8.3)$$

Joint distribution of $(X_{(i_1)}, \ldots, X_{(i_j)})$, where $1 \leq i_1 \leq i_2 \leq \cdots \leq i_j \leq n$, can be found as

$$f_{(i_1), \ldots, (i_j)}(x_{i_1}, \ldots, x_{i_j})$$

$$= \frac{n!}{\displaystyle\prod_{k=1}^{j+1}(i_k - i_{k-1})!} \left[\prod_{k=1}^{j+1} \left\{ F(x_{i_k}) - F(x_{i_{k-1}}) \right\}^{i_k - i_{k-1} - 1} \right] \times \prod_{k=1}^{j} f(x_{i_k})$$

$$(1.8.4)$$

with $i_0 = 0, i_{j+1} = n, F(x_{i_0}) = 0, F(x_{i_{j+1}}) = 1$ and $x_{i_1} \leq \cdots \leq x_{i_j}$.

As a special case of the above formula (1.8.4), one can find the joint *pdf* of $X_{(i_1)}$ and $X_{(i_2)}$, $1 \leq i_1 < i_2 \leq n$, as

$$f_{(i_1),(i_2)}(x_{i_1}, x_{i_2}) = \frac{n!}{(i_1 - 1)!(i_2 - i_1 - 1)!(n - i_2)!} f(x_{i_1})f(x_{i_2})$$

$$\times \left(F(x_{i_1}) \right)^{i_1 - 1} \left[F(x_{i_2}) - F(x_{i_1}) \right]^{i_2 - i_1 - 1}$$

$$\times \left[1 - F(x_{i_2}) \right]^{n - i_2} \qquad (1.8.5)$$

for $-\infty < x_{i_1} < x_{i_2} < \infty$. Further, by taking $i_1 = 1$ and $i_2 = n$ we get the joint *pdf* of $X_{(1)}$ and $X_{(n)}$ (the smallest and the largest of the observations, respectively) as

$$f_{(1),(n)}(x_1, x_n) = n(n-1)f(x_1)f(x_n)\left[F(x_n) - F(x_1) \right]^{n-2} \qquad (1.8.6)$$

The joint *pdf* (1.8.6) can be used to find the probability distribution of, say $R = X_{(n)} - X_{(1)}$, the **sample range** (a measure of dispersion), or $V = (X_{(1)} + X_{(n)})/2$, the **mid-range** (a measure of location).

Order statistics play a major role in experiments where one can obtain *censored* observations. If an experiment is stopped after the first k ordered observations are obtained out of n test units, then it is called Type-II censoring on the right. Thus one obtains only $X_{(1)} \leq \cdots \leq X_{(k)}$. Similarly, Type-II censoring on the left would happen if the smallest k observations are ignored and one ends up with $X_{(k+1)} \leq \cdots \leq X_{(n)}$.

Analogously, if an experiment is discontinued after a fixed length of time t_* then this is called Type-I censoring on the right. For this kind of censoring the length of the experiment is fixed, but the number of observations (say, N) before time t_* is a discrete random variable, and one observes $X_{(1)} \leq \ldots \leq X_{(N)}$ where $N = \#$ observations before the fixed time t_*.

Suppose we are interested in the lifespan of a particular brand of electric bulbs and X_i denote the lifespan of the ith randomly selected bulb ($1 \leq i \leq n$). If we decide to stop the experiment as soon as the kth ordered observation is obtained (i.e., Type-II right censoring is employed), then the expected length of the experiment is $E(X_{(k)})$. On the other hand, if Type-I censoring with fixed time t_* is used, then a bulb fails before time t_* with probability $P(X_i \leq t_*) = F(t_*)$ (F being the common cdf of X_i's). Further, N follows a binomial probability distribution (discussed in Chapter 2) $P(N = k) = \binom{n}{k}(F(t_*))^k(1 - F(t_*))^{n-k}$. This can be used to find the joint distribution of $(X_{(1)}, \ldots, X_{(N)}, N)$ as

$$
\begin{aligned}
g(x_{(1)}, \ldots, x_{(N)}, N = k) &= g(x_{(1)}, \ldots, x_{(N)}|N = k)P(N = k) \\
&= k! \prod_{i=1}^{k} \{f(x_{(i)})/F(t_*)\} \binom{n}{k} (F(t_*))^k \\
&\qquad \times (1 - F(t_*))^{n-k} \\
&= \frac{n!}{(n-k)!}(1 - F(t_*))^{n-k} \prod_{i=1}^{k} f(x_{(i)})
\end{aligned}
$$

where $x_{(1)} \leq \ldots \leq x_{(k)} < t_*, k = 1, 2, \ldots, n$ and $f(\cdot)$ is the pdf of X_i's. The quantity $E(N)$ gives an idea about the number of failures one expects to see within the fixed time t_*.

1.9   Characteristic Function

In Section 1.7 we have discussed the mgf of a $r.v.$ X which may or may not exist. But the most important function which characterizes the probability distribution of a random variable is the characteristic function (cf) which is defined at a real value t as

$$\Psi_X(t) = E(e^{itX}) \tag{1.9.1}$$

where $i = \sqrt{-1}$, the complex number. The above expectation exists always and requires complex (or contour) integration. The cf uniquely determines the probability distribution associated with the $r.v.$ X (i.e., given a cf expression, it is possible to extract the cdf of X). Also, if the moments of a $r.v.$ X exist, then they can be generated from $\Psi_X(t)$ just the way it can be done for the mgf, assuming that the mgf exists, (i.e., by taking derivative(s) of $\Psi_X(t)$ w.r.t. t and then setting $t = 0$).

Chapter 2

Some Common Probability Distributions

In this chapter, we discuss some commonly used probability distributions. We will see that, while discussing a particular probability distribution, we essentially deal with a family of probability distributions indexed by one or more parameters. This family of distributions gives us some liberty to vary the parameter value(s) while staying with a particular functional form. As a result, when it comes to model the outcomes of an experiment (i.e., approximating the actual probability distribution of a random variable), we can choose a suitable parameter value that can give the resultant, commonly used probability distribution a good fit to the actual one. Hence, to emphasize the role of the parameter(s) and keep track of the parameter(s), a typical *pdf* or *pmf* will be denoted by, for example, $f(x|\theta)$ instead of $f(x)$, where θ is a relevant parameter involved in the probability distribution. If there is a single parameter involved, then θ is taken as scalar valued, otherwise θ is taken as vector valued. The collection of all possible values of θ, denoted by Θ, which makes $f(x|\theta)$ a *pdf* or *pmf*, is called the (natural) parameter space of the probability distribution $f(x|\theta)$. For the sake of simplicity, we take θ as scalar valued only, but if a situation demands, then θ can be vector valued too.

Notation: If a random variable has probability distribution (*pmf* or *pdf*) $f(x|\theta)$, then it is written as '$X \sim f(x|\theta)$'.

2.1   Discrete Distributions

Discrete Uniform Distribution

A random variable X has a Discrete Uniform $(1, \theta)$ distribution (or $DU(1, \theta)$ distribution) if

$$f(x|\theta) = P(X = x|\theta) = \frac{1}{\theta}, \quad x = 1, 2, \ldots, \theta \qquad (2.1.1)$$

where θ is a specified integer. Note that the parameter space $\Theta = \{1, 2, 3, \ldots\}$, the set of all positive integers.

Using the identities

$$\sum_{i=1}^{k} i = \frac{k(k+1)}{2} \quad \text{and} \quad \sum_{i=1}^{k} i^2 = \frac{k(k+1)(2k+1)}{6}$$

we can find the mean and variance of X as

$$\mu_X = E(X) = \sum_{x=1}^{\theta} x \frac{1}{\theta} = \frac{(\theta+1)}{2}$$

and

$$E(X^2) = \sum_{x=1}^{\theta} x^2 \frac{1}{\theta} = \frac{(\theta+1)(2\theta+1)}{6} \qquad (2.1.2)$$

which gives

$$\sigma_X^2 = Var(X) = \frac{(\theta+1)(\theta-1)}{12} \qquad (2.1.3)$$

The *mgf* of $DU(1, \theta)$ is

$$\phi_X(t) = \left(\sum_{x=1}^{\theta} e^{xt} \right) \Big/ \theta \qquad (2.1.4)$$

The above distribution can be generalized to Discrete Uniform (θ_1, θ_2) distribution (or $DU(\theta_1, \theta_2)$ distribution) with *pmf*

$$f(x|\theta_1, \theta_2) = \frac{1}{(\theta_2 - \theta_1 + 1)}, \quad x = \theta_1, \theta_1 + 1, \ldots, \theta_2$$

and

$$\Theta = \{(\theta_1, \theta_2)|\theta_1 \leq \theta_2, \text{ and both } \theta_1 \text{ and } \theta_2 \text{ are integers}\}$$

Moments of $DU(\theta_1, \theta_2)$ can be derived easily and hence omitted.

Binomial Distribution

The binomial distribution is based on the idea of a Bernoulli trial.

A Bernoulli trial is an experiment with only two possible outcomes, say S (for 'success') and F (for 'failure'). Let $\theta = P(S) =$ the probability of 'success' in a Bernoulli trial. Define a random variable X as $X = \#$ *successes in a Bernoulli trial.* The range of X is $R_X = \{0, 1\}$ and

$$X = \begin{cases} 1 & \text{with probability } \theta \\ 0 & \text{with probability } (1 - \theta) \end{cases} \qquad (2.1.5)$$

where $0 \le \theta \le 1$. The probability distribution of X given by the *pmf*

$$f(x \mid \theta) = \theta^x (1 - \theta)^{1-x}, \quad x = 0, 1 \qquad (2.1.6)$$

where $\theta \in \Theta = [0, 1]$, is said to be a Bernoulli(θ) distribution.

Now consider a larger experiment which consists of a fixed number of independent and identical Bernoulli trials with θ being probability of success in a single trial. Let $n =$ *the total number (fixed) of Bernoulli trials.* Define the random variable X as $X = \#$ *successes in those n trials.* The range of X is then $R_X = \{0, 1, 2, \ldots, n\}$, and the probability distribution of X is

$$f(x \mid \theta) = P(X = x \mid \theta) = \binom{n}{x} \theta^x (1 - \theta)^{n-x}, \quad x = 0, 1, \ldots, n \quad (2.1.7)$$

where $\theta \in \Theta = [0, 1]$. A random variable having the *pmf* (2.1.7) is said to have a Binomial(n, θ) or B(n, θ) distribution. Thus, a Bernoulli(θ) distribution is a B($1, \theta$) distribution.

Note that $\theta = 0$ implies $P(X = 0) = 1$ and $\theta = 1$ implies $P(X = 1) = 1$, and both happen rarely or only in extreme cases. Therefore, we ignore those cases by restricting $\Theta = (0, 1)$ (open interval).

It can be shown that for a B(n, θ) distribution,

$$E(X) = n\theta, \quad Var(X) = n\theta(1 - \theta) \qquad (2.1.8)$$

and the *mgf* of X is

$$\phi_X(t) = \left(\theta e^t + (1 - \theta)\right)^n \qquad (2.1.9)$$

A B(n, θ) random variable X (i.e., the random variable X having a B(n, θ) distribution) can be characterized as a sum of n independent and identically distributed (*iid*) random variables each having a Bernoulli(θ) distribution. In other words, let $X_1, X_2, \ldots, X_n$ be *iid* Bernoulli(θ) (or $B(1, \theta)$) random variables, then $X = X_1 + X_2 + \ldots + X_n$ has a $B(n, \theta)$ distribution.

Geometric Distribution

Suppose we have a sequence of independent and identical Bernoulli(θ) trials. We define a random variable X which counts the number of trails to get the very first success. The range of X is then $R_X = \{1, 2, 3, \ldots\}$. The probability distribution of X given by the *pmf*

$$f(x|\theta) = P(X = x|\theta) = \theta(1 - \theta)^{x-1}, \quad x = 1, 2, 3, \ldots \quad (2.1.10)$$

where $\theta \in \Theta = (0, 1)$, is called a Geometric (G(θ)) distribution with parameter θ. A generalization of G(θ) distribution is presented below.

Negative Binomial Distribution

Suppose we count the number of Bernoulli trials required to obtain a fixed number of successes (say, r successes). Let the random variable X denote the trial number at which the rth success occurs (r is a fixed positive integer). The range of X is $R_X = \{r, r + 1, \ldots\}$ with *pmf*

$$f(x|r, \theta) = P(X = x|r, \theta) = \binom{x-1}{r-1}\theta^r(1 - \theta)^{x-r} \quad (2.1.11)$$

where $\theta \in \Theta = (0, 1)$, and X is said to have a Negative Binomial(r, θ) (or NB(r, θ)) distribution. For NB(r, θ) distribution, we have

$$E(X) = \frac{r(1 - \theta)}{\theta}, \quad Var(X) = \frac{r(1 - \theta)}{\theta^2} \quad (2.1.12)$$

$$\phi_X(t) = \left(\frac{\theta}{1 - (1 - \theta)e^t}\right)^r, \quad t < -ln(1 - \theta) \quad (2.1.13)$$

It can be seen that NB($1, \theta$)= G(θ). Hence, one can find the $E(X)$, $Var(X)$, or the *mgf* of G(θ) distribution from (2.1.13) by taking $r = 1$.

Poisson Distribution

A random variable X with range $R_X = \{0, 1, 2, \ldots\}$ is said to have a Poisson distribution with parameter λ (Poi(λ)) if the *pmf* of X is

$$f(x|\lambda) = P(X = x|\lambda) = \frac{e^{-\lambda}\lambda^x}{x!}, \quad x = 0, 1, 2, \ldots \quad (2.1.14)$$

where $\lambda \in \Lambda = (0, \infty)$. The parameter λ of Poi(λ) distribution is often called the intensity parameter. Mean, variance, and the *mgf* of Poi(λ) are given by

$$E(X) = Var(X) = \lambda \quad \text{and} \quad \phi_X(t) = e^{\lambda(e^t - 1)} \qquad (2.1.15)$$

Poisson distribution may be looked upon as a limiting form of a binomial distribution. Suppose that in a binomial distribution, the number of trials (n) tends to infinity and the probability of success (θ) in a single trial tends to 0 in such a way that

$$n\theta = (\# \text{ trials}) \times (\text{probability of success}) \rightarrow \lambda \quad (\text{finite})$$

then

$$\lim \binom{n}{x} \theta^x (1 - \theta)^{n-x} = \frac{e^{-\lambda}\lambda^x}{x!}, \quad x = 0, 1, 2, \ldots$$

Hypergeometric Distribution

The probability distribution of the hypergeometric random variable X, the number of successes in a random sample of size n selected from N items of which K are labeled 'success' and $N - K$ labeled 'failure', is

$$f(x \,|\, N, K, n) = \binom{K}{x}\binom{N - K}{n - x} / \binom{N}{n}, \quad x = 0, 1, 2, \ldots, n \quad (2.1.16)$$

Note that there is, implicit in (2.1.16), an additional assumption on the range of X. Binomial coefficients of the form $\binom{a}{b}$ are meaningful only if $0 \le b \le a$, and so the range of X is additionally restricted by the pair of inequalities

$$0 \le x \le K \quad \text{and} \quad 0 \le n - x \le N - K$$

which can be combined as

$$n - (N - K) \le x \le K \qquad (2.1.17)$$

In many cases, n is a small value compared to N and K, and hence, the range $0 \le x \le n$ will be contained in the above range (2.1.17). We, therefore, state the formal range of the hypergeometric random variable X as $R_X = \{0, 1, 2, \ldots, n \,|\, n \le N - K \text{ and } n \le K\}$.

Deriving moments of a hypergeometric distribution is a bit tedious. However, its mean and variance are given as

$$E(X) = \frac{nK}{N} \quad \text{and} \quad Var(X) = \frac{nK(N - n)(N - k)}{N^2(N - 1)} \qquad (2.1.18)$$

Unfortunately, no closed form expression of the moment generating function of a hypergeometric distribution is available.

2.2   Continuous Distributions

Continuous Uniform Distribution

The continuous uniform distribution over the range $[\theta_1, \theta_2]$ is defined by spreading mass (or weight) uniformly over the interval $[\theta_1, \theta_2]$ and denoted by $CU[\theta_1, \theta_2]$. A random variable X with the *pdf* given by

$$f(x|\theta_1, \theta_2) = \begin{cases} \dfrac{1}{(\theta_2 - \theta_1)} & \text{if } x \in [\theta_1, \theta_2] \\ 0 & \text{otherwise} \end{cases} \tag{2.2.1}$$

is said to have $CU[\theta_1, \theta_2]$ distribution. It is easy to verify that

$$E(X) = \frac{(\theta_1 + \theta_2)}{2}, \quad Var(X) = \frac{(\theta_2 - \theta_1)^2}{12}$$

$$\text{and} \quad \phi_X(t) = \frac{(e^{\theta_2 t} - e^{\theta_1 t})}{t(\theta_2 - \theta_1)} \tag{2.2.2}$$

Normal Distribution

One of the most important continuous probability distributions in the entire field of statistics is the normal distribution. The graph of a normal *pdf*, called the **normal curve**, is a bell shaped symmetric curve, and it describes (probability-wise) many real life data sets. The normal distribution is often referred to as the Gaussian distribution in honor of Karl Friedrich Gauss who derived the equation of a normal *pdf*.

The normal distribution has two parameters, usually denoted by μ and σ^2, which are its mean and variance. A random variable X is said to have a normal distribution with mean μ and variance σ^2 which is denoted by $N(\mu, \sigma^2)$, if its *pdf* is given by

$$f(x|\mu, \sigma^2) = \frac{1}{\sqrt{2\pi}\sigma} e^{-(x-\mu)^2/(2\sigma^2)}, \quad -\infty < x < \infty \tag{2.2.3}$$

If X has $N(\mu, \sigma^2)$ distribution, then the random variable $Z = (X - \mu)/\sigma$ has a $N(0, 1)$ distribution, known as the **standard normal distribution**. The relationship between the standard normal *cdf* and any other $N(\mu, \sigma^2)$ *cdf* can be established as

$$
\begin{aligned}
P(X \le x) &= P\big((X - \mu)/\sigma \le (x - \mu)/\sigma\big) \\
&= P\big(Z \le z\big),\; z = (x - \mu)/\sigma \\
&= \frac{1}{\sqrt{2\pi}} \int_{-\infty}^{z} e^{-u^2/2}\, du
\end{aligned}
\tag{2.2.4}
$$

Therefore, all normal probabilities can be computed in terms of the standard normal distribution. Throughout this book we will denote the *pdf* and *cdf* of the standard normal distribution by $\varphi(\cdot)$ and $\Phi(\cdot)$ respectively, i.e.,

$$
\varphi(z) = \frac{1}{\sqrt{2\pi}} e^{-z^2/2} \quad \text{and} \quad \Phi(z) = \int_{-\infty}^{z} \varphi(u)\, du
\tag{2.2.5}
$$

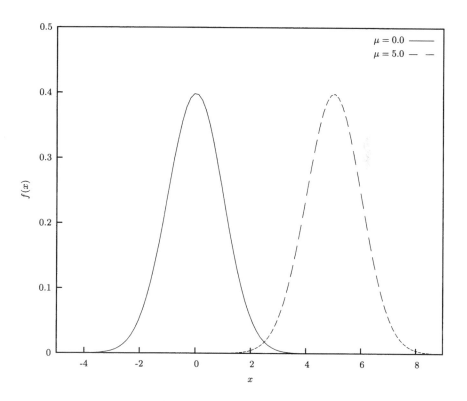

Figure 2.2.1: The *pdf*s of $N(\mu, 1)$ Distribution, $\mu = 0, 5$.

The following simple properties of any $N(\mu, \sigma^2)$ distribution are observed:

(a) The two parameters μ and σ^2 which uniquely determine a $N(\mu, \sigma^2)$ distribution are indeed mean and variance of the distribution, i.e., $E(X) = \mu$ and $Var(X) = \sigma^2$.

(b) The mode, which is the point on the horizontal axis where the normal curve (*pdf*) is a maximum, occurs at $x = \mu$.

(c) The normal curve is symmetric about a vertical axis through the mean μ.

(d) The normal curve has its points of inflexion at $x = \mu \pm \sigma$. In other words, the normal curve is concave if $\mu - \sigma < x < \mu + \sigma$, and is convex otherwise.

(e) The normal curve approaches the horizontal axis asymptotically as we proceed in either direction away from the mean μ.

(f) The *mgf* of $N(\mu, \sigma^2)$ distribution is $\phi_X(t) = e^{\mu t + \sigma^2 t^2/2}$ for $-\infty < t < \infty$.

(g) Probability coverage under the standard normal curve is given in Table A.1 (Appendix-A). As a result, one can easily find the probability content within 1, 2, or 3 standard deviations of mean for $N(\mu, \sigma^2)$ as

$$
\begin{aligned}
P\big(|X - \mu| \leq \sigma\big) &= 0.6826 \\
P\big(|X - \mu| \leq 2\sigma\big) &= 0.9544 \\
P\big(|X - \mu| \leq 3\sigma\big) &= 0.9974
\end{aligned}
\tag{2.2.6}
$$

Gamma Distribution

The gamma family of distributions is a flexible family of distributions on $[0, \infty)$. If α is a positive real value, then the integral

$$
\Gamma(\alpha) = \int_0^\infty t^{\alpha-1} e^{-t}\, dt
\tag{2.2.7}
$$

is called the gamma function evaluated at α. The value of $\Gamma(\alpha)$ is finite for $0 < \alpha < \infty$. It can be shown through integration by parts that

$$
\Gamma(\alpha + 1) = \alpha \Gamma(\alpha), \quad \alpha > 0
\tag{2.2.8}
$$

and for any positive integer k,

$$\Gamma(k+1) = k! \qquad (2.2.9)$$

A random variable X is said to have a (two parameter) gamma distribution with parameters α and β, denoted by $G(\alpha, \beta)$, if the *pdf* of X is given by

$$f(x \,|\, \alpha, \beta) = \frac{1}{\Gamma(\alpha)\beta^\alpha} x^{\alpha-1} e^{-x/\beta}, \;\; 0 < x < \infty \qquad (2.2.10)$$

where $\alpha > 0$ and $\beta > 0$. The parameter α, which influences the peakedness of the distribution, is known as the shape parameter, and β is called the scale parameter.

The mean, variance, and the *mgf* of a $G(\alpha, \beta)$ are respectively

$$E(X) = \alpha\beta, \; Var(X) = \alpha\beta^2, \text{ and } \phi_X(t) = (1 - \beta t)^{-\alpha} \text{ for } t < 1/\beta \qquad (2.2.11)$$

Two important special cases of the gamma distribution are worth mentioning.

If we set $\alpha = k/2$ where k is an integer, and $\beta = 2$, then the *pdf* (2.2.10) becomes

$$f(x \,|\, k) = \frac{1}{\Gamma(k/2)2^{k/2}} x^{k/2-1} e^{-x/2}, \;\; 0 < x < \infty \qquad (2.2.12)$$

and this is known as **chi-square distribution** with k degrees of freedom (*df*), denoted by χ_k^2.

The chi-square distribution plays an important role in statistical inferences, especially when a random sample is taken from a normal population. For example, if Z is a $N(0,1)$ random variable, then $W = Z^2$ follows a χ_1^2 distribution. Also, if $W_1, W_2, \dots, W_k$ are independent and identically distributed as χ_1^2, then $W = \sum_{i=1}^{k} W_i$ follows χ_k^2 distribution.

The other important special case of the gamma distribution is obtained if we set $\alpha = 1$. This gives the special gamma *pdf*

$$f(x \,|\, \beta) = \frac{1}{\beta} e^{-x/\beta}, \;\; x > 0 \qquad (2.2.13)$$

known as an **exponential distribution**   with scale parameter β. Exponential distribution has tremendous applications in modeling lifetimes

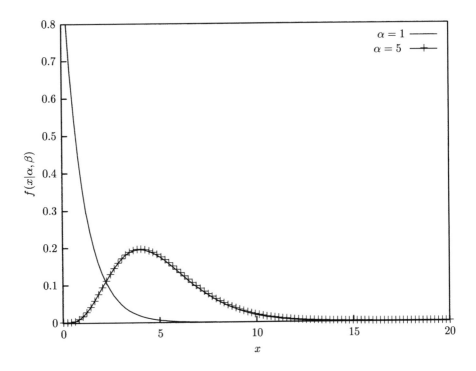

Figure 2.2.2: The *pdfs* of $G(\alpha, 1)$ Distribution, $\alpha = 1, 5$.

due to some interesting properties. Both Chi-square and Exponential distributions will be discussed later (Chapter 6 and Chapter 7) in greater detail in this book.

Lognormal Distribution

A random variable X is said to have a lognormal distribution if (lnX) is normally distributed. Using the fact that (lnX) is a $N(\mu, \sigma^2)$ distribution, one can derive the *pdf* of X as (denoted by $LN(\mu, \sigma^2)$), and

$$f(x \mid \mu, \sigma^2) = \frac{1}{\sqrt{2\pi}\sigma} x^{-1} e^{-(lnx-\mu)^2/(2\sigma^2)}, \quad 0 < x < \infty \qquad (2.2.14)$$

where $-\infty < \mu < \infty$ and $\sigma > 0$. Using the normal *mgf*, we can find the mean and variance of the $LN(\mu, \sigma^2)$ distribution as

$$E(X) = E\left(e^{lnX}\right) = e^{\mu + \sigma^2/2} \quad \text{and} \quad Var(X) = e^{2\mu + \sigma^2}\left(e^{\sigma^2} - 1\right)$$
$$(2.2.15)$$

Interestingly, $LN(\mu, \sigma^2)$ is one such distribution for which the *mgf* does not exist, yet all the moments exist. In fact,

$$E(X^k) = e^{k\mu + k^2\sigma^2/2} \quad \text{for any } k \geq 1 \qquad (2.2.16)$$

The lognormal distribution is very similar in appearance to the gamma distribution as shown in Figure 2.2.2 and Figure 2.2.3.

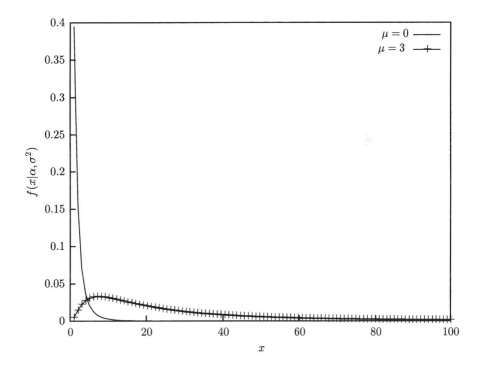

Figure 2.2.3: The *pdfs* of LN $(\mu, 1)$, $\mu = 0, 3$.

Logistic Distribution

A random variable X is said to have a logistic distribution with parameters μ and σ (denoted by $L(\mu, \sigma)$), provided the *pdf* of X is

$$f(x| \mu, \sigma) = \frac{1}{\sigma} \frac{e^{-(x-\mu)/\sigma}}{\left(1 + e^{-(x-\mu)/\sigma}\right)^2}, \quad -\infty < x < \infty \qquad (2.2.17)$$

where $-\infty < \mu < \infty$ and $\sigma > 0$.

The logistic distribution is appealing because of its simple looking *cdf* given as

$$F(x| \mu, \sigma) = P(X \le x) = \left(1 + e^{-(x-\mu)/\sigma}\right)^{-1} \qquad (2.2.18)$$

and it has numerous applications in clinical trials. Mean, variance, and *mgf* of $L(\mu, \sigma)$ are respectively

$$E(X) = \mu, \quad Var(X) = \frac{\pi^2}{3}\sigma^2$$

$$\text{and} \quad \phi_X(t) = \Gamma(1 - \sigma t)\,\Gamma(1 + \sigma t)\,e^{\mu t}, \quad |t| < \frac{1}{\sigma} \qquad (2.2.19)$$

Weibull Distribution

A random variable X has a Weibull distribution with parameters α and β, denoted $W(\alpha, \beta)$, if its *pdf* is given by

$$f(x| \alpha, \beta) = \left(\frac{\alpha}{\beta}\right) x^{\alpha-1} e^{-(x/\beta)^\alpha}, \quad 0 < x < \infty \qquad (2.2.20)$$

where $\alpha > 0$ and $\beta > 0$. Note that if we set $\alpha = 1$, then the *pdf* in (2.2.20) becomes an exponential *pdf* in (2.2.13).

One can easily derive the moments of $W(\alpha, \beta)$ distribution as

$$E(X^k) = \beta^k \Gamma\left(1 + \frac{k}{\alpha}\right) \qquad (2.2.21)$$

from which mean and variance can be obtained as

$$E(X) = \beta \Gamma\left(1 + \frac{1}{\alpha}\right)$$

$$\text{and} \quad Var(X) = \beta^2 \left(\Gamma\left(1 + \frac{2}{\alpha}\right) - \Gamma^2\left(1 + \frac{1}{\alpha}\right)\right) \qquad (2.2.22)$$

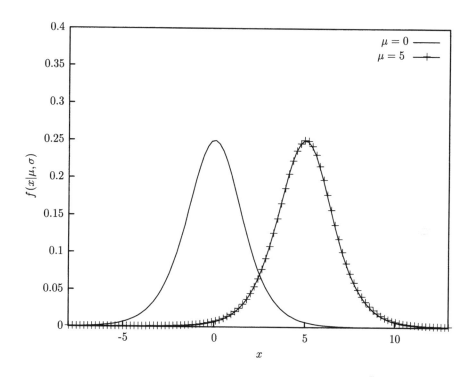

Figure 2.2.4: The *pdfs* of $L(\mu, 1)$, $\mu = 0, 5$.

The *mgf* of $W(\alpha, \beta)$ exists only for $\alpha \geq 1$ and has a very complicated form (not very useful).

The Weibull distribution is often applied to reliability and life testing problems such as the time to failure or life span of a component, measured from some specified time until it fails. For more on this see Chapter 8.

Pareto Distribution

A random variable X is said to have a Pareto distribution with parameters α and β, denoted by $P(\alpha, \beta)$, if its *pdf* is given by

$$f(x \mid \alpha, \beta) = \beta \alpha^{\beta} x^{-(\beta+1)}, \quad \alpha < x < \infty \tag{2.2.23}$$

where $\alpha > 0$ and $\beta > 0$.

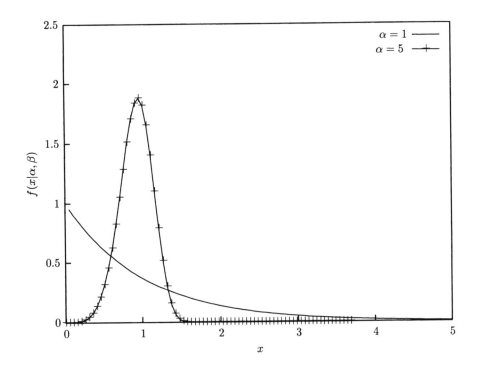

Figure 2.2.5: The *pdfs* of $W(\alpha, 1)$, $\alpha = 1, 5$.

The mean and variance of $P(\alpha, \beta)$ are

$$E(X) = \frac{\alpha\beta}{(\beta - 1)}, \quad \beta > 1 \qquad (2.2.24)$$

and

$$Var(X) = \frac{\alpha^2\beta}{(\beta - 1)^2(\beta - 2)}, \quad \beta > 2 \qquad (2.2.25)$$

If $\beta \leq 2$, then $Var(X)$ does not exist, and $E(X)$ does not exist for $\beta \leq 1$. Any kth moment of X will exist subject to the restriction on $\beta > k$. The *mgf* of $P(\alpha, \beta)$ distribution does not exist.

Student's *t*-Distribution

A distribution which is very similar in shape to the normal distribution is the Student's *t*-distribution with ν degrees of freedom, in short called

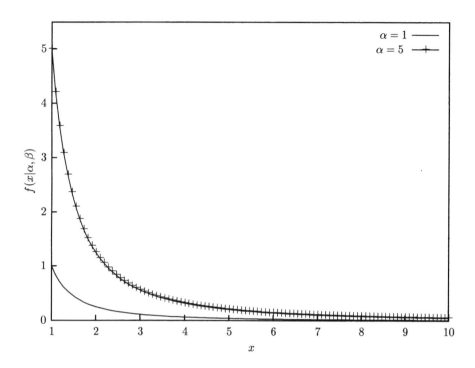

Figure 2.2.6: The *pdfs* of $P(\alpha, 1)$, $\alpha = 1, 5$.

't-distribution with ν *df* and denoted by $t_\nu(\theta)$ distribution. A random variable X has $t_\nu(\theta)$ distribution if its *pdf* is given by

$$f(x \mid \nu, \theta) = \frac{\Gamma\left(\frac{\nu+1}{2}\right)}{\Gamma\left(\frac{\nu}{2}\right)\sqrt{\pi\nu}\left(1 + \frac{(x-\theta)^2}{\nu}\right)^{(\nu+1)/2}}$$

$$-\infty < x < \infty, \ -\infty < \theta < \infty \quad (2.2.26)$$

where $\nu = 1, 2, 3, \ldots$. The *mgf* of $t_\nu(\theta)$ distribution does not exist, and moments exist subject to restrictions on ν. For example,

$$E(X) = \theta, \ \text{for } \nu > 1, \ \text{and} \ Var(X) = \frac{\nu}{(\nu - 2)}, \ \text{for } \nu > 2 \quad (2.2.27)$$

The special case, when $\nu = 1$ is called the Cauchy distribution and

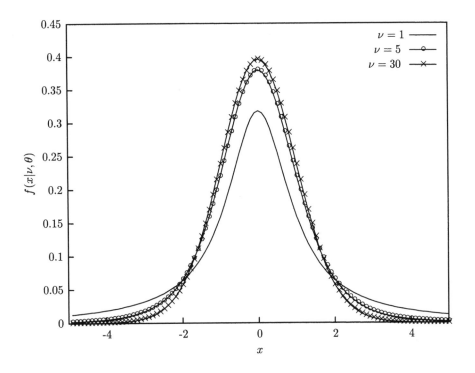

Figure 2.2.7: The *pdfs* of $t_\nu(\theta)$ with $\theta = 0$, $\nu = 1, 5, 30$.

the corresponding *pdf* is

$$f(x \mid \theta) = \frac{1}{\pi\left(1 + (x - \theta)^2\right)}, \quad -\infty < x < \infty, \; -\infty < \theta < \infty \quad (2.2.28)$$

The parameter θ in (2.2.26) measures the center of the distribution and is the median of $t_\nu(\theta)$ distribution. Note that the mean of the Cauchy distribution (2.2.28) does not exist. Also, from the *pdf* (2.2.26) of $t_\nu(\theta)$, as $\nu \to \infty$, $t_\nu(\theta)$ tends to the $N(\theta, 1)$ distribution.

Snedecor's *F*-Distribution

A random variable X is said to have a Snedecor's *F*-distribution (or '*F*-distribution' in short) with ordered degrees of freedom (*dfs*) ν_1 and

ν_2, provided its *pdf* is given as

$$f(x\,|\,\nu_1, \nu_2) = \frac{\Gamma\big((\nu_1+\nu_2)/2\big)}{\Gamma(\nu_1/2)\Gamma(\nu_2/2)} \left(\frac{\nu_1}{\nu_2}\right)^{\nu_1/2} x^{(\nu_1-2)/2}\left(1+\left(\frac{\nu_1}{\nu_2}\right)x\right)^{-(\nu_1+\nu_2)/2}$$

$$x > 0 \qquad (2.2.29)$$

where the *dfs* ν_1 and ν_2 usually take integer values. The above distribution is denoted by F_{ν_1, ν_2}. Mean and variance of the F_{ν_1, ν_2} distribution exist subject to restrictions on ν_1 and ν_2, and they are

$$E(X) = \frac{\nu_2}{\nu_2 - 2}, \quad \nu_2 > 2$$

$$\text{and} \quad Var(X) = 2\left(\frac{\nu_2}{\nu_2 - 2}\right)^2 \frac{(\nu_1+\nu_2-2)}{\nu_1(\nu_2-4)}, \quad \nu_2 > 4 \qquad (2.2.30)$$

The *mgf* of F_{ν_1, ν_2} does not exist, however, the kth moment is given by

$$E(X^k) = \frac{\Gamma\big((\nu_1+2k)/2\big)\Gamma\big((\nu_2-2k)/2\big)}{\Gamma(\nu_1/2)\Gamma(\nu_2/2)}\left(\frac{\nu_2}{\nu_1}\right)^k, \quad \nu_2 > 2k \qquad (2.2.31)$$

Suppose X_1 and X_2 are two independent random variables having $\chi^2_{\nu_1}$ and $\chi^2_{\nu_2}$ distributions, respectively. Then the distribution of $X = (X_1/\nu_1)/(X_2/\nu_2)$ is F_{ν_1, ν_2}. Also, if a random variable Y follows $t_\nu(0)$ distribution, then $X = Y^2$ follows $F_{1, \nu}$ distribution.

Beta Distribution

A random variable X with range (or support) $R_X = [0, 1]$ is said to have a Beta(α, β) distribution, provided the *pdf* is given by

$$f(x\,|\,\alpha, \beta) = \frac{1}{B(\alpha, \beta)} x^{\alpha-1}(1-x)^{\beta-1}, \quad 0 \le x \le 1 \qquad (2.2.32)$$

and $\alpha > 0$, $\beta > 0$. The constant $B(\alpha, \beta) = \Gamma(\alpha)\Gamma(\beta)/\Gamma(\alpha+\beta)$. Mean, variance, and *mgf* of Beta(α, β) distribution are

$$E(X) = \frac{\alpha}{\alpha+\beta}, \quad Var(X) = \frac{\alpha\beta}{(\alpha+\beta)^2(\alpha+\beta+1)} \qquad (2.2.33)$$

and

$$\phi_X(t) = 1 + \sum_{i=1}^{\infty}\left(\prod_{j=0}^{i-1}\frac{\alpha+j}{\alpha+\beta+j}\right)\frac{t^i}{i!} \qquad (2.2.34)$$

For the special case $\alpha = \beta = 1$, the Beta(α, β) distribution reduces to a CU$[0, 1]$ distribution in (2.2.1). Also, if X_1 and X_2 are independent random variables with $G(\alpha_1, \beta)$ and $G(\alpha_2, \beta)$ distributions respectively, then the distribution of $Y = X_1/(X_1 + X_2)$ is $B(\alpha_1, \alpha_2)$.

2.3   Some Limit Theorems

Definition 2.3.1 (Convergence in Distribution) *A sequence* $\left\{F_k(x)\right\}$ *of distribution functions (or cdfs) converges in distribution (or weakly) to a distribution function $F(x)$, if*

$$\lim_{k\to\infty} F_k(x) = F(x)$$

for all continuity points x of $F(x)$.

Let $F_k(x)$ and $F(x)$ be the distribution functions of X_k ($k \geq 1$) and X, respectively. Then, the sequence $\{X_k\}$ of random variables converges in distribution (or weakly) to X if $\left\{F_k(x)\right\}$ converges in distribution to $F(x)$. This is denoted as

$$X_k \xrightarrow{d} X$$

Theorem 2.3.1 *Let $X_k \xrightarrow{d} X$ and g be a continuous function, then $g(X_k) \xrightarrow{d} g(X)$.*

Definition 2.3.2 (Convergence in Probability) *Let $\{X_k\}$ be a sequence of random variables, then $\{X_k\}$ converges in probability to a random variable X if for every $\epsilon > 0$,*

$$P(|X_k - X| > \epsilon) \longrightarrow 0 \qquad as\ k \to \infty$$

It is denoted as

$$X_k \xrightarrow{p} X$$

Definition 2.3.3 (Almost Sure Convergence) *A sequence of random variables $\{X_k\}$ is said to converge almost surely to a random variable X if*

$$P\left(\lim_{k\to\infty} X_k = X\right) = 1$$

and this is denoted as

$$X_k \xrightarrow{a.s.} X$$

or $X_k \to X$ with probability 1.

The next theorem establishes an order among the above three types of convergences.

Theorem 2.3.2

$$X_k \xrightarrow{a.s.} X \quad \Longrightarrow \quad X_k \xrightarrow{p} X \quad \Longrightarrow \quad X_k \xrightarrow{d} X$$

The following theorem provides a result in which convergence in distribution can lead to convergence in probability.

Theorem 2.3.3 *If $X_k \xrightarrow{d} c$ (constant), then $X_k \xrightarrow{p} c$.*

In the following we will see applications of the above three types of convergences.

Theorem 2.3.4 (Central Limit Theorem or CLT) *Let $X_1, X_2, \ldots$ be a sequence of iid random variables with a finite mean μ and a finite variance $\sigma^2 > 0$. Define*

$$Y_k = \frac{1}{k} \sum_{i=1}^{k} X_i \quad and \quad Z_k = \frac{\sqrt{k}(Y_k - \mu)}{\sigma}$$

Then

$$Z_k \xrightarrow{d} Z$$

where Z is a standard normal random variable (i.e., Z follows $N(0,1)$ distribution).

Remark 2.3.1 *Since any real value is a continuity point of the standard normal cdf, the central limit theorem (CLT) says that for any $c \in \mathbb{R}$,*

$$P(Z_k \leq c) \longrightarrow \int_{-\infty}^{c} \frac{1}{\sqrt{2\pi}} e^{-u^2/2} du, \quad as \ k \to \infty$$

Generally speaking, Z_k is approximately a $N(0,1)$ random variable for a sufficiently large k; i.e., the average of sufficiently large number of iid random variables is approximately normally distributed.

Theorem 2.3.5 (Weak Law of Large Numbers or WLLN) *Let $X_1, X_2, \ldots$ be a sequence of iid random variables with a finite mean μ. Define*

$$Y_k = \frac{1}{k} \sum_{i=1}^{k} X_i$$

then

$$Y_k \xrightarrow{p} \mu$$

Theorem 2.3.6 (Strong Law of Large Numbers or SLLN) *Let* $X_1, X_2, \ldots$ *be a sequence of iid random variables with a finite mean μ. Define*

$$Y_k = \frac{1}{k} \sum_{i=1}^{k} X_i$$

then

$$Y_k \xrightarrow{a.s.} \mu$$

The above limit theorems help us study the behavior of certain sample quantities when the sample size approaches infinity. Although the concept of an infinitely large sample size is a hypothetical one, it can often provide us with some useful approximations for the finite sample case. Consider the following applications.

Suppose $X_1, X_2, \ldots, X_n$ be *iid* with a common *cdf* $F(x)$ which is assumed to be strictly monotonic. For $0 < c < 1$, let ξ_c denotes the $100c^{th}$ population percentile and let k be an integer such that $(k/n) \to c$ as $n \to \infty$. Then the asymptotic distribution of $X_{(k)}$ is $N(\xi_c, \sigma_*^2)$ (i.e., as $n \to \infty$, $X_{(k)} \xrightarrow{d} N(\xi_c, \sigma_*^2)$) where $\sigma_*^2 = c(1-c)/\{n(F'(\xi_c))^2\}$.

The above result says that in a large sample, the kth order statistic can be used to estimate the population percentile ξ_c provided $(k/n) \approx c$. Further, the variance σ_*^2 can tell us how fast the convergence is as n grows.

Theorem 2.3.7 (Slutsky's Theorem) *If $X_k \xrightarrow{d} X$ and $Y_k \xrightarrow{p} c$ (where c is a constant), then:*

(i) $X_k Y_k \xrightarrow{d} cX$

(ii) $X_k + Y_k \xrightarrow{d} (c + X)$

Chapter 3

Concepts of Statistical Inference

3.1   Introduction

In any statistical investigation, we collect data which, in many cases, can be described by observable random variables $X_1, X_2, \ldots, X_n$ (all or some of which may as well be multi-dimensional). We then assume a suitable probability distribution, called a probability model or simply a *model*, for the data $\boldsymbol{X} = (X_1, X_2, \ldots, X_n)$. The choice of the model is dictated partly by the nature of the experiment which gives rise to the data and partly by our past experience with a similar data. Often, mathematical simplicity plays a significant role in choosing the model. In general, the model takes the form of a specification of the joint probability distribution of the observable random variables $X_1, X_2, \ldots, X_n$. According to the model, the joint distribution function $F_{\boldsymbol{x}}$ is supposed to be some unspecified member of a suitable class $\mathcal{F}_{\boldsymbol{x}}$ of distributions. For the sake of simplicity, we assume that X_i's are single dimensional if not mentioned otherwise.

In many situations, we assume that the random variables $X_1, X_2, \ldots, X_n$ are independent and identically distributed (*iid*) random variables having a common but unspecified cumulative distribution function (*cdf*) F. Thus, our model states that $F_{\boldsymbol{x}}$ is a member of some class $\mathcal{F}_{\boldsymbol{x}}$ of the form

$$
\begin{aligned}
F_{\mathbf{x}}(x_1, x_2, \ldots, x_n) &= P(X_1 \leq x_1, X_2 \leq x_2, \ldots, X_n \leq x_n) \\
&= P(X_1 \leq x_1)P(X_2 \leq x_2) \ldots P(X_n \leq x_n) \\
&= \prod_{i=1}^{n} F(x_i) \qquad\qquad (3.1.1)
\end{aligned}
$$

where F belongs to some class $\mathcal{F}$ of distribution functions.

As an example, imagine the experiment where n individuals are selected at random (without replacement) from a very large population, and
let the ith random variable X_i represent the height of the ith individual,
$i = 1, 2, \ldots, n$. Since the individuals are selected without replacement,
the probability distribution of X_i, the height of the ith individual, depends on the population composition after excluding the previous $(i-1)$
individuals. But the overall population size (say, N) is so large compared to the overall sample size n that inclusion or exclusion of a few
individuals does not make much difference, and as a result, the probability distribution of X_i is virtually the same as those of $X_1, X_2, \ldots, X_{i-1}$,
$i = 2, 3, \ldots, n$, and they all are independent. F indicates the *cdf* of
a randomly selected individual's height. Because the randomly selected
individual can represent any individual in the population, F therefore is
the *cdf* over the range $R_{\mathbf{x}} =$ collection of 'height' values of all individuals
in the population, which is a subset of the positive side of $\mathbb{R}$.

Besides making the assumption that $X_1, X_2, \ldots, X_n$ are *iid* random
variables with a common distribution F, we may further assume that
the structure of F is a $N(\mu, \sigma^2)$ distribution with μ and σ^2 being left
unspecified. Thus, $F \in \mathcal{F}$, where

$$
\mathcal{F} = \left\{ N(\mu, \sigma^2) \,\middle|\, \mu \in \mathbb{R} \text{ and } \sigma^2 > 0 \right\} \qquad (3.1.2)
$$

With the assumption (3.1.2), the joint distribution function of $\mathbf{X} =
(X_1, X_2, \ldots, X_n)$ given in (3.1.1) becomes

$$
F_{\mathbf{x}}(x_1, x_2, \ldots, x_n) = \prod_{i=1}^{n} \int_{-\infty}^{x_i} \frac{1}{\sqrt{2\pi}\sigma_i} \exp\left(-\frac{1}{2\sigma^2}(u_i - \mu)^2 \right) du_i
$$

$$
(3.1.3)
$$

where $\mu \in \mathbb{R}$ and $\sigma^2 > 0$.

In carrying out the statistical investigation, we then take as our goal
the task of specifying F more completely than is done by the model (say,

(3.1.2)). This task is achieved through the observable random variables $X_1, X_2, \ldots, X_n$ with realizations $X_1 = x_1, X_2 = x_2, \ldots, X_n = x_n$. The observations $(x_1, x_2, \ldots, x_n)$ are used to make an educated guess about the distribution $F_{\mathbf{x}}$ (the model) which is partly unknown since μ and σ^2 are unknown in (3.1.2).

The process of making an educated guess about the probability distribution of the outcome of an experiment is called **statistical inference**. The inference is based on the data $(X_1 = x_1, X_2 = x_2, \ldots, X_n = x_n)$ obtained from one or more experiment(s), and the theory of probability enters into the process in three ways. First, the model used to explain the occurance of the data is probabilistic. Second, certain probabilistic principles are followed in making the inference. Third, the precision or the dependability of our inference is also evaluated in probabilistic terms.

The problem of statistical inference generally takes one of the two following forms:

(**a**) estimation

(**b**) hypothesis testing

(**a**) Sometimes we are interested in a particular characteristic θ of the probability model $F_{\mathbf{x}}$ of $\mathbf{X} = (X_1, X_2, \ldots, X_n)$ where $X_1, X_2, \ldots, X_n$ are *iid* with the common distribution F and θ may be the mean or variance of the distribution F. In order to make a conjecture about this θ, we use some function of the data $\mathbf{X} = (X_1, X_2, \ldots, X_n)$, say $\widehat{\theta} = \widehat{\theta}(\mathbf{X})$. More precisely, if we observe $X_1 = x_1, X_2 = x_2, \ldots, X_n = x_n$, then we put forward the corresponding value of $\widehat{\theta}(\mathbf{X})$, i.e., $\widehat{\theta} = \widehat{\theta}(\mathbf{x}) = \widehat{\theta}(x_1, x_2, \ldots, x_n)$ as a 'likely' value of the characteristic θ. This $\widehat{\theta}$ is called a 'point estimate' of θ, and the functional form $\widehat{\theta}(\mathbf{X})$ is called a **point estimator** of θ. Therefore, in point estimation, the unknown value θ is estimated by a point $\widehat{\theta}$ or a single value $\widehat{\theta}$.

Instead of a single value $\widehat{\theta}$ for θ, one might prefer a collection of values in the form of an interval, say $C(\mathbf{X}) = \left(\widehat{\theta}_{\mathrm{L}}(\mathbf{X}), \widehat{\theta}_{\mathrm{U}}(\mathbf{X})\right)$, such that the actual value θ is likely to belong to this set $C(\mathbf{X})$. Such an estimation is called **set or interval estimation** of θ.

(**b**) On some occasions, we start with some tentative notion about the characteristic θ of the distribution which we are interested in. This idea about θ, called a hypothesis, may come from our past experience in

dealing with similar problems, or may be suggested by some author-
ity (e.g., a leading anthropologist making a guess about the average
height of male members of a lost tribe, or a manufacturer making a
claim about the reliability of an improved product released in the market). We may then like to know how tenable or valid the idea about θ
is in the light of the observations $X_1 = x_1, X_2 = x_2, \ldots, X_n = x_n$. The
inference problem is thus regarded as testing a hypothesis about the unknown characteristic θ of the model. Note that the model used and the
hypothesis being tested are based upon the probability distribution of
$\boldsymbol{X} = (X_1 = x_1, X_2 = x_2, \ldots, X_n = x_n)$. However, while the hypothesis
is an assumption the validity of which is questionable, the validity of the
model is not questioned, but is taken for granted.

In the development of statistical methods, the techniques of inference
that were the first to appear were those which involved a number of
assumptions about the distribution of $\boldsymbol{X} = (X_1, X_2, \ldots, X_n)$. In most
cases, it is assumed that $X_1, X_2, \ldots, X_n$ are *iid* random variables, and in
any case it would be assumed that the joint distribution has a particular
parametric form with some or all of the model parameters being unknown.
Statistical inference in those cases would relate solely to the value(s)
of the unknown parameter(s). This is called **parametric inference**.
On the other hand, statistical inference when no parametric form of the
distribution is assumed is called **nonparametric inference**.

3.2   Sufficiency and Completeness

As mentioned earlier, in parametric statistics, the model, i.e., the probability distribution $F_{\boldsymbol{X}}$ of the data has a known structural form in which
one or more parameters are unknown. For convenience, let us assume that
the parametric form $F_{\boldsymbol{X}}$ depends on the parameter θ, and for simplicity,
we also assume that θ is single dimensional even though multidimensional
θ poses no ambiguity. The parameter θ is indeed a characteristic of the
probability distribution $F_{\boldsymbol{X}}$. **To show the dependence of $F_{\boldsymbol{X}}$ on θ,**
we modify our notation slightly and the distribution of the data
$\boldsymbol{X} = (X_1, X_2, \ldots, X_n)$ **will henceforth be noted by $F_{\boldsymbol{X},\theta}$ or by F_θ.**
Since values of θ over a suitable range Θ constitute the family of possible
probability distribution for our data $\boldsymbol{X}$, θ is thus an index indicating a
suitable class of distributions $\mathcal{F}_{\boldsymbol{X}} = \{F_{\boldsymbol{X},\theta} \mid \theta \in \Theta\}$. In a statistical in-

vestigation, the aim of the investigator is to know which member of the above class $\mathcal{F}_X$ would be most appropriate to his/her study, i.e., to make a decision about the value of the unknown parameter θ.

The only information that helps the investigator in making a decision is supplied by the observed values $x_1, x_2, \ldots, x_n$ of the random variables $X_1, X_2, \ldots, X_n$. However, in most of cases, the observations would be too numerous and too complicated a set of numbers to be directly dealt with, and so a condensation or summarization of data would be most desirable. It is this motivation that leads one to the **principle of sufficiency** or the notion of **sufficient statistics**.

A **sufficient statistic** for a parameter θ is a statistic (or a function of the data) that retains all the essential information about θ. Any additional information in the data , besides the value of the sufficient statistic, is not relevant for θ.

Definition 3.2.1 *A k-dimensional statistic[1] $T(X) = \big(T_1(X), T_2(X), \ldots, T_k(X)\big)$ is called a sufficient statistic for θ (or rather for the family of distributions $\mathcal{F}_X = \{F_{X,\theta} \mid \theta \in \Theta\}$) if the conditional probability distribution of $X = (X_1, X_2, \ldots, X_n)$, given the value of T, does not depend on θ.*

It should be noted that in the above definition, the components $T_1, T_2, \ldots, T_k$ of T need to be functionally unrelated, otherwise, one or more components can be dropped from T without compromising any relevant information. Also, in most of the subsequent examples, the parameter θ is single dimensional and so will be a sufficient statistic. In such a case, we use the notation, $T(X) = T$, to denote a sufficient statistic.

Example 3.2.1: Suppose our observations $X_1, X_2, \ldots, X_n$ are independent, and X_i follows a $B(m_i, \theta)$ distribution, $i = 1, 2, \ldots, n;\ 0 < \theta < 1$.

Consider the statistic $T = \sum_{i=1}^{n} X_i$ which has a $B(m, \theta)$ distribution with $m = m_1 + m_2 + \ldots + m_n$. Now, the *pmf* of T is

$$g(t \mid \theta) = \begin{cases} \binom{m}{t} \theta^t (1-\theta)^{m-t}, & t = 0, 1, 2, \ldots, m \\ 0 & \text{otherwise} \end{cases}$$

For $t \in \{0, 1, 2, \ldots, m\} =$ the range (or support) of T,

$$P\big(X_1 = x_1, X_2 = x_2, \ldots, X_n = x_n \big| T = t\big),$$

---

[1] Any function of the data is called a *statistic*.

$$x_i \in \{1, 2, \ldots, m_i\}, \ i = 1, 2, \ldots, n$$

$$= \ P(X_1 = x_1, \ldots, X_n = x_n, T = t)/P(T = t)$$

$$= \ \begin{cases} P(X_1 = x_1, \ldots, X_n = x_n)/P(T = t) & \text{if } \sum_{i=1}^{n} x_i = t \\ 0 & \text{otherwise} \end{cases}$$

$$= \ \begin{cases} \left(\prod_{i=1}^{n} \binom{n_i}{x_i}\right) \theta^t (1-\theta)^{m-t} / \left(\binom{m}{t} \theta^t (1-\theta)^{m-t}\right) & \text{if } \sum_{i=1}^{n} x_i = t \\ 0 & \text{otherwise} \end{cases}$$

$$= \ \begin{cases} \prod_{i=1}^{n} \binom{n_i}{x_i} / \binom{m}{t} & \text{if } \sum_{i=1}^{n} x_i = t \\ 0 & \text{otherwise} \end{cases}$$

Thus, when T is given, the conditional distribution of $\boldsymbol{X}$ is independent of θ. Hence, T is a sufficient statistic for θ.

When it comes to the continuous case, direct application of the definition becomes difficult since $P(\boldsymbol{T} = \boldsymbol{t}) = 0$ for any $\boldsymbol{t} \in R_T$, the range of T. The following result can be used to determine whether a statistic $\boldsymbol{T}$ is sufficient for θ or not.

Theorem 3.2.1 *If $f_{\boldsymbol{X}}(\boldsymbol{x}|\theta)$ is the joint pdf or pmf of $\boldsymbol{X}$, and $g_{\boldsymbol{T}}(\boldsymbol{t}|\theta)$ is the pdf or pmf of $\boldsymbol{T}(\boldsymbol{X})$, then $\boldsymbol{T}(\boldsymbol{X})$ is a sufficient statistic for θ if, and only if, for every $\boldsymbol{x} \in R_{\boldsymbol{X}}$, the range of $\boldsymbol{X}$, the ratio $f_{\boldsymbol{X}}(\boldsymbol{x}|\theta)/g_{\boldsymbol{T}}(\boldsymbol{T}(\boldsymbol{x})|\theta)$ is free from θ.*

Example 3.2.2: Let $X_1, X_2, \ldots, X_n$ be *iid* $N(\mu, \sigma^2)$ where σ^2 is known. We would like to verify that $T(\boldsymbol{X}) = \overline{X} = \sum_{i=1}^{n} /n$ is a sufficient statistic for μ.

The joint *pdf* of $\boldsymbol{X}$ is

$$f_{\boldsymbol{X}}(\boldsymbol{x}|\mu) \ = \ \prod_{i=1}^{n} (2\pi\sigma^2)^{-1/2} \exp\left(-(x_i - \mu)^2/(2\sigma^2)\right)$$

$$= \ (2\pi\sigma^2)^{-n/2} \exp\left(-\frac{1}{2\sigma^2} \sum_{i=1}^{n} (x_i - \overline{x})^2 - \frac{n}{2\sigma^2}(\overline{x} - \mu)^2\right)$$

Also, note that $\overline{X}$ follows a $N(\mu, \sigma^2/n)$ distribution (which can be obtained from the *mgf* of $\overline{X}$), i.e., the *pdf* of $\overline{X}$ given by $g_{\bar{x}}(\bar{x}|\mu)$ is

$$g_{\bar{x}}(\bar{x}|\mu) = (2\pi\sigma^2/n)^{-1/2} \exp\left(-\frac{n}{2\sigma^2}(\overline{x} - \mu)^2\right)$$

Thus,

$$\frac{f_{\mathbf{X}}(\mathbf{x}\,|\,\mu)}{g_{\bar{x}}(\bar{x}\,|\,\mu)} = n^{-1/2}\left(2\pi\sigma^2\right)^{-(n-1)/2}\exp\left(-\frac{1}{2\sigma^2}\sum_{i=1}^{n}\left(x_i - \bar{x}\right)^2\right)$$

which is free from μ, and hence, $\overline{X}$ is a sufficient statistic for μ.

While dealing with sufficiency, we encounter two questions: first, how to verify whether a statistic is sufficient or not; and second, how to find a sufficient statistic. Theorem 3.2.1 answers the first question, whereas the following theorem settles the second one.

Theorem 3.2.2 (Fisher-Neyman Factorization Theorem) *Let* $f_{\mathbf{X}}(\mathbf{x}\,|\,\theta)$ *be the joint pdf or pmf of the data* $\mathbf{X}$*. A statistic* $\mathbf{T}(\mathbf{X})$ *is a sufficient statistic for* θ *if, and only if, there exists functions* $g_{\mathbf{T}}(\mathbf{t}\,|\,\theta)$ *and* $h(\mathbf{x})$ *such that for all* $\mathbf{x}$ *and all* $\theta \in \Theta$,

$$f_{\mathbf{X}}(\mathbf{x}\,|\,\theta) = g_{\mathbf{T}}(\mathbf{T}(\mathbf{x})\,|\,\theta) \cdot h(\mathbf{x}) \tag{3.2.1}$$

Basically, we factor the joint *pdf* or *pmf* of the data into two components. The component that is free from θ constitutes the function $h(\mathbf{x})$. The other component that depends on θ also depends on the observed data $\mathbf{x}$ only through some function $\mathbf{T}(\mathbf{x})$, and this function $\mathbf{T}$ is a sufficient statistic for θ.

Example 3.2.3: Let $X_1, X_2, \ldots, X_n$ be *iid* random variables from the continuous uniform $\mathrm{CU}(\theta_1, \theta_2)$ with common *pdf* [2]

$$f_{\mathbf{X}}(x\,|\,\theta) = (\theta_2 - \theta_1)^{-1}I_{(\theta_1,\theta_2)}(x)$$

where $\theta = (\theta_1, \theta_2)$ and both θ_1 and θ_2 are unknown. The parameter space is $\Theta = \left\{(\theta_1, \theta_2)\,\middle|\, -\infty < \theta_1 < \theta_2 < \infty\right\}$. The joint *pdf* of $(X_1, X_2, \ldots, X_n)$ is

$$f_{\mathbf{X}}(\mathbf{x}\,|\,\theta) = \begin{cases} (\theta_2 - \theta_1)^{-n} & \text{if } \theta_1 < x_i < \theta_2 \ \forall i \\ 0 & \text{otherwise} \end{cases}$$

$$= \begin{cases} (\theta_2 - \theta_1)^{-n} & \text{if } \theta_1 < x_{(1)} \leq x_{(n)} < \theta_2 \\ 0 & \text{otherwise} \end{cases}$$

where $x_{(1)}$ and $x_{(n)}$ are the smallest and the largest of $x_1, x_2, \ldots, x_n$, respectively. We can further write the joint *pdf* as

$$f_{\mathbf{X}}(\mathbf{x}\,|\,\theta) = (\theta_2 - \theta_1)^{-n}I_{(\theta_1,\infty)}(x_{(1)}) \cdot I_{(-\infty,\theta_2)}(x_{(n)})$$

---

[2] $I_{(\theta_1,\theta_2)}(x) = 1$, when $\theta_1 < x < \theta_2$; $= 0$, otherwise.

Define $t_1 = T_1(\boldsymbol{x}) = x_{(1)}$, $t_2 = T_2(\boldsymbol{x}) = x_{(n)}$, $h(\boldsymbol{x}) = 1$, and
$g_{\boldsymbol{T}}(\boldsymbol{t}\,|\,\theta) = (\theta_2 - \theta_1)^{-n} I_{(\theta_1, \infty)}(x_{(1)}) \cdot I_{(-\infty, \theta_2)}(x_{(n)})$. Thus, by applying
Theorem 3.2.2, $\boldsymbol{T} = (T_1, T_2) = (X_{(1)}, X_{(n)})$ is a sufficient statistic for
$\theta = (\theta_1, \theta_2)$.

Example 3.2.4: Assume that $X_1, X_2, \ldots, X_n$ are *iid* $N(\mu, \sigma^2)$ where
both the parameters μ and σ^2 are unknown, and can be written as a
two-dimensional parameter $\theta = (\mu, \sigma^2)$. It is easy to see that the joint
pdf of $\boldsymbol{X} = (X_1, X_2, \ldots, X_n)$ depends on the sample $\boldsymbol{X}$ only through the
following two values:

$$T_1 = \overline{X} = \frac{1}{n}\sum_{i=1}^{n} X_i \quad \text{and} \quad T_2 = \frac{1}{(n-1)}S = \frac{1}{(n-1)}\sum_{i=1}^{n}(X_i - \overline{X})^2$$

It is seen that the joint *pdf* $f_{\boldsymbol{X}}(\boldsymbol{x}\,|\,\theta)$ of $\boldsymbol{X}$ can be written as

$$f_{\boldsymbol{X}}(\boldsymbol{x}\,|\,\theta) = g_{\boldsymbol{T}}(T_1(\boldsymbol{x}), T_2(\boldsymbol{x})\,|\,\theta) \cdot h(\boldsymbol{x})$$

where

$$g_{\boldsymbol{T}}(t_1, t_2\,|\,\theta) = (2\pi\sigma^2)^{-n/2} \exp\left(-(n(t_1 - \mu)^2 + (n-1)t_2)/(2\sigma^2)\right)$$

and $h(\boldsymbol{x}) = 1$. Therefore, by the Factorization Theorem, $\boldsymbol{T}(\boldsymbol{X}) = (T_1(\boldsymbol{X}),$
$T_2(\boldsymbol{X})) = (\overline{X}, S/(n-1))$ is a sufficient statistic for (μ, σ^2).

The sufficient statistic $(\overline{X}, S/(n-1))$, in Example 3.2.4, contains all
the relevant information about (μ, σ^2) for a normal model when both
μ and σ^2 are unknown. But this may not be the case if the model is
other than normal. For example, if our *iid* observations have a Cauchy
distribution with location parameter μ and scale parameter σ, then the
common *pdf* is given by

$$f(x\,|\,\mu, \sigma) = \frac{1}{\pi\sigma}\left(1 + \left(\frac{x - \mu}{\sigma}\right)^2\right)^{-1}, \quad x \in \mathbb{R}, \ \mu \in \mathbb{R}, \ \text{and} \ \sigma > 0$$

$$(3.2.2)$$

This *pdf*, when plotted, looks very similar to the *pdf* of $N(\mu, \sigma^2)$, but the
statistic $(\overline{X}, S/(n-1))$ is **not** sufficient for (μ, σ^2). One can verify that
with $(T_1, T_2) = (\overline{X}, S/(n-1))$, the decomposition (3.2.1) of the joint *pdf*
of $\boldsymbol{X}$ under a Cauchy model (3.2.2) does not hold. Therefore, the choice
of a sufficient statistic is very much model dependent.

So far we have discussed the sufficiency property that a statistic T may have in relation to a parameter θ which essentially is an index identifying a family of probability distributions. Now we shall consider another property, called **completeness**.

Consider a statistic T (may be multidimensional) based on the random variables $X_1, X_2, \ldots, X_n$ having a joint distribution $f_X(x | \theta)$, $\theta \in \Theta$. The distribution of T, in general, depends on θ. Let the probability distribution of T be $g_T(t | \theta)$, $\theta \in \Theta$.

Definition 3.2.2 *A statistic T or, more precisely, the family of distributions $\{g_T(t | \theta), \theta \in \Theta\}$ of T, is called complete if for any function $\psi(T)$ of T, $E\big(\psi(T)\big) = 0$ for all $\theta \in \Theta$ implies $\psi(T) = 0$ with probability 1 for all $\theta \in \Theta$.*

Notice that completeness is a property of a family of probability distributions, not of a particular distribution.

Example 3.2.5: We have already seen that if $X_1, X_2, \ldots, X_n$ are *iid* Bernoulli(θ), then the statistic $T = \sum_{i=1}^{n} X_i$ is sufficient for θ (see Example 3.2.1). T has a B(n, θ) distribution. Let ψ be a function such that $E\big(\psi(T)\big) = 0$ for all $\theta \in (0, 1)$. Then,

$$
\begin{aligned}
0 = E\big(\psi(T)\big) &= \sum_{t=0}^{n} \psi(t) \binom{n}{t} \theta^t (1 - \theta)^{n-t} \\
&= (1 - \theta)^n \sum_{t=0}^{n} \psi(t) \binom{n}{t} \left(\frac{\theta}{1 - \theta} \right)^t \quad (3.2.3)
\end{aligned}
$$

for all $\theta \in (0, 1)$. Since $(1 - \theta) \neq 0$, (3.2.3) implies

$$
\sum_{t=0}^{n} \psi(t) \binom{n}{t} \eta^t = 0 \quad (3.2.4)
$$

for all $\eta = \theta/(1 - \theta) \in (0, \infty)$. By taking limit η tending to 0 (*i.e.*, $\eta \to 0$), we can get $\psi(0) = 0$. Next, by differentiating both sides of (3.2.4) with respect to η, and then taking $\eta \to 0$, we can obtain $\psi(1) = 0$. Repeat the same procedure by differentiating further and taking $\eta \to 0$ and we eventually have $\psi(2) = 0, \psi(3) = 0, \ldots, \psi(n) = 0$. Thus, $\psi(T) \equiv 0$. Hence, T is a complete statistic.

Example 3.2.6: Let $X_1, X_2, \ldots, X_n$ be a random sample from Poisson(θ) (Poi(θ)) distribution with common *pmf* $f(x | \theta) = \exp(-\theta) \theta^x / x!$, $x =$

$0, 1, 2, \ldots$, where $\theta \in \Theta = (0, \infty)$. By Factorization Theorem, it is seen that $T = \sum_{i=1}^{n} X_i$ is a sufficient statistic for θ. Also, T has a Poisson$(n\theta)$ distribution (easily seen from the mgf of T). Let $\psi(T)$ be a function such that $E(\psi(T)) = 0$ for all $\theta > 0$, i.e.,

$$0 = E(\psi(T)) = \exp(-n\theta) \sum_{t=0}^{\infty} k(t)\theta^t \qquad (3.2.5)$$

where $k(t) = \psi(t)n^t/t!$ and $\theta > 0$. The equation (3.2.5) implies that

$$\sum_{t=0}^{\infty} k(t)\theta^t = 0 \quad \text{for all } \theta \in (0, \infty) \qquad (3.2.6)$$

Differentiating both sides of (3.2.6) j times with respect to θ ($j = 1, 2, 3, \ldots$), and then taking $\theta \to 0$, we can get $k(j) = 0$, $j = 1, 2, 3, \ldots$, and it then provides $k(0) = 0$. Therefore, $k(T) = 0$ with probability 1; i.e., $\psi(T) = 0$ with probability 1 implying that T is complete.

Example 3.2.7: Let $X_1, X_2, \ldots, X_n$ be $iid\,\mathrm{N}(\mu, \sigma^2)$ where μ is unknown but σ^2 is known. Earlier we have seen that $T = \overline{X}$ is sufficient for θ and has $\mathrm{N}(\mu, \sigma^2/n)$ distribution. Let $\psi(T)$ be a function such that $E(\psi(T)) = 0$ for all $\mu \in \mathbb{R}$ and it implies

$$\int_{-\infty}^{\infty} \psi(t) \exp\left(-nt^2/(2\sigma^2) + n\mu t/\sigma^2\right) dt = 0 \qquad (3.2.7)$$

for $\mu \in \mathbb{R}$. The left hand side of (3.2.7) is the bilateral Laplace transform of the function $\psi(t) \exp\left(-nt^2/(2\sigma^2)\right)$. From the unicity theorem of Laplace transform, it follows that

$$\psi(t) \exp\left(-nt^2/(2\sigma^2)\right) = 0 \quad \text{for all } \mu \in \mathbb{R}$$

i.e., $\psi(t) = 0$ for all $\mu \in \mathbb{R}$. Hence, $T = \overline{X}$ is a complete statistic.

As we have seen earlier, a sufficient statistic retains only the information relevant for the unknown parameter θ. The remaining information in the original sample (i.e., total information in the sample minus the information in a sufficient statistic) is redundant for θ, and this gives rise to the concept of an **ancillary statistic**.

Definition 3.2.3 *A statistic $S(\boldsymbol{X})$ whose distribution does not depend on the parameter θ is called an ancillary statistic for θ.*

Notice that the distribution of an ancillary statistic is free from θ, and hence, carries no information about θ. It is therefore natural to expect that an ancillary statistic should have nothing to do with a sufficient statistic, i.e., an ancillary statistic should be independent of a sufficient statistic. The following result tells exactly when this is going to happen.

Theorem 3.2.3 (Basu's Theorem) *If $T(X)$ is sufficient for θ and is complete, then $T(X)$ is independent of any ancillary statistic $S(X)$.*

Basu's Theorem is useful in the sense that it allows us to prove the independence of two statistics without finding the joint probability distribution of the two statistics.

Example 3.2.8: Let $X_1, X_2, \ldots, X_n$ be *iid* with *pdf*

$$f(x\,|\,\theta) = e^{-(x-\theta)}, \quad -\infty < \theta < x < \infty \qquad (3.2.8)$$

The joint *pdf* of $X = (X_1, X_2, \ldots, X_n)$ is

$$
\begin{aligned}
f_X(x\,|\,\theta) &= \prod_{i=1}^{n} f(x_i\,|\,\theta) \\
&= e^{-\Sigma(x_i-\theta)} I_{(\theta,\infty)}(x_1) \ldots I_{(\theta,\infty)}(x_n) \\
&= e^{-\Sigma(x_{(i)}-\theta)} I_{(\theta,\infty)}(x_{(1)}) \qquad (3.2.9)
\end{aligned}
$$

where $x_{(1)} \le x_{(2)} \le \cdots \le x_{(n)}$.[3] Using Factorization Theorem, it is easy to see that a sufficient statistic for θ is $T(X) = X_{(1)}$. The *pdf* of $X_{(1)}$ is given by

$$g_{X_{(1)}}(x_{(1)}\,|\,\theta) = n\, e^{-n(x_{(1)}-\theta)} I_{(\theta,\infty)}(x_{(1)}) \qquad (3.2.10)$$

Also, if for some function $\psi(T) = \psi(X_{(1)})$ we have $E\big(\psi(T)\big) = 0$ for all $\theta \in \mathbb{R}$, then we can get

$$\int_{\theta}^{\infty} \psi(t) e^{-nt}\, dt = 0 \quad \text{for all } \theta \in \mathbb{R} \qquad (3.2.11)$$

Differentiating both sides of (3.2.11) with respect to θ yields

$$-\psi(\theta) e^{-n\theta} = 0 \quad \text{for all } \theta \in \mathbb{R}$$

i.e., $\psi(T) = 0$ with probability 1. Hence $T = X_{(1)}$ is a complete sufficient statistic for θ. Is $T = X_{(1)}$ independent of $S = \sum_{i=1}^{n}(X_i - \overline{X})^2$? Note that S can be also written as $S = \sum_{i=1}^{n}(Y_i - \overline{Y})^2$ where $Y_i = X_i - \theta$ and Y_i's are *iid* having *pdf* $f(y) = e^{-y}$, $y > 0$. Clearly, Y_i's are ancillary and so are $\overline{Y}$ and S. By Basu's Theorem, $X_{(1)}$ is independent of S.

---

[3] $X_{(1)} \le X_{(2)} \le \cdots \le X_{(n)}$ are called order statistics of $X = (X_1, X_2, \ldots, X_n)$.

3.3   Methods of Estimation

In this section, we will consider two popular methods of estimating a parameter, namely — (i) method of moments estimation, and (ii) maximum likelihood estimation.

Method of Moments Estimators (MMEs)

Let $X_1, X_2, \ldots, X_n$ be iid with pdf or pmf $f(x|\boldsymbol{\theta})$, where $\boldsymbol{\theta} = (\theta_1, \theta_2, \ldots, \theta_k)$. Method of moments estimators (MMEs) of $\theta_1, \theta_2, \ldots, \theta_k$ are found by equating the first k sample moments to the corresponding k population moments, and then solving the resultant system of k equations for k unknown parameters. In other words, let $\mu'_{(1)}(\boldsymbol{\theta}), \mu'_{(2)}(\boldsymbol{\theta}), \ldots, \mu'_{(k)}(\boldsymbol{\theta})$ be the first k population raw moments, and $m'_{(j)}(\boldsymbol{X}) = \sum_{i=1}^{n} X_i^j / n$, $j = 1, 2, \ldots, k$, be the first k-sample moments. We can obtain the MMEs of $\theta_1, \theta_2, \ldots, \theta_k$, say $\widehat{\theta}_{1(\mathrm{MM})}, \widehat{\theta}_{2(\mathrm{MM})}, \ldots, \widehat{\theta}_{k(\mathrm{MM})}$, by solving

$$m'_{(j)}(\boldsymbol{X}) = \mu'_{(j)}(\boldsymbol{\theta}), \quad i = 1, 2, \ldots, k \tag{3.3.1}$$

Maximum Likelihood Estimators (MLEs)

Suppose we have a random vector $\boldsymbol{X} = (X_1, X_2, \ldots, X_n)$ with joint pdf or pmf $f_{\boldsymbol{X}}(\boldsymbol{x}|\boldsymbol{\theta})$ where $\boldsymbol{\theta} = (\theta_1, \theta_2, \ldots, \theta_k) \in \boldsymbol{\Theta}$ (k-dimensional). The likelihood function is defined by

$$L(\boldsymbol{\theta}|\boldsymbol{X}) = f_{\boldsymbol{X}}(\boldsymbol{X}|\boldsymbol{\theta}) \tag{3.3.2}$$

For each observed $\boldsymbol{X} = \boldsymbol{x}$, let $\widehat{\boldsymbol{\theta}}(\boldsymbol{x})$ be a parameter value at which $L(\boldsymbol{\theta}|\boldsymbol{X} = \boldsymbol{x})$ attains its supremum (or maximum) as a function of $\boldsymbol{\theta}$. Then the maximum likelihood estimator (MLE) of $\boldsymbol{\theta}$ based on $\boldsymbol{X}$ is $\widehat{\boldsymbol{\theta}}(\boldsymbol{X})$.

If the likelihood function $L(\boldsymbol{\theta}|\boldsymbol{X})$ is differentiable in each θ_i, then the MLE of θ_i's are found by solving the following equations

$$\frac{\partial}{\partial \theta_i} L(\boldsymbol{\theta}|\boldsymbol{X}) = 0, \quad i = 1, 2, \ldots, k \tag{3.3.3}$$

and the solution $\widehat{\boldsymbol{\theta}}_{\mathrm{ML}} = \left(\widehat{\theta}_{1(\mathrm{ML})}, \widehat{\theta}_{2(\mathrm{ML})}, \ldots, \widehat{\theta}_{k(\mathrm{ML})} \right)$ satisfies

$$\left(\left(\frac{\partial^2}{\partial \theta_i \partial \theta_j} L(\boldsymbol{\theta}|\boldsymbol{X}) \right) \right) \Bigg|_{\boldsymbol{\theta} = \widehat{\boldsymbol{\theta}}_{\mathrm{ML}}} \leq 0 \tag{3.3.4}$$

The left hand side of (3.3.4) is a $k \times k$ matrix and '≤ 0' implies non-positive definite.

Note that quite often $L(\boldsymbol{\theta}| \boldsymbol{X})$ is too complicated to differentiate. Hence, it is more convenient to use $lnL(\boldsymbol{\theta}| \boldsymbol{X})$ since 'logarithm' preserves monotonicity of $L(\boldsymbol{\theta}| \boldsymbol{X})$.

Example 3.3.1: Let $X_1, X_2, \ldots, X_n$ be *iid* Bernoulli(θ). Find the MME and MLE of θ.

Since there is only one unknown parameter involved, we find the MME of θ by equating the first sample moment with the population analogue as $\overline{X} = E(X_1) = \theta$. Hence, $\widehat{\theta}_{\text{MM}} = \overline{X}$.

To find the MLE of θ, first note that the likelihood function is

$$L(\theta| \boldsymbol{X}) = \prod_{i=1}^{n} \theta^{X_i}(1-\theta)^{1-X_i}$$
$$= \theta^{\Sigma X_i}(1-\theta)^{n-\Sigma X_i}$$

Let $X = \sum_{i=1}^{n} X_i$, then the log likelihood function can be written as

$$lnL(\theta| \boldsymbol{X}) = Xln\theta + (n-X)ln(1-\theta)$$

It is seen that $\widehat{\theta}_{\text{ML}} = X/n = \overline{X}$ which maximizes $lnL(\theta| \boldsymbol{X})$.

Example 3.3.2: Let $X_1, X_2, \ldots, X_n$ be *iid* CU$(0, \theta)$ (continuous uniform over the interval $(0, \theta)$). Find the MME and MLE of θ.

The *pdf* of X_i is

$$f(x_i| \theta) = \frac{1}{\theta}I_{(0,\theta)}(x_i), \quad i = 1, 2, \ldots, n$$

Hence, the likelihood function is

$$L(\theta| X_1, \ldots, X_n) = \frac{1}{\theta^n}\prod_{i=1}^{n} I_{(0,\theta)}(X_i)$$
$$= \frac{1}{\theta^n}I_{(X_{(n)},\infty)}(\theta) \qquad (3.3.5)$$

where $X_{(n)} = \max_i X_i$. Note that the equation, $\frac{\partial}{\partial \theta}lnL(\theta| \boldsymbol{X}) = 0$, has no solution. But we observe that the likelihood function $L(\theta| \boldsymbol{X})$ is monotonically decreasing for $\theta \geq X_{(n)}$. Therefore, $L(\theta| \boldsymbol{X})$ is maximized at $\theta = \widehat{\theta}_{\text{ML}} = X_{(n)}$.

On the other hand, if we equate the first sample moment with the first population moment, we get $\overline{X} = \theta/2$, i.e., $\widehat{\theta}_{\text{MM}} = 2\overline{X}$. It can be seen that the MME and MLE of θ above are vastly different.

Both the estimation methods described above have advantages and disadvantages. For example, one may not obtain a closed form expression of the MLE always, whereas MMEs may be easier to derive and have simpler closed form expressions. A specific example of this situation is seen when *iid* observations are available from a Gamma distribution with both scale and shape parameters unknown (see Chapter 7 for details). On the other hand, MLE is always a function of a sufficient statistic and invariant under transformations, as mentioned in the next two theorems.

Theorem 3.3.1 *If T is a sufficient statistic for θ, then the MLE of θ is a function of T.*

Theorem 3.3.2 *If $\widehat{\theta}_{ML}$ is the MLE of θ, and $\eta = \psi(\theta)$ is a function of θ, then $\widehat{\eta}_{ML} = \psi(\widehat{\theta}_{ML})$.*

The above two theorems are useful tools in finding estimators of a parameter θ (scalar or vector valued). Theorem 3.3.1 says that it is enough to focus on a sufficient statistic as far as maximum likelihood estimation is concerned. Reducing the original data set by a sufficient statistic makes the problem less cumbersome and more tractable. Also, quite often we are not interested in estimating the original model parameter, rather our focus may be on some function of the original model parameter. Theorem 3.3.2 helps in deriving MLE's of such functions of the model parameter.

Example 3.3.3: As an application of Theorem 3.3.2, we consider a random variable X representing the number of daily auto accidents in a small town. It is assumed that X follows a Poisson(θ) distribution where $\theta > 0$ is the average number of auto accidents per day. To estimate $\eta = P(X = 0)$, the probability of not having any auto accident in a day, we take a random sample of n distinct days and record the number of auto accidents on those days as $X_1, X_2, \ldots, X_n$. Obviously, $X_1, X_2, \ldots, X_n$ are *iid* Poisson(θ), and a sufficient statistic is $T = \sum_{i=1}^{n} X_i$ following a Poisson($n\theta$) distribution. Also, it can be seen that the MLE of θ is $\widehat{\theta}_{\mathrm{ML}} = \overline{X}$. But we are only interested in estimating the parameter, $\eta = P(X = 0) = e^{-\theta}$. By Theorem 3.3.2, the MLE of η is $\widehat{\eta}_{\mathrm{ML}} = \exp(-\widehat{\theta}_{\mathrm{ML}}) = e^{-\overline{X}}$.

3.4   Methods of Evaluating Estimators

Based on our data $X = (X_1, X_2, \dots, X_n)$ following a joint *pdf* or *pmf* $f(x\,|\,\theta)$,[4] we wish to estimate a real valued function $\eta = \eta(\theta)$ of θ. Let $\widehat{\eta} = \widehat{\eta}(X)$ be an estimator of η. Note that $\widehat{\eta}$ is a random variable. When $X = x$ is observed, $\widehat{\eta}(X) = \widehat{\eta}(x)$ is used to estimate the value of η, or just an 'estimate' of η. But how do we judge whether $\widehat{\eta}$ is a 'good' or a 'bad' estimator of η?

Intuitively, $\widehat{\eta} = \widehat{\eta}(X)$ may be looked upon as a 'good' estimator if and only if it gives values $\widehat{\eta}(x)$ that deviate from $\eta = \eta(\theta)$ only by a 'small' margin, i.e., if the probability distribution of $\widehat{\eta}$ has a high degree of concentration around $\eta = \eta(\theta)$, whatever the true value of θ ($\in \Theta$) may be. More precisely, we may say that $\widehat{\eta}$ is the best estimator if, for every $d > 0$,

$$P\left[|\widehat{\eta} - \eta| < d\right] \geq P\left[|\widehat{\eta}^* - \eta| < d\right] \quad \text{for all } \theta \in \Theta \qquad (3.4.1)$$

whatever the rival estimator $\widehat{\eta}^* = \widehat{\eta}^*(X)$ may be.

Even though the condition (3.4.1) is a desirable one in finding the 'best' estimator of η, it is difficult to apply due to the mathematical complexities involved. Given two estimators $\widehat{\eta}$ and $\widehat{\eta}^*$ of η, it may not be easy to derive the probability terms on two sides of the inequality in (3.4.1).

A simpler and mathematically convenient criterion, known as **mean squared error**, is thus adopted to judge the performance of an estimator.

Definition 3.4.1 *The mean squared error (MSE) of an estimator $\widehat{\eta}$ of a parameter $\eta = \eta(\theta)$ is defined by $MSE(\widehat{\eta}) = E\left[(\widehat{\eta} - \eta)^2\right]$ (which is a function of θ and will be assumed to exist for each $\theta \in \Theta$).*

Note that the MSE has two components, of which one is due to **bias** and the other is due to **variance**. For $MSE(\widehat{\eta})$,

$$
\begin{aligned}
E\left[(\widehat{\eta} - \eta)^2\right] &= E\left[\left((\widehat{\eta} - E(\widehat{\eta})) + (E(\widehat{\eta}) - \eta)\right)^2\right] \\
&= (E(\widehat{\eta}) - \eta)^2 + E\left[(\widehat{\eta} - E(\widehat{\eta}))^2\right] \\
&= \left(B_\theta(\widehat{\eta})\right)^2 + Var_\theta(\widehat{\eta}) \qquad (3.4.2)
\end{aligned}
$$

---

[4]We use the notation $f(x\,|\,\theta)$ to denote the joint *pdf* or *pmf* of $X = (X_1, X_2, \dots, X_n)$ instead of $f_X(x\,|\,\theta)$ since the subscript is obvious from the context.

where $B_\theta(\widehat{\eta}) = E(\widehat{\eta}) - \eta$ is called the **bias** of $\widehat{\eta}$. Note that both the bias and variance of $\widehat{\eta}$ depend, in general, on θ; but we will drop this subscript often when it is clear from the context.

Definition 3.4.2 *An estimator $\widehat{\eta}$ of $\eta = \eta(\theta)$ is called unbiased if it has zero bias for all θ, or $E(\widehat{\eta}) = \eta$ for all $\theta \in \Theta$.*

From (3.4.2), it is obvious that for an unbiased estimator, its MSE is same as its variance. Therefore, a good amount of attention has been paid, while developing the theory of statistical inference, to find estimators which are not only unbiased but also have minimum variance.

Definition 3.4.3 *An estimator is called a (uniformly) minimum variance unbiased estimator (MVUE) of $\eta(\theta)$ if it is unbiased and has the smallest variance (for each $\theta \in \Theta$) among all unbiased estimators of $\eta(\theta)$. Thus, $\widehat{\eta}$ is an MVUE of $\eta(\theta)$ if*

$$E(\widehat{\eta}) \;=\; \eta(\theta) \quad \text{for all } \theta \in \Theta$$
$$\text{and} \quad Var(\widehat{\eta}) \;\leq\; Var(\widehat{\eta}^*) \quad \text{for all } \theta \in \Theta$$

for any other unbiased estimator $\widehat{\eta}^*$ of $\eta(\theta)$.

In the following, we now present some useful theorems on MVUEs and ordinary unbiased estimators.

Theorem 3.4.1 *An MVUE, if it exists, is unique, in the sense that if both $\widehat{\eta}_1$ and $\widehat{\eta}_2$ are MVUEs, then $\widehat{\eta}_1 = \widehat{\eta}_2$ with probability 1.*

According to Theorem 3.4.1, since an MVUE is unique, henceforth we shall refer to it as UMVUE (unique minimum variance unbiased estimator).

While one can check directly whether an estimator is unbiased or not, it is not immediately apparent how to satisfy oneself that an estimator has the smallest variance among all unbiased estimators. Two approaches are available to achieve this task. One method is based on the use of the Cramér-Rao inequality, and other one is based on the use of Rao-Blackwell inequality.

Theorem 3.4.2 (Cramér-Rao Inequality) *Let $\boldsymbol{X} = (X_1, X_2, \dots, X_n)$ be a sample with pdf or pmf $f(\boldsymbol{x}|\theta)$, and let $\widehat{\eta}(\boldsymbol{X})$ be any estimator where*

$E\big[\widehat{\eta}(\boldsymbol{X})\big]$ is differentiable with respect to θ. Suppose the joint pdf or pmf $f(\boldsymbol{x}\,|\,\theta)$ satisfies

(a) $\dfrac{\partial}{\partial\theta}\displaystyle\int\cdots\int h(\boldsymbol{x})f(\boldsymbol{x}\,|\,\theta)\,dx_1\ldots dx_n =$

$$\int\cdots\int h(\boldsymbol{x})\dfrac{\partial}{\partial\theta}f(\boldsymbol{x}\,|\,\theta)\,dx_1\ldots dx_n$$

if $\boldsymbol{X}$ is continuous; or

(b) $\dfrac{\partial}{\partial\theta}\displaystyle\sum_{x_n}\cdots\sum_{x_1}h(\boldsymbol{x})f(\boldsymbol{x}\,|\,\theta) = \sum_{x_n}\cdots\sum_{x_1}h(\boldsymbol{x})\dfrac{\partial}{\partial\theta}f(\boldsymbol{x}\,|\,\theta)$

if $\boldsymbol{X}$ is discrete;

for any function $h(\boldsymbol{x})$ such that $E|h(\boldsymbol{x})| < \infty$. Then,

$$Var\big(\widehat{\eta}(\boldsymbol{X})\big) \geq \frac{\left\{\frac{d}{d\theta}E\big[\widehat{\eta}(\boldsymbol{X})\big]\right\}^2}{E\left[\left(\frac{\partial}{\partial\theta}lnf(\boldsymbol{X}\,|\,\theta)\right)^2\right]} \tag{3.4.3}$$

The quantity in the denominator on the right hand side of (3.4.3) is called **Fisher information**, or the **information number**, and is denoted by

$$I(\theta) = E\left[\left(\frac{\partial}{\partial\theta}lnf(\boldsymbol{X}\,|\,\theta)\right)^2\right] \tag{3.4.4}$$

If we have *iid* observations $X_1, X_2, \ldots, X_n$ with common *pdf* or *pmf* $f(x\,|\,\theta)$, then,

$$I(\theta) = nE\left[\left(\frac{\partial}{\partial\theta}lnf(X\,|\,\theta)\right)^2\right] = n\cdot i(\theta) \quad \text{(say)} \tag{3.4.5}$$

where $i(\theta)$ is called Fisher information per observation.

Example 3.4.1: Suppose $X_1, X_2, \ldots, X_n$ are *iid* from a $N(\theta, \sigma^2)$ population where σ^2 is known. Consider the problem of estimation of $\eta(\theta) = \theta,\ \theta \in \Theta = \mathbb{R} =$ the set of real numbers.

Here,

$$f(x\,|\,\theta) = \frac{1}{\sqrt{2\pi}\sigma}\exp\left(-\frac{(x-\theta)^2}{2\sigma^2}\right)$$

and so $lnf(x\,|\,\theta) = (\text{constant}) - (x-\theta)^2/(2\sigma^2)$ where the 'constant' is free from θ. It is seen that $i(\theta) = 1/\sigma^2$. Using Theorem 3.4.2,

$$Var\big(\widehat{\theta}(\boldsymbol{X})\big) \geq \frac{\left\{\frac{d}{d\theta}E\big[\widehat{\theta}(\boldsymbol{X})\big]\right\}^2}{\left(n/\sigma^2\right)} \tag{3.4.6}$$

Let $E\left[\widehat{\theta}(\boldsymbol{X})\right] = \theta + B(\widehat{\theta})$, where $B(\widehat{\theta})$ is the bias of $\widehat{\theta}$, and then (3.4.6) can be written as

$$Var\left(\widehat{\theta}(\boldsymbol{X})\right) \geq \frac{\sigma^2}{n}\left(1 + B'(\widehat{\theta})\right)^2 \qquad (3.4.7)$$

If $\widehat{\theta}(\boldsymbol{X})$ is an unbiased estimator of θ, or $B(\widehat{\theta}) = 0$ for all θ, then (3.4.7) reduces further to

$$Var\left(\widehat{\theta}(\boldsymbol{X})\right) \geq \frac{\sigma^2}{n} \qquad (3.4.8)$$

Note that the estimator $\widehat{\theta} = \overline{X}$ has variance of σ^2/n. Hence, $\overline{X}$ is the UMVUE of θ. Also, the regularity condition of Theorem 3.4.2 is satisfied in this example.

Example 3.4.2: Suppose we have *iid* observations $X_1, X_2, \ldots, X_n$ following $N(\mu, \sigma^2)$ where σ^2 is unknown, but μ is known. Call $\theta = \sigma^2$, the parameter space being $\Theta = (0, \infty)$. We want to estimate (i) $\eta(\theta) = \theta = \sigma^2$, and (ii) $\eta(\theta) = \sqrt{\theta} = \sigma$.

(i) Estimation of σ^2 (Normal Variance)

The *pdf* of each observation is

$$f(x \mid \theta) = \frac{1}{\sqrt{2\pi\theta}} \exp\left(-\frac{(x - \mu)^2}{2\theta}\right)$$

and then,

$$lnf(x \mid \theta) = c - \frac{1}{2}ln\theta - \frac{(x - \mu)^2}{2\theta}$$

where 'c' is a constant free from θ. It is now easy to see that

$$i(\theta) = E\left[\left(\frac{\partial}{\partial \theta}lnf(X \mid \theta)\right)^2\right] = \frac{1}{2\theta^2} \qquad (3.4.9)$$

Therefore, by Theorem 3.4.2, for any estimator $\widehat{\theta} = \widehat{\sigma^2}$ of $\eta(\theta) = \sigma^2 = \theta$ with bias function $B(\widehat{\theta})$,

$$Var(\widehat{\theta}) \geq \frac{2\theta^2}{n}\left(1 + B'(\widehat{\theta})\right)^2 \qquad (3.4.10)$$

where $B'(\widehat{\theta}) = \frac{d}{d\theta}B(\widehat{\theta})$.

Now, we consider the MLE of θ given by

$$\widehat{\theta}_{\text{ML}} = \frac{1}{n}\sum_{i=1}^{n}(X_i - \mu)^2 \qquad (3.4.11)$$

Since $\sum_{i=1}^{n}(X_i - \mu)^2/\theta$ follows a χ_n^2 distribution, it can be verified that $\widehat{\theta}_{\mathrm{ML}}$ is unbiased for θ and $Var(\widehat{\theta}_{\mathrm{ML}}) = 2\theta^2/n$. Therefore, the variance of $\widehat{\theta}_{\mathrm{ML}}$ attains the Cramér-Rao lower bound, hence, $\widehat{\theta}_{\mathrm{ML}}$ is the UMVUE for $\theta = \sigma^2$.

(ii) Estimation of σ (Normal Standard Deviation)

The parametric function to be estimated now is $\eta(\theta) = \sqrt{\theta} = \sigma$. If $\widehat{\eta}(\boldsymbol{X})$ is an estimator of $\eta(\theta) = \sqrt{\theta}$, then by Theorem 3.4.2,

$$Var(\widehat{\eta}(\boldsymbol{X})) \geq \frac{2\theta^2}{n}\Big(\frac{1}{2\sqrt{\theta}} + B'(\widehat{\eta})\Big)^2 \tag{3.4.12}$$

If $\widehat{\eta}(\boldsymbol{X})$ is an unbiased estimator of $\eta(\theta) = \sqrt{\theta}$, then

$$Var(\widehat{\eta}(\boldsymbol{X})) \geq \frac{\theta}{2n} \tag{3.4.13}$$

Now, we consider the MLE of $\eta(\theta)$ given by

$$\widehat{\eta}_{\mathrm{ML}} = \sqrt{\sum_{i=1}^{n}(X_i - \mu)^2/n} \tag{3.4.14}$$

which is *not* an unbiased estimator, since

$$E(\widehat{\eta}_{\mathrm{ML}}) = \frac{\sqrt{2}\,\Gamma\big((n+1)/2\big)}{\sqrt{n}\,\Gamma(n/2)}\sqrt{\theta} \tag{3.4.15}$$

An unbiased estimator of $\eta(\theta) = \sqrt{\theta}$ is, thus,

$$\widehat{\eta}_{\mathrm{U}}(\boldsymbol{X}) = \sqrt{\frac{n}{2}}\,\frac{\Gamma(n/2)}{\Gamma\big((n+1)/2\big)}\widehat{\eta}_{\mathrm{ML}} \tag{3.4.16}$$

A direct derivation of the variance of $\widehat{\eta}_{\mathrm{U}}(\boldsymbol{X})$ shows

$$Var(\widehat{\eta}_{\mathrm{U}}(\boldsymbol{X})) = \theta\left[\frac{n}{2}\left(\frac{\Gamma(n/2)}{\Gamma\big((n+1)/2\big)}\right)^2 - 1\right] \tag{3.4.17}$$

which does not match the right hand side of (3.4.13). As a result, one may wonder whether it is possible to obtain another unbiased estimator with variance smaller than that of $\widehat{\eta}_{\mathrm{U}}$ in (3.4.17). We will address this question in Theorem 3.4.4.

Before we go to the next theorem, it should be pointed out that the regularity condition in Theorem 3.4.2 prohibits certain distributions to take advantage of the Cramér-Rao inequality, as illustrated by the following example.

Example 3.4.3: Let $X_1, X_2, \ldots, X_n$ be *iid* $CU(0, \theta)$ (continuous uniform), $\theta \in \Theta = (0, \infty)$. Our goal is to estimate $\eta(\theta) = \theta$.

Note that the *pdf* of a single observation is

$$f(x \mid \theta) = \frac{1}{\theta} I_{(0, \theta)}(x)$$

As a result,

$$\frac{d}{d\theta} \int_0^\theta h(x) f(x \mid \theta) \, dx \neq \int_0^\theta h(x) \frac{\partial}{\partial \theta} f(x \mid \theta) \, dx$$

The above inequality is true in general if the support of the distribution depends on the parameter we intend to estimate. The regularity condition of Theorem 3.4.2 is not satisfied, and hence, we cannot apply the Cramér-Rao inequality to evaluate the performance of an estimator of θ.

A generalization of Cramér-Rao inequality, known as Chapman-Robbins inequality, helps us handle the cases like the one in Example 3.4.3.

Theorem 3.4.3 (Chapman-Robbins Inequality) *Let X have pmf or pdf $f(x \mid \theta)$. $\widehat{\eta}_v(X)$ is an unbiased estimator of $\eta(\theta)$. If θ_0 is a fixed point in Θ, and $\epsilon \neq 0$ is such that $\theta_0 + \epsilon \in \Theta$ and $f(x \mid \theta_0) = 0 \Rightarrow f(x \mid \theta_0 + \epsilon) = 0$, then*

$$Var_{\theta_0}\left(\widehat{\eta}_v(X)\right) \geq \sup_\epsilon \frac{\left(\eta(\theta_0 + \epsilon) - \eta(\theta_0)\right)^2}{E_{\theta_0}\left[\left(\frac{f(X \mid \theta_0 + \epsilon)}{f(X \mid \theta_0)} - 1\right)^2\right]} \qquad (3.4.18)$$

Example 3.4.4: (Continuation of Example 3.4.3).
Note that (from Example 3.4.3) $f(X \mid \theta) = 0 \Rightarrow f(x \mid \theta + \epsilon) = 0$ when $-\infty < \epsilon < 0$. Hence, Chapman-Robbins inequality can be applied and the lower bound to the variance of an unbiased estimator of $\eta(\theta) = \theta$ is

$$\sup_{-\theta < \epsilon < 0} \frac{\epsilon^2}{E_\theta\left[\left(\frac{f(X \mid \theta + \epsilon)}{f(X \mid \theta)} - 1\right)^2\right]} = \sup_{-\theta < \epsilon < 0} \left(\frac{\epsilon^2 (\theta + \epsilon)^n}{\theta^n - (\theta + \epsilon)^n}\right)$$

If $n = 1$, the above lower bound is $\theta^2/4$.

The Cramér-Rao inequality, or more generally, the Chapman-Robbins inequality can help us in obtaining UMVUE estimators provided the lower bound to the variance is attained. But if an unbiased estimator does not attain the variance lower bound, then still it does not mean that the estimator is not UMVUE. The following theorem, coupled with the concept of completeness discussed earlier, helps us in settling the question whether an unbiased estimator is UMVUE or not even if the estimator does not attain the Cramér-Rao lower bound.

Theorem 3.4.4 (Rao-Blackwell Theorem) *Let $\widehat{\eta}_U^*(X)$ be any unbiased estimator of $\eta(\theta)$, and let T be a sufficient statistic for θ. Define $\widehat{\eta}_U(T) = E\big[\widehat{\eta}_U^* \big| T\big]$. Then $\widehat{\eta}_U(T)$ is unbiased for $\eta(\theta)$ and $Var_\theta\big(\widehat{\eta}_U(T)\big) \leq Var_\theta\big(\widehat{\eta}_U^*(X)\big)$ for all $\theta \in \Theta$.*

Corollary 3.4.1 *If the sufficient statistic T is complete, then the unbiased estimator $\widehat{\eta}_U(T)$, as defined in Theorem 3.4.4, is the UMVUE.*

Example 3.4.5: (Continuation of Example 3.4.2, part (ii)).
The statistic $T = \sum_{i=1}^{n}(X_i - \mu)^2$ is sufficient for $\theta = \sigma^2$. Also, T has a complete family of distributions. Therefore, by Corollary 3.4.1 and Theorem 3.4.4,

$$\widehat{\eta}_U(T) = \sqrt{\frac{n}{2}}\,\frac{\Gamma(n/2)}{\Gamma((n+1)/2)}\sqrt{\frac{T}{n}}$$

is the UMVUE of $\eta(\theta) = \sqrt{\theta} = \sigma$.

The real utility of Theorem 3.4.4 and Corollary 3.4.1 lies in problems where $\eta(\theta)$ is complicated.

Example 3.4.6: Let $X_1, X_2, \ldots, X_n$ be *iid* Poisson(θ). We want to estimate

$$\eta(\theta) = P(X = k) = \frac{\theta^k e^{-\theta}}{k!}, \quad \theta \in \Theta = (0, \infty)$$

Define a random variable Y as $Y = I(X_1 = k)$.[5] Then,

$$E_\theta(Y) = 1 \cdot P(X_1 = k) + 0 \cdot P(X_1 \neq k) = \eta(\theta), \quad \forall \theta \in \Theta$$

So, Y is an unbiased estimator of $\eta(\theta)$. Now, recall that $T = \sum_{i=1}^{n} X_i$ is sufficient for θ, and T, following a Poisson($n\theta$), has a complete family of distributions. Let $\widehat{\eta}_U(T) = E(Y|T)$. It can be seen that for $T = t$, $\widehat{\eta}_U(t) = P(X_1 = k \mid T = t) = \binom{t}{k}(n-1)^{t-k}/n^t$. Thus,

$$\widehat{\eta}_U(T) = \binom{T}{k}(n-1)^{T-k}/n^T \tag{3.4.19}$$

is the UMVUE of $\eta(\theta) = \theta^k e^{-\theta}/k!$.

---

[5] $I(X_1 = k) = 1$, when $X_1 = k$; $= 0$, otherwise.

3.5   Elements of Hypothesis Testing

Consider the random vector $X = (X_1, X_2, \ldots, X_n)'$ having a joint probability distribution (*pmf* or *pdf*) $f(x|\theta)$ where $\theta \in \Theta$, the parameter space. It is assumed that f is a known function except for the parameter θ. Our inference problem is to have more specific knowledge about the true value of θ within Θ. This problem is to test a **statement** (called a **hypothesis**) when we have before us a suggestion in the form

$$H_0 : \theta \in \Theta_0 \qquad (3.5.1)$$

where Θ_0 is a specified proper subset of Θ. A statement such as (3.5.1), whose validity needs to be verified on the basis of a set of observations on X, is called a statistical hypothesis and is generally denoted, as in (3.5.1), by H_0.

In this section, we will discuss the procedure that one needs to follow in testing (or verifying) a hypothesis.

The formulation of criteria that are used in constructing a test for any given hypothesis is the subject-matter of the theory of hypothesis testing. For this purpose, it will be useful to make a distinction between a **simple** hypothesis and a **composite** hypothesis.

Definition 3.5.1 *If a hypothesis H_0 specifies the joint distribution of $X = (X_1, X_2, \ldots, X_n)'$ completely, then H_0 is called a simple hypothesis. Otherwise, H_0 is said to be a composite hypothesis.*

Consider the case where we have observations $X_1, X_2, \ldots, X_n$ *iid* $N(\mu, 1)$, $\mu \in \mathbb{R}$, and we want to test the validity of the statement (hypothesis) "$\mu = 0$". The statement "$\mu = 0$" makes the above probability distribution completely known, and hence it is a **simple** hypothesis. On the other hand, if $X_1, X_2, \ldots, X_n$ are independent with X_i following $N(\mu_i, 1)$, $\mu_i \in \mathbb{R}$, $i = 1, 2, \ldots, n$, and one wants to verify the statement "$\mu_1 = \mu_2 = \ldots = \mu_n$", then under this statement X_i's are supposed to be *iid* $N(\mu, 1)$ for some unknown $\mu \in \mathbb{R}$ (common for all X_i's). As a result, the distribution of X_i's still remains unspecified (nonunique), and hence the statement "$\mu_1 = \mu_2 = \ldots = \mu_n$" is termed as a **composite** hypothesis.

The validity of a hypothesis H_0 regarding the distribution of $X = (X_1, X_2, \ldots, X_n)'$ is judged on the basis of the realization (observation) $x = (x_1, x_2, \ldots, x_n)'$ of X. The value of x will dictate whether we should consider H_0 to be valid or not. Suppose $R_{\mathbf{x}}$ denotes the range of $\mathbf{X}$ (i.e., the set of all possible values that $\mathbf{X}$ can assume). The range $R_{\mathbf{x}}$ is partitioned into two regions, say, A and R, called **acceptance region** and **rejection region**, respectively. If the observation $\mathbf{x}$ falls in the region A, then H_0 is considered to be valid, and H_0 is said to be accepted. On the other hand, if $\mathbf{x}$ belongs to the region R ($R = A^c$ since $A \cap R = \phi$), then H_0 is not considered to be valid and H_0 is said to be rejected. Often the rejection region R is referred to as **critical region**.

It should be kept in mind that the acceptance (rejection) of H_0 does not mean that H_0 has been proved to be true (untrue); rather it only means that as far as the evidence $\mathbf{x}$ is concerned, H_0 seems to be a plausible (nonplausible) hypothesis.

Testing a hypothesis H_0 is thus a procedure to partition $R_{\mathbf{x}}$ into A and $R(= A^c)$ which will determine whether to accept H_0 or not based on the evidence $\mathbf{x}$. To each test procedure there corresponds a unique partition of $R_{\mathbf{x}}$; and conversely, to each partition there corresponds a unique test procedure. Furthermore, each partition can be characterized by the corresponding critical region (rejection region) only.

Any hypothesis which differs from the null hypothesis is called an alternative hypothesis, and is denoted by H_1. An alternative hypothesis is considered in such a way that it is the one to be accepted when the null hypothesis is rejected.

For a parameter θ, the general format of the null and alternative hypotheses is $H_0 : \theta \in \Theta_0$ and $H_1 : \theta \in \Theta_0^c$ where Θ_0 is some subset of the parameter space Θ and Θ_0^c is its complement. But practicality and common sense can make our alternative hypothesis confined to a smaller subclass (smaller than Θ_0^c).

For example, an automobile manufacturer claims that its new compact models get, on an average, 30 miles per gallon (mpg) on highways. Let θ be the mean of the mileage distribution for these cars. Common sense tells that the manufacturer is not going to underrate the car, but we suspect that the mileage claim might be overstated. We want to see if the manufacturer's claim of $\theta = 30$ mpg can be rejected. Therefore, we set out null hypothesis as $H_0 : \theta = 30$. If H_0 is not accepted, we are sure that θ is less than 30 mpg since we believe that the manufacturer's claim is too high. Hence, our alternative hypothesis is $H_1 : \theta < 30$.

If we reject the null hypothesis when it is in fact *true*, we have made

an error that is called a **type-I error**. On the other hand, if we accept
the null hypothesis when it is in fact *false*, we have made another error
that is called a **type-II error**. Table 3.5 indicates how these errors occur.

	Decision Made	
Truth about H_0	Accept H_0	Reject H_0
H_0 is true	Correct decision	Type-I error
H_0 is false	Type-II error	Correct decision

Table 3.5.1: Type-I and Type-II Errors.

Note that we make a decision about H_0 (acceptance or rejection) based
on $\boldsymbol{X}$, and hence, commission of an error of either type is a random event.
Therefore, we can only talk about the probability of making each error.
Usually for a given sample size, an attempt to reduce the probability of
one type of error results in an increase in the probability of the other
type of error. In practical applications, one type of error may be more
serious than the other. In such a case, careful attention is given to the
probability of more serious error. It is possible to reduce both types of
errors simultaneously by increasing the sample size, but this may not
always be possible due to limited resources.

The probability with which we are willing to risk a type-I error is
called the **level of significance** of a test and it is often denoted by α.
The probability of making a type-II error is denoted by β. The quantity
$(1 - \beta)$ represents the probability of rejecting H_0 when it is in fact false,
and this $(1 - \beta)$ is called the **power** of the test.

According to the partition of A (acceptance region) and R (rejection or
critical region) discussed earlier, the probability of type-I error associated
with a given θ is $P_\theta(\boldsymbol{X} \in R) = P_\theta(R)$, provided $\theta \in \Theta_0$ (in (3.5.1)). On
the other hand, the probability of type-II error associated with a given θ
is $P_\theta(\boldsymbol{X} \in A) = P_\theta(A)$, provided $\theta \notin \Theta_0$. Since $R = A^c$, we can express
the probability of type-II error as $(1 - P_\theta(R))$ for $\theta \notin \Theta_0$.

The probability $P_\theta(R)$, regarded as a function of θ, $\theta \in \Theta$, is called
the **power function** (or the **operating characteristic function**) of
the test, i.e.,

$$P_\theta(R) = \begin{cases} \text{Probability of type-I error} & \text{if } \theta \in \Theta_0 \\ 1 - \text{Probability of type-II error} & \text{if } \theta \notin \Theta_0 \end{cases} \qquad (3.5.2)$$

For fixed level of significance α $(0 < \alpha < 1)$ we choose a partition of $R_X = A \bigcup R$ such that $P_\theta(R) \leq \alpha$ for $\theta \in \Theta_0$. The value $\sup_{\theta \in \Theta_0} P_\theta(R)$ is called the **size** of the test and it may not necessarily be same as α.

Likelihood Ratio Tests

The likelihood ratio method of hypothesis testing is based on maximum likelihood estimation. If $\boldsymbol{X} = (X_1, X_2, \ldots, X_n)'$ is a random vector with *pdf* or *pmf* $f(\boldsymbol{x}|\theta)$, then the likelihood function is defined as

$$L(\theta|\boldsymbol{X}) = f(\boldsymbol{X}|\theta)$$

Definition 3.5.2 *The likelihood ratio test statistic for testing* $H_0 : \theta \in \Theta_0$ *against* $H_1 : \theta \in \Theta_0^c$ *is*

$$\lambda(\boldsymbol{X}) = \frac{\sup_{\theta \in \Theta_0} L(\theta|\boldsymbol{X})}{\sup_{\theta \in \Theta} L(\theta|\boldsymbol{X})} \qquad (3.5.3)$$

A likelihood ratio test (LRT) is a test procedure that has a rejection region of the form $R = \{\boldsymbol{x}|\, \lambda(\boldsymbol{x}) \leq c\}$, where c is a suitable value between 0 and 1.

The rationale behind LRT principle is that if H_0 is really true then it would be supported by the data, and hence the restricted maximization of $L(\theta|\boldsymbol{X})$ in the numerator of (3.5.3) would come close to the unrestricted maximization in the denominator of (3.5.3) making $\lambda(\boldsymbol{X})$ come close to 1. Otherwise, if H_0 is really false, $\lambda(\boldsymbol{X})$ would be close to 0. The value of c which determines acceptance or rejection of H_0 is determined by the level of significance, i.e., α. Also, if $T = T(\boldsymbol{X})$ is a sufficient statistic for θ, then for convenience, one can use the likelihood function based on T instead of $\boldsymbol{X}$.

Example 3.5.1: Let $X_1, X_2, \ldots, X_n$ be *iid* $N(\theta, 1)$ random variables. Consider testing $H_0 : \theta = \theta_0$ against $H_1 : \theta \neq \theta_0$, where θ_0 is a prefixed value. It can be shown that the LRT statistic is

$$\lambda(\boldsymbol{X}) = \exp\left\{ -n(\overline{X} - \theta_0)^2/2 \right\} \qquad (3.5.4)$$

where $\overline{X} = \sum_1^n X_i/n$. The rejection region R is $R = \{\boldsymbol{x}|\, \lambda(\boldsymbol{x}) \leq c\}$ (with c strictly between 0 and 1) which can be rewritten as

$$R = \left\{ \boldsymbol{x}|\, |\overline{x} - \theta_0| \geq \sqrt{-(2/n)lnc} \right\}$$

Call $\sqrt{-(2/n)lnc} = c_*$, which ranges between 0 and ∞ as c ranges between 0 and 1. How do we choose c, and hence c_*? Note that there is a one-to-one correspondence between c and c_*. If we set our probability of type-I error at level α, then

$$\begin{aligned} \alpha &= P_{\theta_0}(|\overline{X} - \theta_0| \geq c_*) \\ &= P_{\theta_0}(|\sqrt{n}(\overline{X} - \theta_0)| \geq \sqrt{n}\,c_*) \end{aligned}$$

Since $\sqrt{n}(\overline{X} - \theta_0)$ follows $N(0,1)$ distribution, $\sqrt{n}\,c_*$ must be equal to $z_{(\alpha/2)}$, the upper $(\alpha/2)$-cutoff point of a standard normal curve; i.e., $c_* = \frac{1}{\sqrt{n}} z_{(\alpha/2)}$. The rejection region thus becomes

$$\begin{aligned} R &= \left\{x \mid |\overline{x} - \theta_0| \geq \frac{1}{\sqrt{n}} z_{(\alpha/2)}\right\} \\ &= \left\{x \mid \overline{x} \leq \theta_0 - \frac{1}{\sqrt{n}} z_{(\alpha/2)} \quad \text{or} \quad \overline{x} \geq \theta_0 + \frac{1}{\sqrt{n}} z_{(\alpha/2)}\right\} \end{aligned}$$

The power function $\beta(\theta)$ is obtained as

$$\begin{aligned} \beta(\theta) &= P_\theta\left(|\overline{X} - \theta_0| \geq \frac{1}{\sqrt{n}} z_{(\alpha/2)}\right) \\ &= 1 - P_\theta\left(\theta_0 - \frac{1}{\sqrt{n}} z_{(\alpha/2)} \leq \overline{X} \leq \theta_0 + \frac{1}{\sqrt{n}} z_{(\alpha/2)}\right) \\ &= 1 - \left[\Phi\left(\sqrt{n}(\theta_0 - \theta) + z_{(\alpha/2)}\right) - \Phi\left(\sqrt{n}(\theta_0 - \theta) - z_{(\alpha/2)}\right)\right] \end{aligned}$$

$$(3.5.5)$$

where $\Phi(\cdot)$ is the standard normal *cdf*. Figure 3.5.1 gives the plot of $\beta(\theta)$ in (3.5.5) for $\theta_0 = 0$, $\alpha = 0.05$, and $n = 5$.

Definition 3.5.3 *A test with power function $\beta(\theta)$ is unbiased if $\beta(\theta^*) \geq \beta(\theta_0^*)$ for any $\theta_0^* \in \Theta_0$ and $\theta^* \in \Theta_0^c$.*

Note that in Example 3.5.1, the power function $\beta(\theta)$ is always greater than α for $\theta \neq \theta_0$ (which can be proved analytically by using the symmetry and unimodality of a normal curve). Therefore, the LRT test for the normal model does have an unbiased power function. Also, unbiasedness of a test procedure is a reasonable requirement which essentially says that the probability of making a correct decision is greater than or equal to the probability of making a wrong decision.

Definition 3.5.4 *Let $\mathfrak{C}$ be a class of tests for testing $H_0 : \theta \in \Theta_0$ against $H_1 : \theta \in \Theta_0^c$. A test in class $\mathfrak{C}$ with power function $\beta_0(\theta)$, is a uniformly*

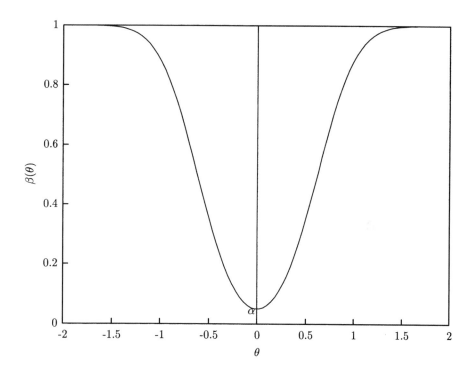

Figure 3.5.1: The Power Function $\beta(\theta)$.

most powerful (UMP) test in class $\mathfrak{C}$ provided $\beta_0(\theta) \geq \beta(\theta)$ for all $\theta \in \Theta_0^c$, where $\beta(\theta)$ is the power function of any other test in class $\mathfrak{C}$.

Usually, the class $\mathfrak{C}$, mentioned in the above definition, is taken as the collection of all tests with significance level α. For obvious notational simplicity, we can thus denote such a class by $\mathfrak{C}_\alpha$ instead of $\mathfrak{C}$.

In many problems with standard probability distributions, the UMP tests do not exist since $\mathfrak{C}_\alpha$ is too large, and consists of tests with power function very good ($\beta(\theta)$ value is closed to 1) in a subregion of Θ_0^c, but very poor ($\beta(\theta)$ value is closed to 0) in another subregion of Θ_0^c. In a situation like this, we look for an optimal test in a smaller class of tests like $\mathfrak{C}_\alpha^U$, the class of all unbiased tests with significance level α. An UMP test in $\mathfrak{C}_\alpha^U$ is called a uniformly most powerful unbiased (UMPU) test with significance level α.

One of the major goals in hypothesis testing is to identify UMP tests within suitable classes. The following results, which clearly describes UMP level α tests for testing a simple null against a simple alternative, is useful for more general situations.

Theorem 3.5.1 (Neyman-Pearson Lemma) *For testing $H_0 : \theta = \theta_0$ against $H_1 : \theta = \theta_1$, consider a rejection region of the form*

$$R_{NP} = \left\{ x \middle|\ f(x|\theta_1) > kf(x|\theta_0) \right\}$$

where $k \geq 0$ is such that $\alpha = P_{\theta_0}(X \in R)$. A test is UMP (in $\mathfrak{C}_\alpha$) with significance level α if and only if it has the above rejection region, R_{NP}.

Example 3.5.2: For *iid* observations $X_1, X_2, \ldots, X_n$ with $N(\theta, 1)$ distribution, the rejection region for testing $H_0 : \theta = \theta_0$ against $H_1 : \theta = \theta_1$ $(\theta_1 > \theta_0)$, is

$$\begin{aligned}
R_{\mathrm{NP}} &= \left\{ x \middle|\ f(x|\theta_1) > kf(x|\theta_0) \right\} \\
&= \left\{ x \middle|\ \bar{x} - \theta_0 > k_*(\theta_0, \theta_1, n) \right\} \qquad (3.5.6)
\end{aligned}$$

where $k_* = k_*(\theta_0, \theta_1, n) = ((\theta_1 - \theta_0)/2) + \ln k/(n(\theta_1 - \theta_0))$. Since θ_0, θ_1, and n are all known, finding k is equivalent to finding k_*, and this is done subject to the level condition, i.e.,

$$\begin{aligned}
\alpha &= P_{\theta_0}(\bar{X} - \theta_0 > k_*) \\
&= P_{\theta_0}\left(\sqrt{n}(\bar{X} - \theta_0) > \sqrt{n}k_* \right) \qquad (3.5.7)
\end{aligned}$$

Since $\sqrt{n}(\bar{X} - \theta_0)$ follows a standard normal distribution , from (3.5.7) we can obtain

$$k_* = \frac{1}{\sqrt{n}} z_\alpha$$

Therefore, the rejection region for testing $H_0 : \theta = \theta_0$ against $H_1 : \theta = \theta_1$ $(\theta_1 > \theta_0)$ is

$$R_{\mathrm{NP}} = \left\{ x \middle|\ \bar{x} - \theta_0 > z_\alpha/\sqrt{n} \right\} \qquad (3.5.8)$$

which is free from θ_1. The test with this rejection region has maximum power at θ_1. But since the rejection region is free from θ_1, it does not really matter which θ_1 $(> \theta_0)$ is being used. As long as θ_1 is greater than θ_0, the test with rejection region (3.5.8) is the most powerful in the class of all tests with the level α for testing $H_0 : \theta = \theta_0$ against $H_1 : \theta > \theta_0$.

Using the monotone likelihood ratio (MLR) property of the distribution of $\sqrt{n}(\overline{X} - \theta_0)$ (which is $N(0, 1)$), it can be shown that the test with the critical region (3.5.8) is indeed UMP for testing $H_0 : \theta \leq \theta_0$ against $H_1 : \theta > \theta_0$. The MLR property and its usage are discussed in the following.

Definition 3.5.5 *A family of pdfs or pmfs $\{h(s| \theta), \theta \in \Theta\}$ of a univariate random variable S with real valued parameter θ has a monotone likelihood ratio (MLR) if, for every $\theta_2 > \theta_1$, $r(s) = \left(h(s| \theta_2)/h(s| \theta_1)\right)$ is a nondecreasing function of s on the set $\left\{s\mid h(s| \theta_1) > 0 \text{ or } h(s| \theta_2) > 0\right\}$, where $r(s)$ takes the value ∞ if $h(s| \theta_1) = 0$.*

For a nonnegative random variable S we have the following useful results which essentially use the MLR property. For notational simplicity we may use $h_\theta(s)$ instead of $h(s|\theta)$ as the *pdf* or *pmf* of S.

Lemma 3.5.1 *Let S be a nonnegative random variable with pdf or pmf $h_\theta(s)$, $\theta \in \Theta$. If*

$$\pi(s) = \frac{h_{\theta_1}(s)}{h_{\theta_2}(s)}$$

is increasing in s for $\theta_1, \theta_2 \in \Theta$, then $P_{\theta_1}(S > c) \geq P_{\theta_2}(S > c)$ for all $c > 0$.

Lemma 3.5.2 *$P_{\theta_1}(S > c) \geq P_{\theta_2}(S > c)$ for all $c > 0$ implies that*

$$E_{\theta_1}[S] \geq E_{\theta_2}[S]$$

Theorem 3.5.2 *Let S be a nonnegative random variable with pdf or pmf $h_\theta(s)$, $\theta \in \Theta$. If*

$$\pi(s) = \frac{h_{\theta_1}(s)}{h_{\theta_2}(s)}$$

is increasing in s for $\theta_1, \theta_2 \in \Theta$, then

$$E_{\theta_1}[\phi(S)] \leq E_{\theta_2}[\phi(S)]$$

for any nonincreasing function ϕ.

Large Sample Tests

Recall Example 3.5.1 where we had to test $H_0 : \theta = \theta_0$ against $H_1 : \theta \neq \theta_0$ for *iid* observations $X_1, X_2, \ldots, X_n$ from a $N(\theta, 1)$ distribution. The LRT test procedure is characterized by the rejection region

$$R = \left\{ \boldsymbol{x} \mid |\overline{x} - \theta_0| \geq \frac{1}{\sqrt{n}} z_{(\alpha/2)} \right\}$$

which is actually determined by $(\overline{x} - \theta_0)$ or just by $\overline{x}$. In fact, the boundary of the rejection region, i.e., $(\theta_0 \pm z_{(\alpha/2)}/\sqrt{n})$ is determined by the probability distribution of the random variable $\overline{X}$ under the null hypothesis (or $\theta = \theta_0$). The random variable, in this case $\overline{X}$, which determines the boundary of the rejection region, is called the test statistic of the test procedure.

In many problems, it is not easy to derive the boundary of a rejection region due to the complicated probability distribution of the test statistic under the null hypothesis. In such cases, it may be possible to derive approximate rejection regions by using large sample (asymptotic) properties of useful yet common statistics (functions of the sample $\boldsymbol{X}$).

Suppose we wish to test a hypothesis about a real valued parameter θ, and $T(\boldsymbol{X})$ is a point estimator of θ based on the random sample $\boldsymbol{X} = (X_1, X_2, \ldots, X_n)'$ of size n. For example, $T(\boldsymbol{X})$ could be the MLE of θ. Suppose $\sigma_T^2(\theta)$ is the variance of $T(\boldsymbol{X})$, then $Z_* = (T(\boldsymbol{X}) - \theta)/\sigma_T$ follows $N(0, 1)$ distribution asymptotically,[6] and we can use $T(\boldsymbol{X})$ as an approximate test statistic for testing a hypothesis on θ. If σ_T is unknown, then replace it by a suitable estimate $\widehat{\sigma}_T$ so that the asymptotic normality still holds.

The likelihood ratio test procedure discussed earlier gives an explicit definition of the test statistic $\lambda(\boldsymbol{X})$ (see 3.5.3) and an explicit form of the rejection region as $R = \{\boldsymbol{x} \mid \lambda(\boldsymbol{x}) \leq c\}$. The constant c is chosen subject to the level condition,

$$P_\theta\big(\lambda(\boldsymbol{X}) \leq c\big) \leq \alpha \quad \text{for all } \theta \in \Theta_0$$

Hence, we need to derive the probability distribution of $\lambda(\boldsymbol{X})$ under the null hypothesis. If $\lambda(\boldsymbol{X})$ has a complicated probability distribution which might frustrate one from deriving the value of c, then we can look at the

---

[6] For large sample size n, the distribution of Z_* can be approximated by the $N(0, 1)$ distribution.

asymptotic distribution of $\lambda(\boldsymbol{X})$ (or a suitable monotone function of it) and derive an approximate value of c for large sample sizes. This is given in the following theorem.

Theorem 3.5.3 *Let $X_1, X_2, \ldots, X_n$ be iid following a probability distribution with pdf or pmf $f(x \mid \theta), \theta \in \Theta$. For testing $H_0 : \theta \in \Theta_0$ against $H_1 : \theta \in \Theta_0^c$, the distribution of $\left(-2 \ln \lambda(\boldsymbol{X}) \right)$ is asymptotically χ_ν^2 (chi-square distribution with ν degrees of freedom) provided some regularity conditions are satisfied by $f(x \mid \theta)$. Also, ν, the degrees of freedom, is determined by the difference between the number of free parameters specified by $\theta \in \Theta$ and the number of free parameters specified by $\theta \in \Theta_0$.*

Theorem 3.5.3 says that for sufficiently large sample size (and subject to some regularity conditions) the distribution of $(-2ln\lambda(\boldsymbol{X}))$ is approximated by χ_ν^2. Since rejection of $H_0 : \theta \in \Theta_0$ for 'small' values of $\lambda(\boldsymbol{X})$ is equivalent to rejection of H_0 for 'large' values of $(-2ln\lambda(\boldsymbol{X}))$, asymptotically H_0 is rejected if and only if $(-2ln\lambda(\boldsymbol{X})) \geq \chi_{\nu, \alpha}^2$, where $\chi_{\nu, \alpha}^2$ is the right (or upper) tail α-probability cutoff point of the χ_ν^2 distribution.

Example 3.5.3: Consider the random vector $\boldsymbol{X} = (X_1, X_2, \ldots, X_k)'$ having a joint probability mass function (*pmf*):

$$
\begin{aligned}
f(\boldsymbol{x} \mid \boldsymbol{\theta}) &= P(X_1 = x_1, X_2 = x_2, \ldots, X_k = x_k \mid \boldsymbol{\theta}) \\
&= \frac{n!}{x_1! x_2! \ldots x_k!} \theta_1^{x_1} \theta_2^{x_2} \ldots \theta_k^{x_k}, \qquad (3.5.9)
\end{aligned}
$$

where n is a fixed positive integer, $\boldsymbol{\theta} = (\theta_1, \theta_2, \ldots, \theta_k)$, $0 \leq \theta_i \leq 1, 1 \leq i \leq k$, $\sum_{i=1}^k \theta_i = 1$, and $x_1 + x_2 + \ldots + x_k = n$. The above probability distribution is called a **multinomial distribution** (a generalization of the binomial distribution), and it is denoted by $M(n, \theta_1, \ldots, \theta_k)$. Assume that θ_i's are all unknown and we want to test

$$ H_0 : (\theta_1, \theta_2, \ldots, \theta_k) = (\theta_1^0, \theta_2^0, \ldots, \theta_k^0) $$

against

$$ H_1 : \theta_i \neq \theta_i^0 \text{ for some } i $$

Such a hypothesis testing problem arises when we want to test a population composition. For example, five years ago, an artificial lake was stocked with fish in the following proportions: 30% bluegill, 25% bass, 25% catfish, and 20% pike. Five years later, a sample of size 400 fish is drawn from the lake, and the number of each type fish was recorded as below:

Type of Fish	Bluegill	Bass	Catfish	Pike	Total
Number	118	90	112	80	400

Do these data indicate that the composition of fish has changed over the last five year?

Note that for testing $H_0 : \boldsymbol{\theta} = (\theta_1, \theta_2, \ldots, \theta_k) = (\theta_1^0, \theta_2^0, \ldots, \theta_k^0) = \boldsymbol{\theta}^0$ (say), the MLEs of θ_i under H_0 are simply θ_i^0. The unrestricted MLEs are $\widehat{\theta}_i = X_i/n, i = 1, 2, \ldots, k$, where $X_i = \#$ fish of type-i. Therefore, the likelihood ratio is

$$\lambda(\boldsymbol{x}) = \frac{(\theta_1^0)^{x_1} \cdots (\theta_k^0)^{x_k}}{(\widehat{\theta}_1)^{x_1} \cdots (\widehat{\theta}_k)^{x_k}} = \prod_{i=1}^{k} \left(\frac{n\theta_i^0}{x_i}\right)^{x_i}$$

$$\text{or} \quad -2ln\lambda(\boldsymbol{x}) \quad = \quad -2\sum_{i=1}^{k} x_i ln\left(\frac{n\theta_i^0}{x_i}\right)$$

$$= \quad 2\sum_{i=1}^{k} x_i ln\left(\frac{x_i}{e_i}\right) \qquad (3.5.10)$$

where $e_i = n\theta_i^0$ is the expected number of items of ith type when H_0 is true. By Theorem 3.5.3, the distribution of $(-2ln\lambda(\boldsymbol{x}))$ is asymptotically (as $n \to \infty$) χ_ν^2. When H_0 is true, all θ_i values are specified, and hence the number of free parameters under H_0 is 0. In the full parameter space Θ, the θ_i values are subject to the restriction: $\theta_1 + \theta_2 + \ldots + \theta_k = 1$. Therefore, the number of free parameters in Θ are $(k-1)$. Hence, for large n, we reject H_0 if $(-2ln\lambda(\boldsymbol{x}))$ exceeds $\chi_{k-1, \alpha}^2$.

For the above mentioned fish example, $k = 4$, and $(\theta_1^0, \theta_2^0, \theta_3^0, \theta_4^0) = (0.30, 0.25, 0.25, 0.20)$. Using (3.5.10), it is found that

$$-2\ln\lambda(\boldsymbol{x}) = 2.4543$$

Taking $\alpha = 0.05$, from the chi-square distribution table in Appendix A, we can obtain $\chi_{k-1, \alpha}^2 = \chi_{3, 0.05}^2 = 7.8147$. Since $(-2ln\lambda(\boldsymbol{x})) = 2.4543 < 7.8147$, we accept the H_0 statement, and conclude that five years later, the proportions of each fish type remain the same as 30% bluegill, 25% bass, 25% catfish, and 20% pike. Moreover, since $n = 400$ is sufficiently large, we can use this asymptotic result easily.

The Notion of P-Value

The level of significance α in a hypothesis testing is the probability of rejecting H_0 when it is true. The value of α is set before the test is done.

However, two different values of α may lead to two different conclusions for the same sample data as seen in the following example.

Example 3.5.4: A lending agency has lowered their auto finance rate by 0.5%. Past records of accounts before the interest rate change showed the mean amount borrowed by new car buyers to be \$11,520. The management believes that the lower rate will encourage new car buyers to borrow more money. A random sample of 81 accounts were examined one year after the interest rate reduction. The average amount borrowed was $\bar{x} = \$12,300$ with standard deviation $s = \$3,000$. Test the management's belief at 0.05 and 0.01 levels of significance. Assume that amount borrowed by a new car buyer follows a normal distribution.

In this problem, we would like to test $H_0 : \mu = \$11,520$ against $H_1 : \mu > \$11,520$, where μ represents the mean amount borrowed by new car buyers after the rate change. The test statistic $T(\boldsymbol{X}) = \sqrt{n}(\bar{X} - 11520)/s$ follows t_{n-1}, i.e., t_{80}, distribution under H_0. The sample of size $n = 81$ gives $\bar{x} = \$12,300$ with $s = \$3,000$ resulting the value of T as $t = \sqrt{81}(12300 - 11520)/3000 = 2.34$. If α is taken as 0.05, then the critical value is $t_{n-1,\alpha} = t_{80,0.05} = 1.6641 < t = 2.34$, and hence H_0 is rejected at 5% level. On the other hand, if α is used as 0.01, then $t_{80,0.01} = 2.374 > t = 2.34$ and H_0 is accepted at 1% level.

In general, if the data $\boldsymbol{x}$ falls in the rejection region of a test procedure with level α, then it will also be in the rejection region of the same test with level $\alpha_* > \alpha$. A natural question is: "What is the smallest level of significance at which the sample data will tell us to reject H_0?" The answer is the 'p-value' associated with the test statistic value.

The p-value for the data $\boldsymbol{x}$ (test statistic $T(\boldsymbol{x})$) is the smallest value of α for which the data $\boldsymbol{x}$ (test statistic $T(\boldsymbol{x})$) will lead to rejection of H_0.

In Example 3.5.4, the test statistic value $T(\boldsymbol{x})$ is $t = T(\boldsymbol{x}) = 2.34$. The test statistic value $T(\boldsymbol{X}) = \sqrt{n}(\bar{X} - \mu_0)/s$ follows t_{n-1}-distribution under $H_0 : \mu = \mu_0$. In this case, the p-value is computed as

$$p\text{-value} = P_{\mu_0}\left(T(\boldsymbol{X}) \geq t\right) = P(t_{n-1} \geq 2.34) \approx 0.014$$

More generally, if $H_0 : \theta \in \Theta_0$ is a composite hypothesis, then the

p-value is described as

$$p\text{-value} = \sup_{\theta \in \Theta_0} P_\theta \left(\begin{array}{c} \text{test statistic is as or more extreme than} \\ \text{the observed test statistic value} \end{array} \right)$$

$$(3.5.11)$$

Instead of a fixed α value, p-value has more appeal because it presents the evidence against accepting H_0 and in favor of H_1. The smaller the p-value that one can obtain, the stronger the sample evidence that shows H_1 is true.

3.6 Set (or Interval) Estimation

In Section 3.3, we discussed point estimation of a parameter θ based on $\boldsymbol{X} = (X_1, X_2, \dots, X_n)'$ where we had to guess a single value $\widehat{\theta}(\boldsymbol{X})$ as the value of θ. Since $\boldsymbol{X}$ is random (having a probability distribution $f(\boldsymbol{x}|\theta)$, $\theta \in \Theta$) it is unlikely that $\widehat{\theta}(\boldsymbol{x})$, when $\boldsymbol{X} = \boldsymbol{x}$ is observed, would be the exact value of θ. All that we try by taking $\widehat{\theta}(\boldsymbol{x})$ to be the estimate of θ is that $\widehat{\theta}(\boldsymbol{x})$ is likely to differ from θ by a 'small amount'. That is why it is customary to provide standard error (standard deviation of $\widehat{\theta}(\boldsymbol{X})$) or an estimate of the standard error as a measure of variation of $\widehat{\theta}(\boldsymbol{X})$. Armed with the standard error (SE) or its estimate, one expects θ to lie between $\widehat{\theta}(\boldsymbol{X}) \pm \text{SE}$, or more likely to lie between $\widehat{\theta}(\boldsymbol{X}) \pm 2\,\text{SE}$. This idea leads us to the concept of set estimation based on the observed data $\boldsymbol{X} = \boldsymbol{x}$. We propose a set $C = C(\boldsymbol{x})$ such that $C \subseteq \Theta$ and

$$P\big(\theta \in C(\boldsymbol{X})\big) \geq (1 - \alpha) \qquad (3.6.1)$$

for some preassigned probability $(1 - \alpha)$, regardless of the true value of θ. Such a set C is called a set estimate of θ with confidence level $(1 - \alpha)$, or in short, a '$(1 - \alpha)$ level confidence set' of θ.

If θ is real valued (which we will assume throughout this section for simplicity), then we usually prefer the set estimate C to be an interval of the form $C = [L, U]$, where $L = L(\boldsymbol{X})$ and $U = U(\boldsymbol{X})$ are two statistics calculated from $\boldsymbol{X} = (X_1, X_2, \dots, X_n)'$ such that

$$P(L \leq \theta \leq U) \geq (1 - \alpha) \qquad (3.6.2)$$

for a preassigned probability $(1 - \alpha)$ and any $\theta \in \Theta$. Note that being functions of $\boldsymbol{X}$, L and U are random variables and hence, $[L, U]$ is a random interval. From (3.6.2), it is said that the random interval $[L, U]$ will cover the unknown parameter θ with at least probability $(1 - \alpha)$, whatever the true value of θ may be.

For instance, let $\alpha = 0.05$ (or $1 - \alpha = 0.95$), we can interpret the meaning of (3.6.2) as the following: let repeated observations x be taken on the random vector $\boldsymbol{X} = (X_1, X_2, \dots, X_n)'$ (representing the outcome of an experiment), and in each case, $l = L(\boldsymbol{x})$ and $u = U(\boldsymbol{x})$ are determined. Then, in about 95% or more of the cases the interval (l, u) will include θ, and it will fail to do so in about 5% or less of the cases.

Also, for any interval estimator $C = [L, U] = [L(\boldsymbol{X}), U(\boldsymbol{X})]$ of a parameter θ, the **coverage probability** of C is the probability that the random interval covers the true parameter θ. Usually, the coverage probability of C is a function of θ since the probability distribution of $\boldsymbol{X}$ depends on θ. The infimum of the coverage probabilities, i.e.,

$$\inf_{\theta \in \Theta} P\big(L(\boldsymbol{X}) \leq \theta \leq U(\boldsymbol{X})\big)$$

is called the **confidence coefficient** of the interval $C = [L, U]$.

Methods of Finding Set (or Interval) Estimators

Two popular methods of obtaining set or interval estimators are: (a) inverting a test statistic, and (b) making use of pivots.

(a) Inverting a test statistic

There is a strong relationship between set (or interval) estimation and hypothesis testing. Consider testing $H_0 : \theta = \theta_0$ against $H_1 : \theta \neq \theta_0$ for any $\theta_0 \in \Theta$. Let A be the acceptance region (which depends on θ_0 apart from $\boldsymbol{X}$) of a test with level α. For each $\boldsymbol{x} \in R_{\mathbf{x}}$, define a set $C(\boldsymbol{x}) \subseteq \Theta$ such that

$$C(\boldsymbol{x}) = \Big\{\theta_0 \Big|\, \boldsymbol{x} \in A\Big\} \tag{3.6.3}$$

Then, the random set $C(\boldsymbol{X})$ is a $(1 - \alpha)$ level confidence set for θ.

Example 3.6.1: Let $X_1, X_2, \dots, X_n$ be *iid* $N(\theta, 1)$ and we would like to derive a $(1 - \alpha)$ level confidence interval (CI) for θ.

First, note that for testing $H_0 : \theta = \theta_0$ against $H_1 : \theta \neq \theta_0$, the UMPU test after observing $\boldsymbol{X} = \boldsymbol{x}$ gives the acceptance region

$$A = \Big\{\boldsymbol{x} \Big|\, |\bar{x} - \theta_0| \leq \frac{1}{\sqrt{n}} z_{(\alpha/2)}\Big\}$$

Rewrite A for θ_0 as

$$A = \left\{ x \middle| \ \bar{x} - \frac{1}{\sqrt{n}} z_{(\alpha/2)} \leq \theta_0 \leq \bar{x} + \frac{1}{\sqrt{n}} z_{(\alpha/2)} \right\}$$

which gives

$$C(x) = \left\{ \theta_0 \middle| \ x \in A \right\} = \left[\bar{x} - \frac{1}{\sqrt{n}} z_{(\alpha/2)}, \ \bar{x} + \frac{1}{\sqrt{n}} z_{(\alpha/2)} \right]$$

Thus, $C(X) = \left[\bar{X} - \frac{1}{\sqrt{n}} Z_{(\alpha/2)}, \bar{X} + \frac{1}{\sqrt{n}} Z_{(\alpha/2)} \right]$ is a $(1-\alpha)$ level CI for θ.

(b) Making use of pivots

A functional $Q(X, \theta)$ (comprising both data and parameters) is called a pivot if the probability distribution of $Q(X, \theta)$ is completely known (or free of all parameters).

Since the distribution of $Q(X, \theta)$ is completely known, it is possible to find a set (known) B such that $(1 - \alpha) = P\big(Q(X, \theta) \in B\big)$. We then find a set estimate $C(X)$ of θ defined as

$$C(x) = \left\{ \theta \middle| \ Q(x, \theta) \in B \right\} \tag{3.6.4}$$

Example 3.6.2: Let $X_1, X_2, \ldots, X_n$ be *iid* $N(0, \sigma^2)$. Find a $(1 - \alpha)$ level confidence interval (CI) for σ^2.

The sufficient statistic for σ^2 is $S = \sum_{i=1}^{n} X_i^2$, and $Q(S, \sigma^2) = S/\sigma^2$ follows a χ_n^2-distribution which is completely known since the sample size n is known. By denoting the lower and upper $(\alpha/2)$-cutoff points of a χ_n^2 distribution by $\chi_{n,\,(1-\alpha/2)}^2$ and $\chi_{n,\,\alpha/2}^2$, respectively, we have

$$
\begin{aligned}
(1 - \alpha) &= P\left(\chi_{n,\,(1-\alpha/2)}^2 \leq \frac{S}{\sigma^2} \leq \chi_{n,\,\alpha/2}^2 \right) \\
&= P\left(\frac{S}{\chi_{n,\,\alpha/2}^2} \leq \sigma^2 \leq \frac{S}{\chi_{n,\,(1-\alpha/2)}^2} \right)
\end{aligned}
$$

Hence, $\left[S/\chi_{n,\,\alpha/2}^2, \ S/\chi_{n,\,(1-\alpha/2)}^2 \right]$ is a $(1 - \alpha)$ level CI for σ^2.

Remark 3.6.1 *There are problems where pivots do not exist, and in such cases, one must invert the acceptance region for testing a hypothesis to derive a set estimate of θ. For instance, if X follows Binomial(n, θ) or Poisson(θ), it is impossible to find a pivot which can allow us to follow (3.6.4).*

Chapter 4

Elements of Decision Theory

4.1 Introduction

In Chapter 3, we have discussed how to estimate a parameter θ given a set of observations $\boldsymbol{X} = (X_1, X_2, \ldots, X_n)'$ following a joint probability distribution, $f(\boldsymbol{x}|\theta), \theta \in \Theta$. (For the sake of simplicity we assume that the distribution of $\boldsymbol{X}$ depends only on a single, scalar valued parameter θ. Multiparameter cases, i.e., vector valued parameter $\boldsymbol{\theta}$ can be considered suitably if the situation demands.) A point estimate of θ was denoted by $\widehat{\theta}$ or $\widehat{\theta}(\boldsymbol{X})$ (with suitable subscripts to indicate the type of estimator or the method used to derive it). On the other hand, we used the notation C or $C(\boldsymbol{X}) = [L(\boldsymbol{X}), U(\boldsymbol{X})]$ to denote an interval estimate of θ. But whether we consider point or interval estimation of θ, it is always a tricky issue to evaluate the performance of an estimator. Traditionally, if $\widehat{\theta} = \widehat{\theta}(\boldsymbol{X})$ is a point estimate of θ, then we study the bias of $\widehat{\theta}$ given as $\mathrm{B}(\widehat{\theta}) = (E(\widehat{\theta}) - \theta)$, and the mean squared error (MSE) of $\widehat{\theta}$ given as $\mathrm{MSE}(\widehat{\theta}) = E[(\widehat{\theta} - \theta)^2]$. Both $\mathrm{B}(\widehat{\theta})$ and $\mathrm{MSE}(\widehat{\theta})$ are functions of θ. Similarly, if $C = [L(\boldsymbol{X}), U(\boldsymbol{X})]$ is an interval estimate of θ, then we study **Probability Coverage** (PC) of C given as $P(C) = P(L(\boldsymbol{X}) \leq \theta \leq U(\boldsymbol{X}))$, and **Expected Length** (EL) of C given as $\mathrm{EL}(C) = E[U(\boldsymbol{X}) - L(\boldsymbol{X})]$. (More generally, when we consider a set estimate $C = C(\boldsymbol{X})$ for a parameter vector $\boldsymbol{\theta}$, EL is replaced by **Expected Volume** (EV) of the set $C(\boldsymbol{X})$ given as $\mathrm{EV}(C) =$

$E[\text{volume of } C]$.) Again, both $P(C)$ and $\text{EL}(C)$ are functions of $\theta, \theta \in \Theta$.

In decision theory, not only do we unify both the estimation types (point and interval estimations) under a single framework, but also we are allowed to consider other ways to evaluate estimators and identify the optimal ones. The general framework of decision theory is described below in simple terms.

Based on the data $\boldsymbol{X} = (X_1, X_2, \dots, X_n)'$ following a probability distribution $f(\boldsymbol{x}|\theta), \theta \in \Theta$, we take a decision $\delta(\boldsymbol{X})$ about the parameter θ. The decision $\delta(\boldsymbol{X})$ could be a point estimate, or an interval estimate of θ, or a procedure to test a hypothesis on θ. Roughly speaking, the decision $\delta(\boldsymbol{X})$ indicates our 'understanding' about the parameter θ based on the data $\boldsymbol{X}$.

Since the data $\boldsymbol{X}$ gives only a partial information about θ, our decision $\delta(\boldsymbol{X})$ is never perfect (in most of the real life problems), and as a result, we incur a loss (due to our imperfect decision) 'L' which depends on $\delta(\boldsymbol{X})$ as well as θ and can be expressed as $L = L\big(\delta(\boldsymbol{X}), \theta\big)$. Note that the loss depends on three things: (i) the parameter value θ (unknown to us), (ii) the random data $\boldsymbol{X}$, and (iii) the structural form of δ. It is assumed that L is bounded below by a constant M, and without loss of generality we can take $M = 0,$[1] i.e., $L\big(\delta(\boldsymbol{X}), \theta\big) \geq 0, \ \forall \ \delta(\boldsymbol{X}), \theta$.

Since our decision $\delta(\boldsymbol{X})$ is data dependent, so is the loss $L\big(\delta(\boldsymbol{X}), \theta\big)$. Therefore, to evaluate the performance of our **decision making process** δ (the functional form), we look at the **risk of** δ defined as

$$R(\delta, \theta) = E\Big[L\big(\delta(\boldsymbol{X}), \theta\big)\Big] = \int L\big(\delta(\boldsymbol{x}), \theta\big) f(\boldsymbol{x}|\theta) \, d\boldsymbol{x} \qquad (4.1.1)$$

The expression $R(\delta, \theta)$ is called the risk function of the decision making process δ, provided the expectation in (4.1.1) exists. Note that $R(\delta, \theta)$ is nonnegative because the loss is assumed to be so. Suppose that $\mathcal{D}$ denotes the class of all possible decisions about θ, then it is seen that $R : \mathcal{D} \times \Theta \to \mathbb{R}^+$, where $\mathbb{R}^+$ is the positive part of the real line. The class $\mathcal{D}$ has uncountably many elements since it contains all possible real functions of $\boldsymbol{X}$.

---

[1] If $M \neq 0$, then we can look at a new loss L^* defined as $L^* = L - M \geq 0$.

4.2   Optimality Criteria

Using the risk $R(\delta, \theta)$, we can partially order the elements of $\mathcal{D}$. An element δ of $\mathcal{D}$ is called a 'decision rule' (or just a 'rule').

Definition 4.2.1 *Given two decision rules δ_1 and δ_2,*
(a) δ_1 is said to be as good as δ_2 if

$$R(\delta_1, \theta) \leq R(\delta_2, \theta) \quad \forall \, \theta \in \Theta \tag{4.2.1}$$

(b) δ_1 is said to be better (or uniformly better) than δ_2 if (4.2.1) holds and

$$R(\delta_1, \theta) < R(\delta_2, \theta) \quad for \ some \ \theta \in \Theta$$

Definition 4.2.2 *A decision rule δ_0 is called admissible if there does not exist any other rule δ which is better than δ_0. A decision rule is called inadmissible if it is not admissible.*

Even though admissibility is an optimality criterion, it is rather weak. In order to be admissible all a decision rule δ needs to possess is 'very small' risk at some $\theta \in \Theta$. The Figure 4.2.1 gives a hypothetical example with $\mathcal{D} = \{\delta_1, \delta_2, \delta_3, \delta_4\}$ where δ_3 is admissible yet it has 'high' risk (poor performance) in most of the parameter space. On the other hand, inadmissibility of a decision rule δ_0 is a strong result since it implies that there exists another decision rule (say, δ^*) which is better than δ_0. In the Figure 4.2.1, δ_2 is inadmissible since δ_3 (also δ_4) is better than δ_2.

Another optimality criterion is derived from **minimax principle** where a rule δ is judged by its supremum risk (or maximum risk), i.e., $\sup_{\theta \in \Theta} R(\delta, \theta)$. Note that we can order $\mathcal{D}$ fully in terms of supremum risk of the rules.

Definition 4.2.3 *A decision rule δ_0 is said to be minimax if it has the smallest supremum risk in the class of all rules, i.e.,*

$$\sup_{\theta \in \Theta} R(\delta_0, \theta) = \inf_{\delta \in \mathcal{D}} \sup_{\theta \in \Theta} R(\delta, \theta) \tag{4.2.2}$$

The minimax approach is like guarding against the worst possibility, but again it can result into an 'overkill' as the Figure 4.2.2 illustrates. Here δ_3 is minimax even though it may not be too appealing.

The other popular criterion which again helps us order $\mathcal{D}$ fully is based on the **Bayes' principle**. Traditionally the parameter θ is believed to be

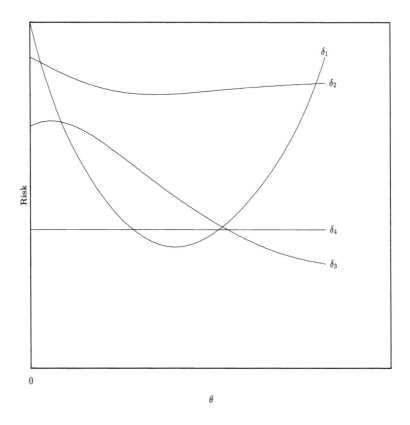

Figure 4.2.1: Risk Curves of δ_1, δ_2, δ_3, and δ_4.

completely unknown. The way it is understood is that the nature selects θ (unknown to us) from Θ following a probability distribution $\pi(\theta)$ (on Θ), and the probability distribution of $\boldsymbol{X}$, i.e., $f(\boldsymbol{x}|\theta)$, depends on θ thus selected. The probability distribution $\pi(\theta)$ of θ is called a **prior distribution**.

As θ is looked upon as a random variable, the expected value of $R(\delta, \theta)$, for variation of θ, will be a guiding principle in evaluating a rule δ. For a rule δ, its Bayes risk with respect to (w.r.t.) prior $\pi(\theta)$,

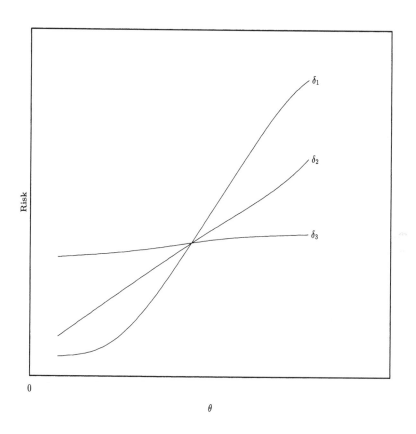

Figure 4.2.2: Risk Curves of δ_1, δ_2, and δ_3.

$\theta \in \Theta$, is defined as

$$r(\delta, \pi) = E\Big[R(\delta, \theta)\Big] = \begin{cases} \displaystyle\int_\Theta R(\delta, \theta)\pi(\theta)\, d\theta, & \text{if } \pi \text{ is continuous} \\[2em] \displaystyle\sum_{\theta \in \Theta} R(\delta, \theta)\pi(\theta), & \text{if } \pi \text{ is discrete} \end{cases}$$

$$(4.2.3)$$

Definition 4.2.4 *A decision rule δ_π is called a Bayes rule w.r.t. prior $\pi(\theta)$ if*

$$r(\delta_\pi, \pi) \leq r(\delta, \pi) \quad \text{for all } \delta \in \mathbf{D} \tag{4.2.4}$$

The concept of a prior distribution of an unknown parameter has always been debatable as whether this could be a practical approach in reality. Example 4.2.1 illustrates the usefulness of the Bayes principle.

Example 4.2.1: Consider the case of lot sampling where an inspector has to decide whether to accept or reject a lot of known size N of items (may be manufactured or agricultural products) by observing the number of defective items in a random sample of size n from the lot. The actual number of defective items in the whole lot, say K, is unknown to the inspector.

The observable random variable X, the number of defective items in a sample of size n, has a hypergeometric distribution with *pmf*

$$f(x\mid K) = \frac{\binom{K}{x}\binom{N-K}{n-x}}{\binom{N}{n}}, \quad x = 0, 1, 2, \ldots, n \qquad (4.2.5)$$

It is assumed that n is a small integer compared to N and K. But if the original lot is formed by taking N items at random from a production line then K, the number of defective items, would itself be a random variable. Assuming that the production process yields a proportion p of defective items, it is reasonable to take a binomial prior distribution of K as

$$\pi(k) = P(K = k) = \binom{N}{k} p^k (1-p)^{N-k}, \quad k = 0, 1, 2, \ldots, N \quad (4.2.6)$$

where p may be known from past experience.

The prior distribution may in turn depend on other parameters, known as super or hyper parameters, which may or may not be known from past experience. In case these super parameters are unknown (i.e., we just know the structural form of the prior distribution), then they can be estimated in the Empirical Bayes setup (not discussed here).

4.3   Loss Functions

In a point estimation problem, the loss function $L(\delta(\boldsymbol{X}), \theta)$ should be 'small' when $\delta(\boldsymbol{X})$ is close to θ, and the loss generally increases as $\delta(\boldsymbol{X})$ deviates more from θ. For a real valued parameter θ, let

$$\Delta = \delta(\boldsymbol{X}) - \theta \qquad (4.3.1)$$

denote the error in estimating θ by $\delta(\boldsymbol{X})$. Usually, $L(\delta(\boldsymbol{X}), \theta)$ is taken as a convex function, say $L(\Delta)$, of Δ. Two commonly used loss functions are

$$\text{Absolute Error Loss (AEL)} : L(\delta(\boldsymbol{X}), \theta) = |\Delta| \qquad (4.3.2)$$

and

$$\text{Squared Error Loss (SEL)} : L(\delta(\boldsymbol{X}), \theta) = \Delta^2 \qquad (4.3.3)$$

For point estimation of a parameter vector $\boldsymbol{\theta} = (\theta_1, \theta_2, \ldots, \theta_k)'$ by a decision rule $\boldsymbol{\delta}(\boldsymbol{X}) = (\delta_1(\boldsymbol{X}), \delta_2(\boldsymbol{X}), \ldots, \delta_k(\boldsymbol{X}))'$, the loss functions AEL and SEL can be generalized as $\sum_{i=1}^{k} |\delta_i(\boldsymbol{X}) - \theta_i|$ and $\sum_{i=1}^{k} (\delta_i(\boldsymbol{X}) - \theta_i)^2$, respectively.

A variation of SEL, one that penalizes over and under estimations differently, is reflected through the loss function

$$L(\delta(\boldsymbol{X}), \theta) = \begin{cases} K_1 \Delta^2 & \text{if } \delta(\boldsymbol{X}) \geq \theta \\ K_2 \Delta^2 & \text{if } \delta(\boldsymbol{X}) < \theta \end{cases} \qquad (4.3.4)$$

where K_1 and K_2 $(K_1 \neq K_2)$ are known constants. Another loss function which penalizes over and under estimations differently and has applications in real estate market is the LINEX[2] (Varian (1975)) loss function given as

$$L(\delta(\boldsymbol{X}), \theta) = b \exp(a\Delta) - c\Delta - b \qquad (4.3.5)$$

where $a, c \neq 0, b > 0$ are known suitable constants. Note that in (4.3.5), $L(\delta(\boldsymbol{X}), \theta) = L(\Delta)$ and $L(0) = 0$. For a minimum to exist at $\Delta = 0$, we must have $ab = c$. Thus, (4.3.5) can be rewritten as

$$L(\delta(\boldsymbol{X}), \theta) = L(\Delta) = b\{\exp(a\Delta) - a\Delta - 1\} \qquad (4.3.6)$$

For $a > 0$, the loss (4.3.6) is asymmetric with overestimation being more costly than underestimation and vice-versa for $a < 0$.

For a nonnegative scalar valued parameter θ (for example, when θ is a scale parameter), two popular loss functions are

$$\text{Squared Log Error (SLE)} : L(\delta(\boldsymbol{X}), \theta) = \left(ln\delta(\boldsymbol{X}) - ln\theta\right)^2 \qquad (4.3.7)$$

and

$$\text{Entropy Loss (EL)} : L(\delta(\boldsymbol{X}), \theta) = ln\left(\delta(\boldsymbol{X})/\theta\right) - \left(\delta(\boldsymbol{X})/\theta\right) - 1 \qquad (4.3.8)$$

---

[2]It is abbreviated from 'Linear-Exponential'.

where $\delta(X), \theta > 0$.

In testing a hypothesis $H_0 : \theta \in \Theta_0$ against $H_1 : \theta \in \Theta_0^c$, any decision rule $\delta(X)$ can take two possible values, $a_0 =$ 'accept H_0' and $a_1 =$ 'accept H_1'. The sets $A = \{x|\ \delta(x) = a_0\}$ and $R = \{x|\ \delta(x) = a_1\}$ are respectively the acceptance and rejection regions of a rule $\delta(X)$.

A loss function in a hypothesis testing problem should reflect the errors if $\theta \in \Theta_0$ and $\delta(x) = a_1$, or if $\theta \in \Theta_0^c$ and $\delta(x) = a_0$. The simple loss, known as '0-1 loss', which takes into account such errors (of wrong decisions) is given as

$$L(\delta(x), \theta) = \begin{cases} 0 & \text{if } \theta \in \Theta_0 \text{ and } \delta(x) = a_0 \text{ or } \theta \in \Theta_0^c \text{ and } \delta(x) = a_1 \\ 1 & \text{if } \theta \in \Theta_0 \text{ and } \delta(x) = a_1 \text{ or } \theta \in \Theta_0^c \text{ and } \delta(x) = a_0 \end{cases}$$
$$(4.3.9)$$

The risk of any decision rule $\delta(X)$ under the loss (4.3.9) is given as

$$R(\delta, \theta) = \begin{cases} 0 \cdot P(\delta(X) = a_0) + 1 \cdot P(\delta(X) = a_1) & \text{if } \theta \in \Theta_0 \\ 0 \cdot P(\delta(X) = a_1) + 1 \cdot P(\delta(X) = a_0) & \text{if } \theta \in \Theta_0^c \end{cases}$$

$$= \begin{cases} \text{Probability of type-I error} & \text{if } \theta \in \Theta_0 \\ \text{Probability of type-II error} & \text{if } \theta \in \Theta_0^c \end{cases} \quad (4.3.10)$$

which is closely related to the concept of power function described in (3.5.2) and this is due to the special nature of the 0-1 loss function which judges whether a decision is right or wrong.

A more general loss function for testing H_0 against H_1 can be given as

$$L(\delta(x), a_0) = \begin{cases} 0 & \text{if } \theta \in \Theta_0 \\ L(a_0, \theta) & \text{if } \theta \in \Theta_0^c \end{cases}$$
$$(4.3.11)$$

$$L(\delta(x), a_1) = \begin{cases} L(a_1, \theta) & \text{if } \theta \in \Theta_0 \\ 0 & \text{if } \theta \in \Theta_0^c \end{cases}$$

where both $L(a_0, \theta)$ and $L(a_1, \theta)$ are suitable nonnegative scalar values. The risk of a rule $\delta(X)$ under (4.3.11) is

$$R(\delta, \theta) = L(a_0, \theta)P(\delta(X) = a_0) + L(a_1, \theta)P(\delta(X) = a_1)$$
$$= L(a_0, \theta)(1 - \beta(\theta)) + L(a_1, \theta)\beta(\theta) \quad (4.3.12)$$

where $\beta(\theta) = P(\delta(X) = a_1)$ is the power function based on the decision rule δ.

The loss function in an interval estimation problem usually has two components: (i) a measure of whether the interval estimate correctly includes the true value θ, and (ii) a measure of the size of the interval estimate. Let $C = [L(\boldsymbol{X}), U(\boldsymbol{X})]$ be an interval estimate of a scalar valued parameter θ. Then the length of C, say $l(C)$, is called a measure of size, and the indicator function,

$$I_C(\theta) = \begin{cases} 1 & \text{if } \theta \in C \\ 0 & \text{if } \theta \notin C \end{cases}$$

expresses the correctness measure of C in containing θ.

The loss function should reflect the fact that a good interval estimate $C = [L(\boldsymbol{X}), U(\boldsymbol{X})]$ should have smaller $l(C)$ and large correctness $I_C(\theta)$. One such loss function for interval estimation of θ by C is given as

$$L(C, \theta) = a\, l(C) - b\, I_C(\theta) \qquad (4.3.13)$$

where a and b are suitable positive constants reflecting the relative weights of the two components of the loss function. Using the loss (4.3.13) one can get the associated risk function as

$$R(C, \theta) = a\, E\big[l(C)\big] - b\, P(\theta \in C) \qquad (4.3.14)$$

which also has two components: (i) the expected length of the interval, and (ii) the coverage probability of the interval.

In the multiparameter case, when θ is vector valued and C is a set estimate of θ, we replace $l(C)$ by $v(C)$, the volume of the set $C = C(\boldsymbol{X})$.

4.4 Admissible, Minimax, and Bayes Rules

We start this section by studying the structure of a Bayes rule first. The risk function of a decision rule δ is

$$R(\delta, \theta) = \int_{R_{\mathbf{x}}} L\big(\delta(\boldsymbol{x}), \theta\big) f(\boldsymbol{x} | \theta) \, d\boldsymbol{x}$$

and the Bayes risk of δ *w.r.t.* a prior distribution $\pi(\theta)$ of θ over Θ is

$$r(\delta, \pi) = \int_{\Theta} \int_{R_{\mathbf{x}}} L\big(\delta(\boldsymbol{x}), \theta\big) f(\boldsymbol{x} | \theta) \pi(\theta) \, d\boldsymbol{x} d\theta \qquad (4.4.1)$$

We assume that the functions $L(\delta(x), \theta)$ and $f(x|\theta)$ are such that $r(\delta, \pi)$ admits a change in the order of integration in (4.4.1), so that we have

$$r(\delta, \pi) = \int_{R_X} \int_\Theta L(\delta(x), \theta) f(x|\theta) \pi(\theta) \, d\theta dx \qquad (4.4.2)$$

In a Bayesian setup where θ is treated as a variable with a (marginal) distribution $\pi(\theta)$, the probability distribution $f(x|\theta)$, in true sense, is the conditional distribution of $X|\theta$. Thus the joint distribution of (X, θ) is

$$m(x, \theta) = f(x|\theta) \pi(\theta) \qquad (4.4.3)$$

and the conditional distribution of θ given $X = x$, called the **posterior distribution** of θ given $X = x$, is

$$\pi(\theta|x) = \frac{m(x, \theta)}{m(x)} \qquad (4.4.4)$$

where $m(x) = \int_\Theta m(x, \theta) \, d\theta$. Using (4.4.3) and (4.4.4), $r(\delta, \pi)$ can be rewritten as

$$\begin{aligned}
r(\delta, \pi) &= \int_{R_X} \int_\Theta L(\delta(x), \theta) \, \pi(\theta|x) m(x) \, d\theta dx \\
&= \int_{R_X} \left[\int_\Theta L(\delta(x), \theta) \, \pi(\theta|x) \, d\theta \right] m(x) \, dx \\
&= \int_{R_X} \left\{ E\left[L(\delta(x), \theta) | \theta \sim \pi(\theta|x) \right] \right\} m(x) \, dx \quad (4.4.5)
\end{aligned}$$

The inner expectation in (4.4.5), or

$$E\left[L(\delta(x), \theta) | \theta \sim \pi(\theta|x) \right] \qquad (4.4.6)$$

is called the **Bayes Expected Loss** (BEL) when the posterior distribution of θ is $\pi(\theta|x)$ for given $X = x$.

Suppose for each $X = x$ in R_X there exists an element $\delta_\pi(x)$ such that

$$E\left[L(\delta_\pi(x), \theta) | \theta \sim \pi(\theta|x) \right] = \inf_{\delta(x)} E\left[L(\delta(x), \theta) | \theta \sim \pi(\theta|x) \right] \quad (4.4.7)$$

Then the function $\delta_\pi(x), x \in R_X$, minimizes (4.4.5), and hence δ_π is the Bayes rule w.r.t. π. In other words, for fixed $X = x$, a Bayes rule is a value $\delta_\pi(x)$ of $\delta(x)$ that minimizes the BEL.

Theorem 4.4.1 *For point estimation of a real valued parameter θ, the Bayes rule $\delta_\pi(\boldsymbol{X})$ is:*
(a) $\delta_\pi(\boldsymbol{X}) = E\big[\theta\big|\,\theta \sim \pi(\theta|\,\boldsymbol{X})\big]$ if the loss is SEL (4.3.3)
(b) $\delta_\pi(\boldsymbol{X}) = Median\big(\theta\big|\,\theta \sim \pi(\theta|\,\boldsymbol{X})\big)$ if the loss is AEL (4.3.2)

To prove the above theorem, it is easy to note that the BEL under SEL is the MSE of θ where θ follows $\pi(\theta|\,\boldsymbol{x})$. The MSE of θ is minimized when $\delta(\boldsymbol{x}) = E\big[\theta\big|\,\theta \sim \pi(\theta|\,\boldsymbol{x})\big]$. Similarly, the BEL under AEL is nothing but the $E\big|\theta - \delta(\boldsymbol{X})\big|$ where θ follows $\pi(\theta|\,\boldsymbol{x})$, and is minimized when $\delta(\boldsymbol{x})$ = median of $\pi(\theta|\,\boldsymbol{x})$.

In many real life problems where the loss functions may not be as simple as AEL or SEL, and may reflect complex monetary or other resource losses, the minimization in (4.4.7) may not be possible analytically. But the integration can be evaluated and minimization can be carried out numerically. It is seen that after observing $\boldsymbol{X} = \boldsymbol{x}$, we need to solve (4.4.7) only for the value $\boldsymbol{X} = \boldsymbol{x}$, and need not be concerned with what the value of $\delta_\pi(\boldsymbol{x}^*)$ would be for $\boldsymbol{X} = \boldsymbol{x}^*$ where $\boldsymbol{x}^*$ is different from $\boldsymbol{x}$.

Example 4.4.1: Let $X_1, X_2, \ldots, X_n$ be *iid* $N(\theta, \sigma^2)$ and let the prior distribution of θ, $\pi(\theta)$, be $N(\mu, \tau^2)$, where μ, τ^2 and σ^2 are known. Our goal is to estimate θ.

The problem can be reduced first by sufficiency. It is known that all the relevant information concerning θ is contained in $\overline{X}$ which follows $N(\theta, \sigma^2/n)$.

Following the steps in (4.4.3) and (4.4.4) it can be shown that for $\overline{X} = \bar{x}$ the posterior distribution of θ is $\pi(\theta|\,\bar{x}) = N(\eta, \delta^2)$ where

$$\eta = E\big[\theta\big|\,\bar{x}\big] = \frac{\tau^2}{\tau^2 + (\sigma^2/n)}\bar{x} + \frac{\sigma^2/n}{\tau^2 + (\sigma^2/n)}\mu$$

and

$$\delta^2 = Var\big(\theta\big|\,\bar{x}\big) = \frac{\tau^2\sigma^2/n}{\tau^2 + (\sigma^2/n)}$$

Since the posterior distribution is normal which is symmetric about its mean (= median), the Bayes rule under SEL or AEL is $\delta_\pi(\overline{X}) = \eta$ (as given above).

At this point we need to add a few more words about Bayes rules. Usually the prior distribution $\pi(\theta)$ is a probability distribution with $\int_\Theta \pi(\theta)\,d\theta = 1$. This guarantees that the posterior $\pi(\theta|\,\boldsymbol{x})$ is also a probability distribution since it is a conditional distribution of θ given $\boldsymbol{x}$. But the existence of the posterior distribution is not conditional on $\pi(\theta)$ being a probability distribution. In other words, it is possible to have $\pi(\theta)$ with $\int_\Theta \pi(\theta)\,d\theta = \infty$, yet the posterior $\pi(\theta|\,x)$ is a probability distribution. Such a prior $\pi(\theta)$ is called an improper or generalized prior, and the resultant Bayes rule is called an **improper** or **generalized Bayes rule**. A prior $\pi(\theta)$ with $\int_\Theta \pi(\theta)\,d\theta < \infty$ is called a proper prior and the corresponding Bayes rule is called a **Proper Bayes** or just a **Bayes rule**.

We now turn our attention to the optimality criterion of admissibility. We shall explore the connection between admissibility of a rule and its being Bayes (*w.r.t.* some prior distribution) while answering the following two questions:

(**Q.1**) For a given problem, how to find an admissible rule? and

(**Q.2**) Given a rule, how to check its admissibility?

The following theorem which answers above (**Q.1**) states that Bayes rules are generally admissible.

Theorem 4.4.2 *A proper Bayes rule is admissible if it is unique except for risk equivalence. (Two rules are called equivalent if they have identical risk functions.)*

Theorem 4.4.2 can be used as a tool to generate admissible decision rules. As long as the proper Bayes rule δ_π (*w.r.t.* prior $\pi(\theta)$) is unique, it is admissible irrespective of the fact that the prior distribution $\pi(\theta)$ may or may not reflect the true nature of the parameter θ.

Example 4.4.2: Let $X_1, X_2, \ldots, X_n$ be *iid* Bernoulli(θ), $\theta \in (0,1)$. $X = \sum_{i=1}^n X_i$ is the minimal sufficient statistic for θ, and X follows $B(n,\theta)$, $R_X = \{0, 1, \ldots, n\}$. Suppose that we would like to find an admissible rule for θ. Take a prior $\pi(\theta)$ of θ as a Beta(α, β) $(\alpha > 0, \beta > 0)$ distribution. It can be shown that the posterior distribution $\pi(\theta|\,X)$ is Beta($X + \alpha, n - X + \beta$) distribution. Therefore, the unique Bayes rule under SEL is $\delta_\pi^{\alpha,\beta}(X) = E(\theta|\,X) = (X + \alpha)/(\alpha + \beta + n)$. By Theorem 4.4.2, $\delta_\pi^{\alpha,\beta}(X)$ is admissible for θ under SEL. In fact, we have achieved more; we have just produced a class of admissible rules,

$$\mathcal{D}_{\alpha,\beta} = \left\{ \delta_\pi^{\alpha,\beta}(X) = (X + \alpha)/(\alpha + \beta + n)\big|\,\alpha > 0, \beta > 0 \right\}$$

since $\delta_\pi^{\alpha,\beta}(X)$ is admissible for every fixed pair of $\alpha > 0, \beta > 0$.

Interestingly, in Example 4.4.2, the class of admissible rules, $\mathcal{D}_{\alpha,\beta}$, does not contain the traditional estimator $\delta_0 = X/n$ which is UMVUE as well as MLE. The only way $\delta_\pi^{\alpha,\beta}(X) = \delta_0$ if and only if $\alpha = \beta = 0$. But Beta(α, β) is not a probability distribution for $\alpha = \beta = 0$. So the question is: "Does there exist a prior probability distribution for which $\delta_0(X)$ is a Bayes rule?"

More generally, this brings out our question (**Q.2**) that is "Given a rule, how do we check its admissibility?" We first shall look at a more special case that deals with unbiased estimators and SEL.

Theorem 4.4.3 *Under SEL, no unbiased estimator can be a proper Bayes rule.*

Theorem 4.4.3 states that there does not exist any prior (proper) $\pi(\theta)$ for which an unbiased estimator can be made as a proper Bayes rule under SEL. Therefore, in Example 4.4.2, no proper prior can make $\delta_0(X) = X/n$ a proper Bayes rule under SEL, and hence the admissibility of $\delta_0(X)$ under SEL can not be answered by Theorem 4.4.2.

However, note that, if we consider an improper prior $\pi_*(\theta) \propto \left(\theta(\theta - 1)\right)^{-1}$,[3] then $\delta_0(X) = X/n$ is improper (or generalized Bayes) w.r.t. $\pi_*(\theta)$. But admissibility of improper or generalized rules is not guaranteed always. Also, by choosing α and β very close to 0 (yet > 0), we can make the proper prior Beta(α, β) arbitrarily close to above $\pi_*(\theta) \propto \left(\theta(\theta - 1)\right)^{-1}$. At the same time, the proper Bayes rule $\delta_\pi^{\alpha,\beta}(X)$ can be made arbitrarily close to $\delta_0(X)$.

Definition 4.4.1 *Let* $\left\{\pi_k(\theta)\right\}_{k \geq 1}$ *be a sequence of proper priors and* $\left\{\delta_\pi^k(X)\right\}$ *be the corresponding proper Bayes rules (under some loss function).*

(a) *A rule* $\delta(X)$ *is said to be a limit of Bayes rules* $\left\{\delta_\pi^k(X)\right\}$, *hence* $\delta(X)$ *is called a 'limiting Bayes' rule, if for almost all* x, $\delta_\pi^k(x) \to \delta(x)$ *as* $k \to \infty$.

(b) *A rule* $\delta(X)$ *is said to be Bayes 'in the wide sense' provided*

$$\lim_{k \to \infty} \left[r(\delta_\pi^k, \pi_k) - r(\delta, \pi_k) \right] = 0$$

---

[3] $\int_0^1 \left(\theta(\theta - 1)\right)^{-1} d\theta = \infty.$

(Note that a proper Bayes rule is necessarily a Bayes rule in the wide sense.)

Theorem 4.4.4 *Under some mild regularity conditions, a decision rule δ is admissible if and only if there exists a sequence $\{\pi_k\}$ of prior distributions (possibly improper) such that:*

(i) *Each π_k gives mass only to a closed and bounded subset of Θ (possibly different for each k), and hence has finite total mass*

(ii) *There is a closed and bounded set $\Theta_0 \subset \Theta$ to which each π_k gives mass 1*

(iii) $\lim_{k \to \infty} \left[r(\delta_\pi^k, \pi_k) - r(\delta, \pi_k) \right] = 0$, *where δ_π^k is the Bayes rule w.r.t π_k*

The regularity conditions under which Theorem 4.4.4 holds include the continuity of the risk function and those needed for Cramér-Rao inequality. Also, the above condition (ii) says that an admissible rule is essentially Bayes 'in the wide sense'. The next theorem is a weaker version of Theorem 4.4.4 but easy to implement for checking the admissibility of a given rule.

Theorem 4.4.5 *Consider a decision problem where Θ is a nondegenerate convex subset. Assuming that risk is continuous in θ, a decision rule δ is admissible if there exists a sequence $\{\pi_k\}$ of improper priors such that:*

(i) *The Bayes risks $r(\delta, \pi_k)$ and $r(\delta_\pi^k, \pi_k)$ are finite for all k, where δ_π^k is the Bayes rule w.r.t π_k*

(ii) *For any nondegenerate convex subset $\Theta_0 \subset \Theta$, there exists a $K > 0$ and an integer k_0 such that for $k \geq k_0$, $P_{\pi_k}(\theta \in \Theta_0) \geq K$*

(iii) $\lim_{k \to \infty} \left[r(\delta_\pi^k, \pi_k) - r(\delta, \pi_k) \right] = 0$

While one can use the last two general theorems to prove admissibility of a given rule for a specific problem, there are other techniques available for special cases. One such technique is the use of Cramér-Rao inequality to prove admissibility only under squared error loss (SEL) as shown in the next example.

Example 4.4.3: Without loss of generality, assume that X follows $N(\theta, 1)$, $\theta \in \Theta = \mathbb{R}$. Then, under SEL, $\delta_0(X) = X$ is an admissible rule for θ.

We prove the admissibility of $\delta_0(X) = X$ by the method of contradiction using Cramér-Rao inequality.

Suppose δ_0 is not admissible (i.e., inadmissible), then there exists another rule $\delta_1(X) \neq \delta_0(X)$ such that

$$1 = E\Big[\big(\delta_0(X) - \theta\big)^2\Big] \geq E\Big[\big(\delta_1(X) - \theta\big)^2\Big], \quad \forall \, \theta \in \mathbb{R} \qquad (4.4.8)$$

with strict inequality for some θ. But by Cramér-Rao inequality,

$$E\Big[\big(\delta_1(X) - \theta\big)^2\Big] \geq b^2(\theta) + \big(1 + b'(\theta)\big)^2 \qquad (4.4.9)$$

Since Fisher information $I(\theta) = 1$, and $b(\theta) = E\Big[\delta_1(X) - \theta\Big]$ = the bias of $\delta_1(X)$, combining (4.4.8) and (4.4.9), we get

$$b^2(\theta) + \big(1 + b'(\theta)\big)^2 \leq 1, \quad \forall \, \theta \in \mathbb{R} \qquad (4.4.10)$$

with strict inequality for some θ.   It is seen from (4.4.10) that $b'(\theta) \leq 0 \;\; \forall \; \theta$. In other words, the bias function $b(\theta)$ of $\delta_1(X)$ is decreasing in θ. Note that $b(\theta)$ is not identically equal to 0. (Otherwise, $\delta_1(X)$ would be an unbiased estimator of θ, and by completeness of $N(\theta, 1)$ distribution, $\delta_1(X)$ would be identical to $\delta_0(X)$ which would lead to a contradiction.) So, there must exist $\theta_0 \in \mathbb{R}$ such that $b(\theta_0) \neq 0$.

Case-(i): Let $b(\theta_0) > 0$. So $b(\theta) > 0$, $\forall \; \theta < \theta_0$, since $b(\theta)$ is a decreasing function of θ. From (4.4.10), we have that

$$b^2(\theta) + 2\,b'(\theta) \leq 0, \quad \forall \, \theta \in \mathbb{R}$$

then

$$1 \;\; \leq \;\; -2\,\frac{b'(\theta)}{b^2(\theta)}, \quad \forall \, \theta < \theta_0$$

$$\int_{\underline{\theta}}^{\theta_0} d\theta \;\; \leq \;\; -2\int_{\underline{\theta}}^{\theta_0} \frac{b'(\theta)}{b^2(\theta)} d\theta \quad \text{for any } \underline{\theta} < \theta_0$$

$$\text{or} \quad (\theta_0 - \underline{\theta}) \;\; \leq \;\; 2\Big(\frac{1}{b(\theta_0)} - \frac{1}{b(\underline{\theta})}\Big) \quad \text{for any } \underline{\theta} < \theta_0 \quad (4.4.11)$$

By letting $\underline{\theta} \to -\infty$, we have the left hand side of (4.4.11) approaching to ∞ whereas the right hand side is finite which leads to a contradiction. Hence, we must obtain that $b(\theta_0) \leq 0$.

Case-(ii): Let $b(\theta_0) < 0$. Then $b(\theta) < 0$, $\forall \, \theta > \theta_0$, because $b(\theta)$ is a decreasing function of θ. Similar to the argument of Case-(i), it can be shown that $b(\theta_0) \geq 0$.

Combining the both cases (i) and (ii), it is true that $b(\theta) \equiv 0$, or $E\Big[\delta_1(X) - \delta_0(X)\Big] = 0$, leading to a contradiction since $N(\theta, 1)$, $\theta \in \mathbb{R}$ is a complete family of distributions. Hence, $\delta_0(X) = X$ is admissible under SEL.

Remark 4.4.1 *Using the technique of Cramér-Rao lower bound, one can similarly prove admissibility of δ_0 under SEL for the following distributions: (a) X follows Poisson(λ) and $\delta_0 = X$, (b) X follows $B(n, \theta)$ and $\delta_0(X) = X/n$.*

Now we turn our attention to minimaxity. While we explore the connection between minimaxity and Bayes principle, we address the following two questions:

(**Q.3**) For a given problem, how to find a minimax rule?

(**Q.4**) Given a decision rule, how to check its minimaxity?

The following theorem is widely used to construct minimax rules and it partially answers (**Q.3**).

Theorem 4.4.6 *An admissible equalizer rule is minimax.*

Note that a unique proper Bayes rule is admissible, and hence a unique proper Bayes equalizer rule is minimax.

Recall that an equalizer rule is the one with constant risk. To construct an admissible rule, we first take a suitable (preferably conjugate[4]) family of proper prior distributions, and derive the Bayes rule and its risk. We then choose super parameters (parameters involved in the prior distributions) such that the risk of the Bayes rule turns out to be a constant. The next example illustrates the application of Theorem 4.4.6.

Example 4.4.4: Let X follow $B(n, \theta)$. Find a minimax rule for θ under SEL.

Consider a Beta prior family of distributions,

$$\pi(\theta) = \theta^{a-1}(1 - \theta)^{b-1}\Big/B(a, b), \quad a > 0, b > 0$$

---

[4]A family of prior distributions is called a **conjugate family** for which the posterior distribution has the same structure as the prior.

The posterior distribution of θ, given $X = x$, is

$$\pi(\theta|x) \propto \theta^{a+x-1}(1-\theta)^{n-x+b-1}$$

and hence the Bayes rule under SEL is

$$\delta_\pi(X) = \frac{a+X}{a+b+X}$$

We want to choose a and b (both are positive) such that δ_π has a constant risk. First we derive the regular risk of δ_π under SEL.

$$
\begin{aligned}
R(\delta_\pi, \theta) &= E\left[(\delta_\pi - \theta)^2\right] \\
&= E\left[\left(\frac{a+X}{a+b+n}\right)^2\right] - 2\theta E\left[\frac{a+X}{a+b+n}\right] + \theta^2 \\
&= \frac{a^2 + 2na\theta + n\theta(1-\theta) + n^2\theta^2}{(a+b+n)^2} - 2\frac{\theta(a+n\theta)}{a+b+n} + \theta^2 \\
&= K_0 + K_1\theta + K_2\theta^2 \qquad\qquad (4.4.12)
\end{aligned}
$$

where $K_0, K_1,$ and K_2 are suitable constants depending only on $a, b,$ and n. So it is seen that the risk of δ_π is a constant when $K_1 = K_2 = 0$. Equating the explicit expressions of K_1 and K_2 with 0 and then solving for a and b, we get $a = b = \sqrt{n}/2$. Therefore, a minimax estimator of θ under SEL is δ_π^* (say), where

$$\delta_\pi^*(X) = \frac{X + \sqrt{n}/2}{n + \sqrt{n}} \qquad\qquad (4.4.13)$$

A natural question thus arises is that ' Is the usual estimator, $\delta_0(X) = X/n$, which is UMVUE, MLE, and admissible under SEL, minimax under SEL?'

By plugging in $a = b = \sqrt{n}/2$ in the expression (4.4.12), we get the risk of δ_π^* as

$$R(\delta_\pi^*, \theta) = \frac{1}{4(1+\sqrt{n})^2} \qquad\qquad (4.4.14)$$

On the other hand,

$$R(\delta_0, \theta) = \frac{\theta(1-\theta)}{n} \qquad\qquad (4.4.15)$$

and it is seen that (Figure 4.4.1)

$$\frac{1}{4n} = \sup_\theta R(\delta_0, \theta) > \sup_\theta R(\delta_\pi^*, \theta) = \frac{1}{4(1+\sqrt{n})^2}$$

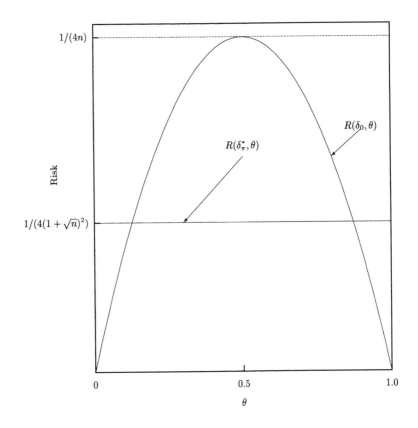

Figure 4.4.1: Risk Curves of δ_0 and δ_π^*.

So, δ_0 is not minimax for θ under SEL. Even though the difference between the supremum risks in (4.4.14) and (4.4.15) is negligible for large n, it still can be substantial for small n.

Remark 4.4.2 *For the above binomial example, if one uses the weighted squared error loss $L_*(\delta, \theta) = (\delta - \theta)^2 / \big(\theta(1 - \theta)\big)$, then $\delta_0 = X/n$ is still admissible under L_*. Under L_*, the risk of δ_0 is a constant $= 1/n$, hence, δ_0 is an admissible equalizer rule under L_*. By Theorem 4.4.6, this rule, δ_0, is minimax under L_*.*

Even though in the Binomial case, the usual estimator $\delta_0 = X/n$ (admissible, MLE, UMVUE) is not minimax under SEL, many times the standard estimators are indeed minimax under common loss functions. The following theorem is useful to prove minimaxity of a given rule and partially answers the above question (**Q.4**).

Theorem 4.4.7 *Let $\{\pi_k(\theta)\}_{k\geq 1}$ be a sequence of proper priors, and $\{\delta^k_\pi(\boldsymbol{X})\}$ be the corresponding proper Bayes rules (under some loss function). Suppose the Bayes risk $r(\delta^k_\pi, \pi_k)$ of δ^k_π w.r.t. π_k converges to a constant c as $k \to \infty$. If $\delta_0(\boldsymbol{X})$ is a decision rule with $R(\delta_0, \theta) \leq c$ for all $\theta \in \Theta$, then δ_0 is minimax.*

Example 4.4.5: As an application of Theorem 4.4.7, we can prove that the sample mean of *iid* observations with a normal distribution is minimax under SEL.

For the sake of simplicity, let us assume that we have a single observation X following $N(\theta, 1)$. The usual estimator of θ is $\delta_0(X) = X$ (MLE, UMVUE, and admissible under SEL). Suppose now we take a sequence of priors $\{\pi_k\}$ such that $\pi_k(\theta) = N(0, k)$ distribution. Under SEL, the Bayes rule corresponding to π_k is $\delta^k_\pi(X) = \frac{k}{k+1}X$ with Bayes risk $r(\delta^k_\pi, \pi_k) = \frac{k}{k+1}$. The limiting Bayes risk is thus

$$\lim_{k\to\infty} r(\delta^k_\pi, \pi_k) = 1 = R(\delta_0, \theta) \quad \forall \theta$$

By Theorem 4.4.7, δ_0 is minimax under SEL.

Using Theorem 4.4.7, one can find a minimax interval estimator for θ under a loss function when X follows $N(\theta, 1)$ as illustrated below.

For an interval estimate $C = C(X)$ of θ under a loss $L(C, \theta) = al(C) - bI_C(\theta)$ with suitable a and b, we consider the proper prior $\pi(\theta) = N(0, k)$. The posterior distribution $\pi(\theta| X)$ is $N\left(\frac{k}{k+1}X, \frac{k}{k+1}\right)$. The Bayes rule w.r.t. $\pi(\theta)$ which minimizes the Bayes expected loss (BEL) has the BEL of the form

$$\text{BEL} = aE_{\pi(\theta|X)}[l(C)] - bE_{\pi(\theta|X)}[I_C(\theta)]$$

Since $l(C)$, the length of C, is a constant in θ, we can obtain

$$\text{BEL} = a \cdot l(C) - b \cdot P_{\pi(\theta|X)}(\theta \in C)$$

where the second term in the above BEL indicates a posterior probability coverage of C. The above BEL is minimized if and only if the posterior probability coverage of C is maximized and that too happens when C has the form: $C(X) = \frac{k}{k+1} X \pm d$. Thus

$$\text{BEL} = 2ad - 2b \cdot P\left(0 < Z < d/\sqrt{k/(k+1)}\right) \qquad (4.4.16)$$

where Z is a standard normal random variable. One can now minimize the BEL (4.4.16) to obtain an optimal d, say d_π^k. Thus, the Bayes interval estimate for θ (w.r.t. $\pi(\theta)$) is $C_\pi^k(X) = \frac{k}{k+1} X \pm d_\pi^k$. Note that the interval $C_0(X) = X \pm d$ can be approximated by $C_\pi^k(X)$ by taking $k \to \infty$ and d is chosen suitably. Using Theorem 4.4.7, it now can be shown that $C_0(X)$ is minimax for certain values of a, b, and d.

Over the last several decades Statistical Decision Theory has produced some interesting results which have counter intuitive features. The most prominant among these is the famous 'Stein's effect' (Stein (1981)) as described below.

Suppose we have independent observations $X_1, \ldots, X_n$ such that each X_i follows $N(\theta_i, 1)$, $1 \le i \le n$. For estimating each θ_i separately under the SEL $(\delta_i - \theta_i)^2$ the usual estimator $\delta_i^0 = X_i$ is admissible. But for simultaneous estimation of $\boldsymbol{\theta} = (\theta_1, \ldots, \theta_n)'$ under the generalized SEL $\sum_{i=1}^n (\delta_i - \theta_i)^2$, the combined estimator $\boldsymbol{\delta}^0 = (\delta_1^0, \ldots, \delta_n^0)'$ is admissible only for $n \le 2$. For $n \ge 3$, the above $\boldsymbol{\delta}^0$ is inadmissible and can be dominated by a host of 'shrinkage' estimators including the famous 'James-Stein estimator' given as $\boldsymbol{\delta}^{JS} = (1 - (p-2)/||\boldsymbol{\delta}^0||^2)\boldsymbol{\delta}^0$. This feature of inadmissibility of a vector of individual admissible estimators when the dimension is greater than a critical value is called 'Stein's effect'. The Stein's effect is found to be applicable for many other distributions, like Gamma, Poisson, etc.

Another interesting decision-theoretic result derived by Stein is the one for normal variance estimation which is discussed in detail in the next chapter.

Chapter 5

Computational Aspects

5.1 Preliminaries

In earlier chapters, especially in Chapter 3 and Chapter 4, we have seen how to evaluate an estimator in terms of bias, variance, and risk. For instance, given *iid* observations $X_1, X_2, \ldots, X_n$ from a N(μ, σ^2) population where both μ and σ are unknown, the variance σ^2 can be estimated by its MLE

$$\widehat{\sigma}^2_{\mathrm{ML}} = \frac{1}{n} \sum_{i=1}^{n} (X_i - \overline{X})^2 = \frac{1}{n} S_X \text{ (say)}$$

where $\overline{X}$ is the sample mean. Using the fact that S_X/σ^2 follows χ^2_{n-1} distribution, it is now easy to find the bias and variance of $\widehat{\sigma}^2_{\mathrm{ML}}$ as

$$
\begin{aligned}
\text{Bias of } \widehat{\sigma}^2_{\mathrm{ML}} &= E\left[\widehat{\sigma}^2_{\mathrm{ML}}\right] - \sigma^2 \\
&= \left(\frac{n-1}{n}\right)\sigma^2 - \sigma^2 \\
&= -\frac{\sigma^2}{n} \qquad\qquad (5.1.1) \\
\text{Variance of } \widehat{\sigma}^2_{\mathrm{ML}} &= E\left[\left(\widehat{\sigma}^2_{\mathrm{ML}} - E\left[\widehat{\sigma}^2_{\mathrm{ML}}\right]\right)^2\right] \\
&= \frac{2(n-1)}{n^2}\sigma^4
\end{aligned}
$$

Using S_X/σ^2 as a pivot, one can find a $(1 - \alpha)$-level confidence interval for σ^2 as

$$\left[\frac{S_X}{\chi^2_{n-1,\,\alpha/2}}, \ \frac{S_X}{\chi^2_{n-1,\,(1-\alpha/2)}}\right]$$

where $\chi^2_{(n-1),\,\alpha/2}$ and $\chi^2_{(n-1),\,(1-\alpha/2)}$ are respectively the upper and lower tail $(\alpha/2)$-probability cutoff points of χ^2_{n-1} distribution. The length of this confidence interval is

$$S_X \left(\frac{1}{\chi^2_{n-1,\,(1-\alpha/2)}} - \frac{1}{\chi^2_{n-1,\,\alpha/2}} \right)$$

which is a random variable. Since the length plays an important role in evaluating a confidence interval, one can find the average (or expected) length of the above confidence interval which turns out to be

$$(n-1)\left(\frac{1}{\chi^2_{n-1,\,(1-\alpha/2)}} - \frac{1}{\chi^2_{n-1,\,\alpha/2}} \right)\sigma^2 \qquad (5.1.2)$$

and this can be compared with the expected length of any other $(1-\alpha)$-level confidence interval for σ^2.

For testing $H_0 : \sigma^2 = \sigma_0^2$ against $H_1 : \sigma^2 > \sigma_0^2$ (for some given σ_0^2), we can consider a test procedure at α-level of significance as:

$$\text{reject } H_0 \quad \text{if} \quad \frac{S_X}{\sigma_0^2} > \chi^2_{n-1,\,\alpha}$$

$$\text{accept } H_0 \quad \text{if} \quad \frac{S_X}{\sigma_0^2} \leq \chi^2_{n-1,\,\alpha}$$

The power of the above test procedure can be found as:

$$\begin{aligned}
\text{Power} \;&=\; P\big(\text{ reject } H_0 \big|\, H_1 \text{ is true }\big) \\
&=\; P_{\sigma^2}\left(\frac{S_X}{\sigma_0^2} > \chi^2_{n-1,\,\alpha} \,\Big|\, \sigma^2 > \sigma_0^2 \right) \\
&=\; P\left(\chi^2_{n-1} > \left(\frac{\sigma_0^2}{\sigma^2}\right)\chi^2_{n-1,\,\alpha} \,\Big|\, \sigma^2 > \sigma_0^2 \right) \qquad (5.1.3)
\end{aligned}$$

which is an increasing function of σ^2 (increases from α to 1 as σ^2 goes from σ_0^2 to ∞).

The above derivations involving σ^2 (point as well as interval estimation, and hypothesis testing) were relatively easy because of the involvement of chi-square distribution. But there are cases where it is difficult to obtain a closed expression as illustrated by the following example.

For estimating the normal variance σ^2 as described above, the MLE of σ^2 is certainly not the optimal estimator in terms of mean squared error (MSE). The optimal estimator of the form (constant)S_X (i.e., a constant multiple of S_X) having the smallest MSE is given by

$$\widehat{\sigma}^2_{opt} = \frac{S_X}{(n+1)} \tag{5.1.4}$$

In other words, the (optimal) estimator $\widehat{\sigma}^2_{opt}$ has the smallest risk under the squared error loss (SEL) in the class of estimators $\{cS_X, \ c > 0\}$. Similar to (5.1.1), one can find the bias, variance, and risk of $\widehat{\sigma}^2_{opt}$. However, an improved estimator which is better than $\widehat{\sigma}^2_{opt}$ (in terms of lower risk under SEL) is[1]

$$\widehat{\sigma}^2_{\mathrm{s}} = \min\left\{ \frac{S_X}{(n+1)}, \ \frac{S_X + n(\overline{X} - \mu_0)^2}{(n+2)} \right\} \tag{5.1.5}$$

where μ_0 is any known real number. Even though the estimator $\widehat{\sigma}^2_{\mathrm{s}}$ looks a bit uncommon, it has a nice interpretation. The estimator $\widehat{\sigma}^2_{\mathrm{s}}$ can be expressed as

$$\widehat{\sigma}^2_{\mathrm{s}} = \begin{cases} S_X/(n+1) & \text{if } n(\overline{X} - \mu_0)^2/S_X > 1/(n+1) \\ \\ \left(S_X + n(\overline{X} - \mu_0)^2\right)/(n+2) & \text{if } n(\overline{X} - \mu_0)^2/S_X \leq 1/(n+1) \end{cases} \tag{5.1.6}$$

The region of $\{n(\overline{X} - \mu_0)^2/S_X > 1/(n+1)\}$ can be treated as a critical (or rejection) region for testing the null hypothesis $H_0 : \mu = \mu_0$ against $H_1 : \mu \neq \mu_0$. Note that the type-I error of the test is not predetermined unlike the regular hypothesis testing case. The test statistic $n(\overline{X} - \mu_0)^2/S_X$, under $H_0 : \mu = \mu_0$, follows $F_{1,(n-1)}$-distribution, and hence the type-I error of the test is

$$\begin{aligned} \text{type-I error} \ &= \ P(\text{reject } H_0 \,|\, H_0 \text{ is true}) \\ &= \ P\left(F_{1,(n-1)} > \frac{1}{(n+1)} \right) \end{aligned} \tag{5.1.7}$$

which is dependent on the sample size n.

If H_0 is rejected, then μ is treated as completely unknown and σ^2 is estimated by $S_X/(n+1)$ (i.e., $\widehat{\sigma}^2_{opt}$). On the other hand, if H_0 is accepted, then μ is treated as known (and equal to μ_0) and σ^2 is estimated by

$$\frac{S_X + n(\overline{X} - \mu_0)^2}{(n+2)} = \frac{1}{(n+2)} \sum_{i=1}^{n} (X_i - \mu_0)^2$$

---

[1]C. Stein (1964) first proposed this estimator.

which is $\widehat{\sigma}^2_{opt}$ with *iid* observations $X_1, X_2, \ldots, X_n$ from $N(\mu_0, \sigma^2)$ distribution.

Finding the bias or risk expressions for $\widehat{\sigma}^2_s$ is not as easy as it would be for $\widehat{\sigma}^2_{ML}$ or $\widehat{\sigma}^2_{opt}$. In fact, it is not possible to get any closed expression of the bias (or risk) of $\widehat{\sigma}^2_s$. To see this, we use the notations $U = \sqrt{n}(\overline{X} - \mu_0)/\sigma$ (which follows $N(\sqrt{n}(\mu - \mu_0)/\sigma, 1)$), and $V = S_X/\sigma^2$ (which follows χ^2_{n-1}). Let

$$F = \frac{U^2}{V} = \frac{n(\overline{X} - \mu_0)^2}{S_X}$$

we can then obtain

$$E\left[\widehat{\sigma}^2_s\right] = E\left[\frac{S_X}{(n+1)} I\left(F > \frac{1}{(n+1)}\right) + \frac{(S_X + \sigma^2 U^2)}{(n+2)} I\left(F \le \frac{1}{(n+1)}\right)\right]$$
$$(5.1.8)$$

The random variable F follows a noncentral $F_{1,(n-1)}(\lambda)$-distribution with the noncentrality parameter λ (say) $= n(\mu - \mu_0)^2/\sigma^2$. Therefore, the bias of $\widehat{\sigma}^2_s$ is given by

$$\text{Bias of } \widehat{\sigma}^2_s = E\left[\widehat{\sigma}^2_s\right] - \sigma^2 =$$

$$\sigma^2 E\left[\frac{S_X/\sigma^2}{(n+1)} I\left(F > \frac{1}{(n+1)}\right) + \frac{(S_X + \sigma^2 U^2)/\sigma^2}{(n+2)} I\left(F \le \frac{1}{(n+1)}\right) - 1\right]$$
$$(5.1.9)$$

From the above bias expression (5.1.9), it is not clear how the estimator $\widehat{\sigma}^2_s$ fares compared to $\widehat{\sigma}^2_{ML}$ or $\widehat{\sigma}^2_{opt}$ in terms of bias.

Similar to the above bias expression (5.1.9), one can also derive the risk expression under SEL (i.e., MSE) for the estimator $\widehat{\sigma}^2_s$. Although $\widehat{\sigma}^2_s$ is known to be superior to $\widehat{\sigma}^2_{opt}$, it is important to study the risk function of $\widehat{\sigma}^2_s$ and see how much improvement it offers over $\widehat{\sigma}^2_{opt}$. If the risk improvement of $\widehat{\sigma}^2_s$ over $\widehat{\sigma}^2_{opt}$ turns out insignificant (over a large section of the parameter space), then one may ignore $\widehat{\sigma}^2_s$ and use $\widehat{\sigma}^2_{opt}$ instead.

Another example where deriving bias (or variance or risk) is more complicated than the one discussed above involves gamma distribution. Suppose that we have *iid* observations $X_1, X_2, \ldots, X_n$ from $G(\alpha, \beta)$ distribution with the *pdf*

$$f(x \mid \alpha, \beta) = \frac{1}{\beta^\alpha \Gamma(\alpha)} e^{-x/\beta} x^{\alpha-1}, \quad x > 0 \qquad (5.1.10)$$

where both α and β are unknown. To estimate α and β by the maximum likelihood estimation method, we first note the likelihood function

$$L(\alpha, \beta \mid \boldsymbol{X}) = \frac{1}{\beta^{n\alpha} (\Gamma(\alpha))^n} \, e^{-n\overline{X}/\beta} \left(\prod_{i=1}^{n} X_i \right)^{\alpha - 1}$$

where $\overline{X} = \sum_{i=1}^{n} X_i / n$; or the logarithmic likelihood function

$$lnL(\alpha, \beta \mid \boldsymbol{X}) = -n\alpha ln\beta - nln\Gamma(\alpha) - \frac{n\overline{X}}{\beta} + (\alpha - 1) \sum_{i=1}^{n} lnX_i \quad (5.1.11)$$

Differentiating the above log-likelihood function *w.r.t.* α and β and setting them equal to 0 yields

$$ln\beta + \frac{\Gamma'(\alpha)}{\Gamma(\alpha)} = \frac{1}{n} \sum_{i=1}^{n} lnX_i \quad \text{and} \quad \alpha\beta = \overline{X} \qquad (5.1.12)$$

where $\Gamma'(\alpha) = \partial\Gamma(\alpha)/\partial\alpha$. The maximum likelihood estimators $\widehat{\alpha}_{\mathrm{ML}}$ and $\widehat{\beta}_{\mathrm{ML}}$ of α and β, respectively, are the solutions of the system of equations in (5.1.12).

Note that the above MLEs of α and β do not have closed expressions, and hence it is impossible to derive bias, variance, and risk expressions of these estimators. Yet it is important for real life problems to study the performance of $\widehat{\alpha}_{\mathrm{ML}}$ and $\widehat{\beta}_{\mathrm{ML}}$ (in terms of bias, variance, or risk), and evaluate them in the presence of other estimators. For example, method of moments estimators (MMEs) have closed expressions for above α and β estimation.

The two examples discussed above were for point estimation. But similar difficulties can also arise either in set (interval) estimation where one may not have a closed expression for the volume (length) of the confidence set (interval) under consideration, or in hypothesis testing where one is interested in studying the complicated power function of a particular test procedure. Essentially, all these problems can be unified to look as an expectation of a suitable functional, say $G(\boldsymbol{X}, \theta)$, i.e., one needs to study $E[G(\boldsymbol{X}, \theta)]$ where $\boldsymbol{X}$ follows a probability distribution $f(\boldsymbol{x} \mid \theta)$.

In Section 5.2, we first talk about numerical integration technique which is becoming popular with fast computers and handy computing packages. Section 5.3 discusses Monte-Carlo simulation method which can be convenient when numerical integration is either too time consuming or not so accurate. Section 5.4 introduces a popular nonparametric

method based on a resampling technique. Finally, in Section 5.5 we will
see simple simulation techniques useful in testing and interval estimation.

5.2   Numerical Integration

Given the random vector $X = (X_1, X_2, \ldots, X_n)'$ following a joint probability distribution $f(x \mid \theta)$, we would like to study $E\big[G(X, \theta)\big]$ where G is
a suitable functional.[2] Usually, the functional G depends on the data X
through the minimal sufficient statistic, say $T = T(X)$. The dimension
of T is much less than that of X. Therefore, for convenience we can write

$$E\big[G(X, \theta)\big] = E\big[G(T, \theta)\big] \tag{5.2.1}$$

where the distribution of T is $g(t \mid \theta)$. The distribution g of T is assumed
to be known except the parameter θ. The expression (5.2.1) can be
written as

$$E\big[G(T, \theta)\big] = \int_{R_T} G(t, \theta) g(t \mid \theta) \, dt \tag{5.2.2}$$

where R_T is the suitable range for T and $dt = dt_1 dt_2 \ldots dt_k$ with k being
the dimension of T.

To study $E\big[G(X, \theta)\big]$ for $\theta \in \Theta$, we select several possible values of θ
from Θ, and for each given value of θ in Θ the integration in (5.2.2) is done
numerically. Finally, numerical integration values are plotted against θ
to get a better understanding of $E\big[G(X, \theta)\big]$.

As an example, we now compute the bias of $\hat{\sigma}_S^2$ through numerical
integration.

Bias of $\hat{\sigma}_S^2 =$

$$\sigma^2 E\left[\frac{V}{(n+1)} I\left(\frac{W}{V} > \frac{1}{(n+1)}\right) + \frac{(V+W)}{(n+2)} I\left(\frac{W}{V} < \frac{1}{(n+1)}\right) - 1\right] \tag{5.2.3}$$

where W follows a noncentral $\chi_1^2(\lambda)$ distribution with noncentrality parameter $\lambda = n(\mu - \mu_0)^2/\sigma^2$, and W is independent of V. The *pdf*s of V
and W are respectively

$$V \sim g_1(v) \;=\; \frac{1}{2^{(n-1)/2}\Gamma((n-1)/2)} e^{-v/2} v^{(n-1)/2-1}$$

---

[2]Here G is a function of the data and the parameter θ. Such a G is called a
'functional'.

$$W \sim g_2(w \mid \lambda) \;=\; \sum_{j=0}^{\infty} e^{-\lambda/2} \frac{(\lambda/2)^j}{j!} \cdot \frac{1}{2^{j+1/2}\Gamma(j+1/2)} e^{-w/2} w^{(j+1/2)-1}$$

$$(5.2.4)$$

As one can see from above, a noncentral $\chi_k^2(\lambda)$ distribution can be characterized as a Poisson mixture of central χ_{k+2J}^2 distributions with J following Poisson$(\lambda/2)$ distribution.

Let

$$\begin{aligned}
A_1(\lambda) &= E\left[\frac{V}{(n+1)} I\left(\frac{W}{V} > \frac{1}{(n+1)}\right)\right] \\
&= \frac{1}{(n+1)} \int_0^{\infty} \int_0^{(n+1)w} v g_1(v) g_2(w \mid \lambda)\, dv\, dw \quad (5.2.5)
\end{aligned}$$

and

$$\begin{aligned}
A_2(\lambda) &= E\left[\frac{(V+W)}{(n+2)} I\left(\frac{W}{V} < \frac{1}{(n+1)}\right)\right] \\
&= \frac{1}{(n+2)} \int_0^{\infty} \int_{(n+1)w}^{\infty} (v+w) g_1(v) g_2(w \mid \lambda)\, dv\, dw
\end{aligned}$$

$$(5.2.6)$$

From (5.2.3), (5.2.5), and (5.2.6), it is easy to see that the bias of $\widehat{\sigma}_{\mathrm{S}}^2$ is a function of σ^2 as well as $\lambda = n(\mu - \mu_0)^2/\sigma^2$ (apart from n), and hence denoting it by $B_n(\widehat{\sigma}_{\mathrm{S}}^2 \mid \sigma^2, \lambda)$ we get

$$\begin{aligned}
\text{Bias of } \widehat{\sigma}_{\mathrm{S}}^2 &= B_n(\widehat{\sigma}_{\mathrm{S}}^2 \mid \sigma^2, \lambda) \\
&= \sigma^2 \Big\{ A_1(\lambda) + A_2(\lambda) - 1 \Big\} \quad (5.2.7)
\end{aligned}$$

where $A_1(\lambda)$ and $A_2(\lambda)$ are given in (5.2.5) and (5.2.6), respectively. The expression $B_n(\widehat{\sigma}_{\mathrm{S}}^2 \mid \sigma^2, \lambda)$ is directly proportional to $A_1(\lambda) + A_2(\lambda) - 1$ and hence we can plot

$$B_n(\widehat{\sigma}_{\mathrm{S}}^2 \mid \sigma^2, \lambda)/\sigma^2 = B_n(\widehat{\sigma}_{\mathrm{S}}^2 \mid 1, \lambda) \quad (5.2.8)$$

against λ to get an idea about the bias of $\widehat{\sigma}_{\mathrm{S}}^2$. This is illustrated below with specific choices of n and λ.

First, using the computer package *Mathematica*, we compute

$$B_n(\widehat{\sigma}_{\mathrm{S}}^2 \mid 1, \lambda) = A_1(\lambda) + A_2(\lambda) - 1$$

for selected values of λ with $n = 5$ and 10. The numerical results are provided in Table 5.2.1. Figure 5.2.1 shows the plots of

$$B_5(\widehat{\sigma}_{\mathrm{S}}^2 \mid 1, \lambda) = B_5 \quad \text{and} \quad B_{10}(\widehat{\sigma}_{\mathrm{S}}^2 \mid 1, \lambda) = B_{10}$$

λ	n		λ	n	
	5	10		5	10
0.0	$-.377$	$-.214$	1.0	$-.362$	$-.203$
0.1	$-.375$	$-.213$	1.5	$-.357$	$-.199$
0.3	$-.372$	$-.210$	2.5	$-.349$	$-.193$
0.5	$-.369$	$-.208$	5.0	$-.339$	$-.186$
0.7	$-.366$	$-.206$	10.0	$-.334$	$-.182$

Table 5.2.1: Computed Values of $B_n(\widehat{\sigma}_{\mathrm{s}}^2 | 1, \lambda)$.

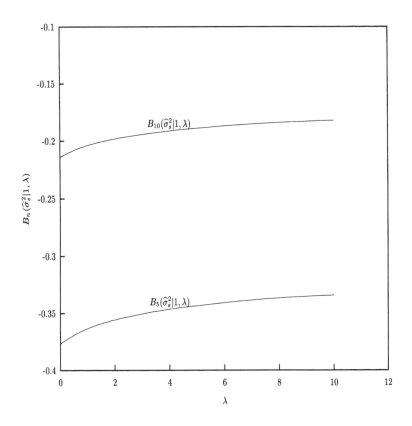

Figure 5.2.1: Plots of $B_5(\widehat{\sigma}_{\mathrm{s}}^2 | 1, \lambda)$ and $B_{10}(\widehat{\sigma}_{\mathrm{s}}^2 | 1, \lambda)$.

Denoting the bias of $\widehat{\sigma}^2_{opt}$ (see (5.1.4)) by $B_n(\widehat{\sigma}^2_{opt}|\sigma^2)$ which depends on the parameter σ^2 only except n, we have

$$B_n(\widehat{\sigma}^2_{opt}|\sigma^2) = \sigma^2\left(\frac{n-1}{n+1}\right) - \sigma^2 = -\frac{2}{(n+1)^2}\sigma^2 \qquad (5.2.9)$$

It is straightforward that

$$B_n(\widehat{\sigma}^2_{opt}|1) = B_n(\widehat{\sigma}^2_{opt}|\sigma^2)/\sigma^2$$

with

$$B_5(\widehat{\sigma}^2_{opt}|1) = -0.33 \quad \text{and} \quad B_{10}(\widehat{\sigma}^2_{opt}|1) = -0.18$$

Remark 5.2.1 *From the above numerical result, it is clear that $\widehat{\sigma}^2_{opt}$ has always larger absolute bias than $\widehat{\sigma}^2_s$ does. Indeed, it is true that $B_n(\widehat{\sigma}^2_s|1,\lambda) \longrightarrow B_n(\widehat{\sigma}^2_{opt}|1)$ as $\lambda \to \infty$ for fixed n. Both $\widehat{\sigma}^2_s$ and $\widehat{\sigma}^2_{opt}$ underestimate σ^2.*

Remark 5.2.2 *The computer package 'Mathematica' works fine for single or double integration(s). But for triple or higher order integrations, it may not be very effective.*

5.3   Monte-Carlo Simulation

Monte-Carlo simulation is an alternative approach (alternative to numerical integration) to evaluate the expectation $E[G(\boldsymbol{X},\theta)]$ where the random variable $\boldsymbol{X}$ follows the distribution $f(\boldsymbol{x}|\theta)$, $\theta \in \Theta$. The distribution $f(\boldsymbol{x}|\theta)$ is assumed to have a known form. The technique of simulation depends heavily on artificial data generated from the distribution $f(\boldsymbol{x}|\theta)$ for selected values of θ. Most of the latest standard statistical software packages have a built-in mechanism to generate data from commonly used probability distributions.

The Monte-Carlo simulation is conducted through a large number (say, M) of steps (or cycles). First, fix a suitable value of $\theta \in \Theta$. Then at the jth step, $1 \leq j \leq M$, generate a random sample $\boldsymbol{X}^{(j)}$ of size n from the known distribution $f(\boldsymbol{x}|\theta)$. For this data $\boldsymbol{X}^{(j)} = \left(X_1^{(j)}, \ldots, X_n^{(j)}\right)$, compute $G(\boldsymbol{X},\theta)$ and call it G_j, i.e., $G_j = G(\boldsymbol{X}^{(j)},\theta)$. After the Mth step, we have generated the simulated (or replicated) values of $G(\boldsymbol{X},\theta)$

as $G_1, G_2, \ldots, G_M$. The expectation of the functional $G(\boldsymbol{X}, \theta)$ is then approximated by the average of $G_1, G_2, \ldots, G_M$, or

$$E\big[G(\boldsymbol{X}, \theta)\big] \approx \frac{1}{M} \sum_{j=1}^{M} G_j = \frac{1}{M} \sum_{j=1}^{M} G(\boldsymbol{X}^{(j)}, \theta) = E_G^*(\theta) \ \text{(say)} \quad (5.3.1)$$

Since the functional $G(\boldsymbol{X}, \theta)$ often depends on the data $\boldsymbol{X}$ through the minimal sufficient statistic $\boldsymbol{T} = \boldsymbol{T}(\boldsymbol{X})$ with dimension $k < n$, it is convenient to conduct the simulation by generating observations on $\boldsymbol{T}$ rather than $\boldsymbol{X}$. To be precise, let (5.2.1) hold and $\boldsymbol{T}$ has distribution $g(\boldsymbol{t} \mid \theta)$, $\theta \in \Theta$, then for a selected value of $\theta \in \Theta$, independently generate $\boldsymbol{T}^{(1)}, \boldsymbol{T}^{(2)}, \ldots, \boldsymbol{T}^{(M)}$ following $g(\boldsymbol{t} \mid \theta)$. The expectation of $G(\boldsymbol{X}, \theta)$ is then approximated by

$$E\big[G(\boldsymbol{X}, \theta)\big] = E\big[G(\boldsymbol{T}, \theta)\big] \approx \frac{1}{M} \sum_{j=1}^{M} G(\boldsymbol{T}^{(j)}, \theta) = E_G^*(\theta) \ \text{(say)} \quad (5.3.2)$$

For several values of θ in Θ, say, $\theta_1, \theta_2, \ldots, \theta_k$, one can thus approximate $E\big[G(\boldsymbol{X}, \theta)\big]$ by using (5.3.2), and obtain $E_G^*(\theta_1), E_G^*(\theta_2), \ldots, E_G^*(\theta_k)$ which can be plotted against θ (for $\theta = \theta_1, \theta_2, \ldots, \theta_k$) to get a better understanding of $E\big[G(\boldsymbol{X}, \theta)\big]$.

To approximate $E\big[G(\boldsymbol{X}, \theta)\big]$, one can write a program code based upon the following steps:

Step-1 : Input θ, (say $\theta = \theta_l$, $l = 1, 2, \ldots, k$)

Step-2 : Generate $\boldsymbol{T}^{(j)}$ following $g(\boldsymbol{t} \mid \theta)$

Step-3 : Compute $G(\boldsymbol{T}^{(j)}, \theta) = G_j$ (say)

Step-4 : Compute $\overline{G}_j = \left(\frac{j-1}{j}\right) \overline{G}_{j-1} + \left(\frac{1}{j}\right) G_j$, where $\overline{G}_0 = 0$

Step-5 : Repeat above steps 2 through 4 for $j = 1, 2, \ldots, M$

Step-6 :

$$E_G^*(\theta) = \overline{G}_M \qquad\qquad\qquad (5.3.3)$$

To demonstrate the above simulation steps, we now compute the bias of $\hat{\sigma}_s^2$. This can be done by using either (5.1.6) or (5.2.3). First, the expression (5.1.6) will be used to demonstrate the bias computation, and then show that the same can also be achieved by (5.2.3) with less complications.

Use of the expression (5.1.6):

For fixed n and μ_0:

Step-1 : Select $\mu \in (-\infty, \infty)$ and $\sigma^2 > 0$

Step-2 : Generate *iid* observations $X_1^{(1)}, X_2^{(1)}, \ldots, X_n^{(1)}$ from $N(\mu, \sigma^2)$

Step-3 : Compute $\widehat{\sigma}_S^2$ for the above set of observations and call it $\widehat{\sigma}_{S(1)}^2$

Step-4 : Repeat above steps 2 and 3 for a large number (say, M) of times and obtain $\widehat{\sigma}_{S(1)}^2, \widehat{\sigma}_{S(2)}^2, \ldots, \widehat{\sigma}_{S(M)}^2$

Step-5 : The bias of $\widehat{\sigma}_S^2$ is approximated as

$$\frac{1}{M} \sum_{j=1}^{M} \widehat{\sigma}_{S(j)}^2 - \sigma^2 = \frac{1}{M} \sum_{j=1}^{M} \left(\widehat{\sigma}_{S(j)}^2 - \sigma^2\right) \qquad (5.3.4)$$

Theoretically (from (5.2.7)), we know that the above simulated bias depends on n, μ, σ^2 only through σ^2 and λ.

Remark 5.3.1 *To avoid computational errors and other difficulties such as 'over-flow' errors, it is suggested that the average in (5.3.4) can be done through the recursive relation (Step-4)in (5.3.3). Also, the idea that the average of a large number of simulated values (i.e., $\sum_{j=1}^{M} \widehat{\sigma}_{S(j)}^2 / M$) approximates the true expectation (i.e., $E\left[\widehat{\sigma}_S^2\right]$) stems from the 'Strong Law of Large Numbers', i.e., $\sum_{i=1}^{M} Y_i / M$ converges to μ_Y almost surely as $M \to \infty$ where Y_i's are iid random variables with mean μ_Y.*

We now show how to use the expression (5.2.3) to approximate the bias of $\widehat{\sigma}_S^2$.

Use of the expression (5.2.3):

Note that in the expression (5.2.3) the random variables V and W are independent, $V \sim \chi_{n-1}^2$ and $W \sim \chi_1^2(\lambda)$ with $\lambda = n(\mu - \mu_0)^2 / \sigma^2$. The random variable W can be characterized as $W = U^2$ where $U \sim N(\pm\sqrt{\lambda}, 1)$. Without loss of generality, we can use $U \sim N(\sqrt{\lambda}, 1)$ to characterize W. Therefore, for fixed n:

Step-1 : Select $\sigma^2 > 0, \lambda > 0$

Step-2 : Generate independent random observations $V^{(1)} \sim \chi^2_{n-1}$ and $U^{(1)} \sim N(\sqrt{\lambda}, 1)$

Step-3 : Compute

$$
\begin{aligned}
g\big(V^{(1)}, W^{(1)}\big) \;=\; & \frac{V^{(1)}}{(n+1)} I\left(\frac{W^{(1)}}{V^{(1)}} > \frac{1}{(n+1)}\right) \\
& + \frac{(V^{(1)} + W^{(1)})}{(n+2)} I\left(\frac{W^{(1)}}{V^{(1)}} < \frac{1}{(n+1)}\right) - 1
\end{aligned}
$$

where $W^{(1)} = \big(U^{(1)}\big)^2$ and call $g\big(V^{(1)}, W^{(1)}\big) = g^{(1)}$

Step-4 : Repeat above steps 2 and 3 for a large number of times (say, M), and obtain $g^{(1)}, g^{(2)}, \ldots, g^{(M)}$

Step-5 : The bias of $\hat{\sigma}^2_s$ is then approximated by

$$
\sigma^2 \left(\frac{1}{M} \sum_{j=1}^{M} g^{(j)}\right) = \sigma^2 \left(\frac{1}{M} \sum_{j=1}^{M} g\big(V^{(j)}, W^{(j)}\big)\right) \qquad (5.3.5)
$$

Remark 5.3.2 *The advantage in using the expression (5.2.3) is that it uses the minimal sufficient statistics and hence uses fewer computational steps. For example, at every data generating step (Step-2) it directly generates $V^{(j)}$ and $U^{(j)}$ instead of $X_1^{(j)}, X_2^{(j)}, \ldots, X_n^{(j)}$ which leads to substantial savings in computational time when n is large.*

Note that a simulated average

$$
\frac{1}{M} \sum_{j=1}^{M} G_j = \frac{1}{M} \sum_{j=1}^{M} G(\boldsymbol{T}^{(j)}, \theta)
$$

which approximates $E\big(G(\boldsymbol{T}, \theta)\big)$ (see (5.3.2)) is based on a very large sample (of size M) of artificially generated observations ($G_j = G(\boldsymbol{T}^{(j)}, \theta)$ values are all *iid*) and therefore subject to sampling variations. As a result, two separate runs of simulations (with same n, M, θ) will not produce exactly the same average. Hence, the magnitude of the fluctuation from one simulation run to another is a key feature in measuring the accuracy of the simulated average. The standard error (SE) which measures the accuracy of the simulated average can be estimated by

$$
\text{SE}_{(M)} = \frac{S_{(M)}}{\sqrt{M}}, \quad \text{where } S_{(M)} = \left(\frac{1}{(M-1)} \sum_{j=1}^{M} \big(G_j - \overline{G}_M\big)^2\right)^{1/2}
$$

$$(5.3.6)$$

with $\overline{G}_M = \sum_{j=1}^{M} G_j/M$. For computational convenience, the value of $S_{(M)}$ can be computed through a recursive relation as shown below.

After generating $G_k = G(\boldsymbol{T}^k, \theta)$, $k = 1, 2, 3, \ldots, j$, the SE is estimated by $\mathrm{SE}_{(j)}$ as

$$\mathrm{SE}_{(j)} = \frac{S_{(j)}}{\sqrt{j}} \quad \text{where}$$

$$S_{(j)} = \left(\frac{(j-2)S_{(j-1)}^2 + (j-1)\overline{G}_{j-1}^2 + G_j^2 - j\overline{G}_j^2}{(j-1)} \right)^{1/2} \tag{5.3.7}$$

One can see how to calculate $\overline{G}_j$ and $\overline{G}_{j-1}$ in Step-4 of (5.3.3). At the end of Mth replication, one gets $\mathrm{SE}_{(M)} = S_{(M)}/\sqrt{M}$, the desired estimated standard error of the result.

Remark 5.3.3 *The estimated standard error $(SE_{(M)})$ of the simulated average plays an important role in determining the value of M, the number of replications, which is to be determined by the investigator. Before undertaking a full fledged simulation study one usually conducts a pilot study with a few possible values of M (typically M ranges from 5,000 to 100,000) to get an idea about $SE_{(M)}$. It is suggested that the value of M should be chosen such that $SE_{(M)}$ should not exceed a fixed percentage (say, 1 or 0.5%) of $\overline{G}_M$ to maintain a high accuracy in approximation of $E\big(G(\boldsymbol{T}, \theta)\big)$ by $\overline{G}_M$.*

As an example of the simulation, the bias of $\widehat{\sigma}_S^2$ is once again computed for $\sigma^2 = 1$ and the selected values of λ with $n = 5$ and 10. The expression (5.2.3) has been used to carry out the simulation with $M = 10,000$ replications. The results are summarized in Table 5.3.1 which can be compared with the corresponding numerical results listed in Table 5.2.1. The estimated standard errors are reported in parentheses.

Remark 5.3.4 *Using the concepts of the central limit theorem, one can expect that the simulated mean $\overline{G}_M$ to be within two standard errors of the actual mean $E(G(\boldsymbol{T}, \theta))$. In Table 5.3.1, the maximum estimated SE is approximately 0.005. Therefore, the reported values of $\overline{G}_M$ are expected to be correct up to the second decimal place.*

λ	n		λ	n	
	5	10		5	10
0.0	$-.3737$	$-.2120$	1.0	$-.3584$	$-.2006$
	(.0043)	(.0037)		(.0045)	(.0037)
0.1	$-.3716$	$-.2106$	1.5	$-.3531$	$-.1966$
	(.0043)	(.0037)		(.0045)	(.0037)
0.3	$-.3681$	$-.2080$	2.5	$-.3453$	$-.1910$
	(.0044)	(.0037)		(.0046)	(.0038)
0.5	$-.3649$	$-.2056$	5.0	$-.3353$	$-.1839$
	(.0044)	(.0037)		(.0047)	(.0038)
0.7	$-.3621$	$-.2035$	10.0	$-.3302$	$-.1803$
	(.0044)	(.0037)		(.0047)	(.0039)

Table 5.3.1: Simulated Bias of $\widehat{\sigma}_{\mathrm{s}}^2$.

5.4   Bootstrap Method of Resampling

In Section 5.2 and Section 5.3, we have discussed methods to evaluate
the performance of an estimator computationally. But both numerical
integration and simulation methods were based on the underlying probability distribution of the data. The assumed probability distribution of
the data is called the **model** of the data. A model is usually selected
based on some expert knowledge or some past experience where similar
situations had been encountered.

For instance, it is well-known that the normal model (or the bell curve)
is widely used in modeling IQ (intelligent quotient) scores of individuals
of a fixed population. To be precise, let *iid* observations $X_1, X_2, \ldots, X_n$
be available where X_i is the IQ score (obtained from a test) of the *i*th
subject in the sample, $1 \leq i \leq n$. It is assumed that $X_1, X_2, \ldots, X_n$ are
iid $\mathrm{N}(\mu, \sigma^2)$ where $\mu \in (-\infty, \infty)$ and $\sigma^2 > 0$. Note that our assumed
normal model is truly a family of distributions. Suppose the variance
σ^2 is estimated by the estimator $\widehat{\sigma}_{\mathrm{s}}^2$ (given in (5.1.5)). The bias or the
risk (under a suitable loss function) of $\widehat{\sigma}_{\mathrm{s}}^2$ depends heavily on the normal
model. Though $\widehat{\sigma}_{\mathrm{s}}^2$ is written in terms of the minimal sufficient statistics, the choice of the minimal sufficient statistics and the corresponding
probability distributions are dependent on the assumed model.

Even though the maximum possible care is taken in a model selection (the applicability of which must be validated by the given data set), there is always plenty of room for making errors which may go undetected. Therefore, one should be very careful in explaining the bias or the risk of an estimator (say, $\widehat{\sigma}_s^2$) of a parameter (σ^2) under the assumed model (say, the normal) which may be misspecified. For example, if an IQ data set really comes from a t-distribution with location parameter μ, scale parameter σ, and 20 degrees of freedom, i.e., if the true *pdf* of each observation is $f_k(x \mid \mu, \sigma)$ with $k = 20$ where

$$f_k(x \mid \mu, \sigma) = \frac{\Gamma\big((k+1)/2\big)}{\sigma \sqrt{k\pi}\, \Gamma\big(k/2\big)} \left(1 + \frac{(x-\mu)^2}{k\sigma^2}\right)^{-(k+1)/2} \tag{5.4.1}$$

then the standard tests for normality may not detect the difference between the t-distribution (5.4.1) and the $N(\mu, \sigma^2)$ distribution unless the sample size is large. This is one of the major drawbacks of parametric inference where the investigator makes a decision based upon the assumed model. To avoid the problem of model misspecification, one can follow a nonparametric approach where the true distribution of the data is assumed to lie in a much larger family, the family of all probability distributions. One then uses large sample (asymptotic) results to evaluate an estimator of a parameter. But for small to moderate sample sizes, the large sample results are hardly applicable and may lead to inappropriate inferences. One, therefore, should keep an open mind in choosing a model and estimating a parameter by carefully employing parametric as well as nonparametric approaches thereby arriving at a meaningful conclusion.

Bootstrap method is a nonparametric resampling technique where the data is allowed to speak for itself. The basic technique is described below in a fairly general setup.

Suppose we have *iid* observations $X_1, X_2, \ldots, X_n$ from a probability distribution $F(x)$ (*cdf*) which is completely unknown. Our parameter of interest is $\theta = \theta(F)$ (i.e., θ is a function of the unknown probability distribution F). The parameter θ can be the population mean, population variance, or population standard deviation, etc. Let θ be estimated by an estimator $\widehat{\theta} = \widehat{\theta}(\boldsymbol{X})$ which may be available from the empirical distribution of the data or may be obtained from a rough parametric consideration. The probability distribution of $\widehat{\theta}$, called the sampling distribution of $\widehat{\theta}$, depends on the distribution of $\boldsymbol{X}$ which is F. Therefore, let $G_F(\cdot)$ be the true probability distribution of the parameter estimator

$\widehat{\theta}$, and bias or risk of $\widehat{\theta}$ will be based upon G_F which is dependent on F. Since F is unknown, so is the distribution $G_F(\cdot)$. But the beauty of the bootstrap method is that it enables us, through computations, to approximate the sampling distribution $G_F(\cdot)$ of $\widehat{\theta}$.

The Bootstrap Procedure:

Let $\{X_1, X_2, \ldots, X_n\}$ be a random sample selected from an unknown probability distribution F. We treat this original sample as a small new population.

Step-1: Draw a random sample of size n from $\{X_1, X_2, \ldots, X_n\}$ with replacement and let the resampled observations be $X_1^{(1)}, X_2^{(1)}, \ldots, X_n^{(1)}$ where the superscript denotes the first replication. Based on $\{X_1^{(1)}, X_2^{(1)}, \ldots, X_n^{(1)}\} = \boldsymbol{X}^{(1)}$ (say), compute $\widehat{\theta}$ and obtain $\widehat{\theta}^{(1)} = \widehat{\theta}(\boldsymbol{X}^{(1)})$.

Step-2: Repeat the above Step-1 independently for a large number of times (say, M). At the jth replication, $1 \leq j \leq M$, we have resampled data set $\boldsymbol{X}^{(j)} = \{X_1^{(j)}, X_2^{(j)}, \ldots, X_n^{(j)}\}$ and the corresponding parameter estimate is given by $\widehat{\theta}^{(j)} = \widehat{\theta}(\boldsymbol{X}^{(j)})$.

Step-3: The computed values $\widehat{\theta}^{(1)}, \widehat{\theta}^{(2)}, \ldots, \widehat{\theta}^{(M)}$ are the replicated values of $\widehat{\theta}$, and can be used to approximate the true distribution G_F of $\widehat{\theta}$. The approximate distribution of $\widehat{\theta}$, say $\widetilde{G}$, is obtained as

$$\widetilde{G}(y) = \frac{\# \, \widehat{\theta}^{(j)}\text{'s } \leq y}{M} = \frac{1}{M} \sum_{j=1}^{M} I_{(-\infty, \, y]}\big(\widehat{\theta}^{(j)}\big) \qquad (5.4.2)$$

where $I_{(-\infty, \, y]}(\cdot)$ is the indicator function defined for the interval $(-\infty, \, y]$.

The reason why $\widetilde{G}(y)$ approximates the true distribution $G_F(y)$ of $\widehat{\theta}$ is not difficult to justify. As mentioned before, the distribution G_F of $\widehat{\theta}$ is completely unknown because F, the distribution of $\boldsymbol{X}$, is assumed (in a nonparametric setup) to be so. However, an estimate of G_F, say $\widehat{G}$, can be obtained as $\widehat{G} = G_{\widehat{F}}$, where $\widehat{F}$ is the empirical distribution based on $\boldsymbol{X}$, i.e.,

$$\widehat{F}(x) = \frac{\# \, X_i\text{'s } \leq x}{n} = \frac{1}{n} \sum_{i=1}^{n} I_{(-\infty, \, x]}(X_i) \qquad (5.4.3)$$

In fact, $\widehat{G} = G_{\widehat{F}}$ is the MLE of G since $\widehat{F}$ is so for F. Now, in the bootstrap method, the jth replicated sample $X_1^{(j)}, X_2^{(j)}, \ldots, X_n^{(j)}$ i.i.d. from $\widehat{F}$ gives $\widehat{\theta}^{(j)}$ which has true probability distribution $G_{\widehat{F}}$, $1 \leq j \leq M$. Thus, after generating $\widehat{\theta}^{(1)}, \widehat{\theta}^{(2)}, \ldots, \widehat{\theta}^{(M)}$ where M is sufficiently large, we are trying to capture the distribution $\widehat{G} = G_{\widehat{F}}$ based on the empirical distribution $\widetilde{G}$ constructed from $\widehat{\theta}^{(1)}, \widehat{\theta}^{(2)}, \ldots, \widehat{\theta}^{(M)}$. When M is sufficiently large and n is not too small,

$$\widetilde{G}(y) \approx \widehat{G}(y) = G_{\widehat{F}}(y) \approx G_F(y) \tag{5.4.4}$$

If $M \to \infty$, then $\widetilde{G}(y) \longrightarrow \widehat{G}(y) = G_{\widehat{F}}(y)$ almost surely; and if further $n \to \infty$, then $\widehat{G}(y) = G_{\widehat{F}}(y) \longrightarrow G_F(y)$ almost surely.

Using the bootstrap method, one can estimate the bias of an estimator $\widehat{\theta}$ by $\widehat{B}_b(\widehat{\theta})$ where

$$\widehat{B}_b(\widehat{\theta}) = \frac{1}{M} \sum_{j=1}^{M} \left(\widehat{\theta}^{(j)} - \widehat{\theta} \right) = \overline{\widehat{\theta}(\cdot)} - \widehat{\theta}$$

with

$$\overline{\widehat{\theta}(\cdot)} = \frac{1}{M} \sum_{j=1}^{M} \widehat{\theta}^{(j)} \tag{5.4.5}$$

The standard error (SE) of $\widehat{\theta}$ is estimated by

$$\widehat{SE}_b(\widehat{\theta}) = \frac{1}{\sqrt{M}} \left(\frac{1}{(M-1)} \sum_{j=1}^{M} \left(\widehat{\theta}^{(j)} - \overline{\widehat{\theta}(\cdot)} \right)^2 \right)^{1/2} \tag{5.4.6}$$

where $\overline{\widehat{\theta}(\cdot)}$ is given by (5.4.5). Similarly, one can estimate the risk of $\widehat{\theta}$ under the squared error loss (SEL) by

$$\widehat{R}_b(\widehat{\theta}) = \frac{1}{M} \sum_{j=1}^{M} \left(\widehat{\theta}^{(j)} - \widehat{\theta} \right)^2 \tag{5.4.7}$$

The subscript '*b*' in above equations indicates that these estimates are obtained by the bootstrap method.

Now we apply the bootstrap method to a real life data set as a demonstration.

Example 5.4.1: The results of the blood tests, showing nearest milligrams of glucose per 100 milliliters of blood, were obtained from 53 nonpregnant women at Boston City Hospital. The data set (Source: *The American Journal of Clinical Nutrition*, Vol. 19, pp. 345–351) is displayed as follows:

The following Figure 5.4.1 is a percentage (or a relative frequency) histogram of the glucose data set.

66	59	69	84	85	83	87	65	90
75	80	73	69	80	87	83	80	86
75	83	87	74	76	79	82	77	93
93	87	70	85	90	85	84	78	86
82	82	84	85	96	91	69	91	70
89	81	74	85	82	77	79	87	

Table 5.4.1: Glucose per 100 Milliliters of Blood (in mg).

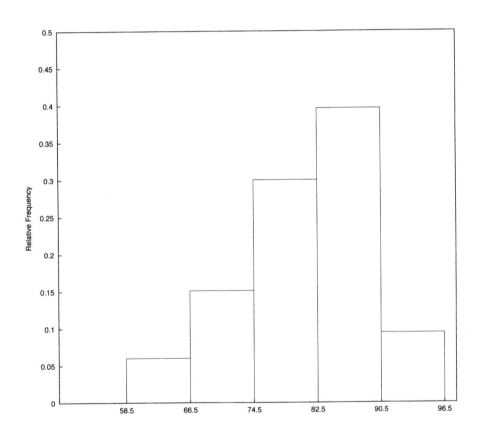

Figure 5.4.1: A Histogram of the Glucose Data Set.

Our goal is to estimate σ^2, the variance of the population. The parameter σ^2 is the variance of blood glucose of the population of all women from which the above sample of 53 was taken. The available estimators of σ^2 are $\widehat{\sigma}^2_{\mathrm{ML}}$, $\widehat{\sigma}^2_{opt}$, and $\widehat{\sigma}^2_{\mathrm{S}}$ (see (5.1.1), (5.1.4), and (5.1.6)). The following table gives bootstrap estimates of bias and SE of the above estimates with $M = 10,000$ replications.

Estimator	Estimated Bias	Estimated SE
$\widehat{\sigma}^2_{\mathrm{ML}}$	-1.2455	0.1171
$\widehat{\sigma}^2_{opt}$	-1.2220	0.1149
$\widehat{\sigma}^2_{\mathrm{S}}$	-1.3601	0.1130

Table 5.4.2: Bootstrap Estimates of Bias and SE.

5.5   Testing and Interval Estimation Based on Computations

Hypothesis Testing

One can consider a hypothesis testing procedure based on statistical simulation and numerical computation. Since fast computing has become convenient, it can be implemented easily at a relatively low cost.

Note that one needs to come up with a suitable test statistic and its cut-off point(s) to test a hypothesis. In this regard the LRT comes in handy due to its asymptotic distribution.

The following test procedure based on statistical simulation and numerical computation is similar to LRT, but does not use the asymptotic chi-square distribution to obtain the cut-off point(s) for the test statistic. This is discussed for two possible cases for testing a hypothesis on a scalar valued parameter.

Test H_0: $\theta = \theta_0$ vs. H_1: $\theta < \theta_0$ or $\theta > \theta_0$ or $\theta \neq \theta_0$.

Case 1 : Nuisance parameter is absent

Step-1 : Obtain $\widehat{\theta}_{\mathrm{ML}}$, the MLE of θ.

Step-2 : Set $\theta = \theta_0$ (specified by H_0). Generate the data $\mathbf{X}$ from $f(x|\theta_0)$ for a large number of (say M) times. For each of these replicated data, recalculate the MLE of θ (pretending that it is unknown). Let these recalculated MLEs be $\widehat{\theta}_{01}, \widehat{\theta}_{02}, \ldots, \widehat{\theta}_{0M}$.

Step-3 : Order the above recalculated MLEs as $\widehat{\theta}_{0(1)} \leq \widehat{\theta}_{0(2)} \leq \cdots \leq \widehat{\theta}_{0(M)}$.

Step-4 :

(a) For testing against H_1: $\theta < \theta_0$, let $\widehat{\theta}_L = \widehat{\theta}_{0(\alpha M)}$.
Reject H_0 if $\widehat{\theta}_{ML} < \widehat{\theta}_L$; accept H_0 if $\widehat{\theta}_{ML} \geq \widehat{\theta}_L$.

(b) For testing against H_1: $\theta > \theta_0$, let $\widehat{\theta}_U = \widehat{\theta}_{0((1-\alpha)M)}$.
Reject H_0 if $\widehat{\theta}_{ML} > \widehat{\theta}_U$; accept H_0 if $\widehat{\theta}_{ML} \leq \widehat{\theta}_U$.

(c) For testing against H_1: $\theta \neq \theta_0$, let $\widehat{\theta}_U = \widehat{\theta}_{0((1-\alpha/2)M)}$ and $\widehat{\theta}_L = \widehat{\theta}_{0((\alpha/2)M)}$.
Reject H_0 if $\widehat{\theta}_{ML} > \widehat{\theta}_U$ or $\widehat{\theta}_{ML} < \widehat{\theta}_L$. Accept otherwise.

As an application of the above method assume that we have *iid* observations $X_1, X_2, \ldots, X_n$ from $N(\theta, 1)$ and we have to test $H_0 : \theta = \theta_0$ against $H_1 : \theta \neq \theta_0$.

1. In Step-1, we have $\widehat{\theta}_{ML} = \overline{X}$.

2. In Step-2, we generate
(1st replication): $X_1^{(1)}, \ldots, X_n^{(1)}$ *iid* $N(\theta_0, 1)$; get $\widehat{\theta}_{01} = \overline{X}^{(1)}$

$\vdots$

(M^{th} replication): $X_1^{(M)}, \ldots, X_n^{(M)}$ *iid* $N(\theta_0, 1)$; get $\widehat{\theta}_{0M} = \overline{X}^{(M)}$
[Note that the values $\widehat{\theta}_{01}, \ldots, \widehat{\theta}_{0M}$ are representing the $N(\theta_0, 1/n)$ distribution; i.e., a histogram of these values clearly follows the $N(\theta_0, 1/n)$ *pdf*.]

3. In Step-3, obtain $\widehat{\theta}_L = \widehat{\theta}_{0((\alpha/2)M)}$ and $\widehat{\theta}_U = \widehat{\theta}_{0((1-\alpha/2)M)}$. $\widehat{\theta}_L$ and $\widehat{\theta}_U$ are approximately $(\theta_0 - z_{(\alpha/2)}/\sqrt{n})$ and $(\theta_0 + z_{(\alpha/2)}/\sqrt{n})$ respectively.

4. In Step-4, reject H_0 if $\widehat{\theta}_{ML} > \widehat{\theta}_U$ or $\widehat{\theta}_{ML} < \widehat{\theta}_L$; accept otherwise. [In classical theory, we reject H_0 if $\widehat{\theta}_{ML} > \theta_0 + z_{(\alpha/2)}/\sqrt{n}$ or $\widehat{\theta}_{ML} < \theta_0 - z_{(\alpha/2)}/\sqrt{n}$.]

Case-2 : Nuisance parameter is present

Suppose the data $\mathbf{X}$ follows a distribution $f(X|\,\boldsymbol{\theta})$ where $\boldsymbol{\theta} = (\theta^{(1)}, \theta^{(2)})$ $\in \boldsymbol{\Theta}$, and $\theta^{(2)}$ can be vector valued. The *pdf* (or *pmf*) $f(X|\,\boldsymbol{\theta})$ is known except $\boldsymbol{\theta}$ and our interest lies in $\theta^{(1)}$ only. Hence $\theta^{(2)}$ is a nuisance parameter (vector). To test H_0: $\theta^{(1)} = \theta_0^{(1)}$ against a suitable alternative H_1: $(\theta^{(1)} < \theta_0^{(1)}$ or $\theta^{(1)} > \theta_0^{(1)}$ or $\theta^{(1)} \neq \theta_0^{(1)})$ at level α, the following steps are followed.

Step-1 : Obtain $\widehat{\boldsymbol{\theta}}_{\mathrm{ML}} = (\widehat{\theta}_{\mathrm{ML}}^{(1)}, \widehat{\theta}_{\mathrm{ML}}^{(2)})$, the MLE of $\boldsymbol{\theta}$.

Step-2 : (a) Set $\theta^{(1)} = \theta_0^{(1)}$, then find the MLE of $\theta^{(2)}$. Call this 're-stricted MLE' of $\theta^{(2)}$, and denote it by $\widehat{\theta}_{\mathrm{RML}}^{(2)}$.

(b) Generate the data $\mathbf{X}$ from $f(x|\,\theta_0^{(1)}, \widehat{\theta}_{\mathrm{RML}}^{(2)})$ for a large number of times (say M). For each of these replicated data, recalculate the MLE of $\boldsymbol{\theta}$ (pretending that it is unknown) and retain the component relevant for $\theta^{(1)}$. Let these recalculated MLEs of $\theta^{(1)}$ be $\widehat{\theta}_{01}^{(1)}, \widehat{\theta}_{02}^{(1)}, \ldots, \widehat{\theta}_{0M}^{(1)}$.

Step-3 : Let $\widehat{\theta}_{0(1)}^{(1)} \leq \widehat{\theta}_{0(2)}^{(1)} \leq \ldots \leq \widehat{\theta}_{0(M)}^{(1)}$ be the ordered values of $\widehat{\theta}_{01}^{(1)}, \widehat{\theta}_{02}^{(1)}, \ldots, \widehat{\theta}_{0M}^{(1)}$.

Step-4 : Similar to that of Case-1 where $\widehat{\theta}_{0(i)}$'s are replaced by $\widehat{\theta}_{0(i)}^{(1)}$'s and $\widehat{\theta}_{\mathrm{ML}}^{(1)}$ is used in place of $\widehat{\theta}_{\mathrm{ML}}$.

Interval Estimation

Earlier we discussed about interval estimation of a parameter based on a pivot. But in many problems a pivot is not readily available. In such a case one tests a mock hypothesis on the parameter, and then inverts the acceptance region to obtain an interval estimate of the parameter. Since we have discussed about hypothesis testing based on simulation and computation, we can take advantage of the same to come up with an interval estimate of the parameter under consideration. In the following we provide the procedure only for the above first case (where there is no nuisance parameter), but can be extended for the second case (where there are nuisance parameters) easily with slight notational changes. Also, the procedure is given for a two sided interval estimate, but can be modified suitably for one sided interval estimates as well. We follow the notations used above for hypothesis testing.

Step-1 : Take several values $\theta_0^1, \theta_0^2, \ldots, \theta_0^K$, over the parameter space of θ, suitably spaced. [It is suggested that these values of θ, numbered somewhere between five and eight, be taken equally spaced over $\widehat{\theta}_{\text{ML}} \pm 2(SE)$, where SE is the standard error or estimated standard deviation of $\widehat{\theta}_{\text{ML}}$.]

Step-2 : Perform a hypothesis testing of H_0^j: $\theta = \theta_0^j$ against H_1^j: $\theta \neq \theta_0^j (j = 1, 2, \ldots, K)$ at level α. For each j, obtain the cut off points $(\widehat{\theta}_{\text{L}}^j, \widehat{\theta}_{\text{U}}^j)$ based on the simulation and computation described earlier, $j = 1, 2, \ldots, K$. Remember that the bounds $(\widehat{\theta}_{\text{L}}^j, \widehat{\theta}_{\text{U}}^j)$ are for $\widehat{\theta}_{\text{ML}}$, the MLE of θ.

Step-3 : Plot the lower bounds $\widehat{\theta}_{\text{L}}^j$, $1 \leq j \leq K$, against θ_0^j and then approximate the plotted curve by a suitable smooth function say, $g_{\text{L}}(\theta_o)$. Similarly, plot the upper bounds $\widehat{\theta}_{\text{U}}^j$, $1 \leq j \leq K$, against θ_0^j and then approximate the plotted curve by a suitable smooth function say, $g_{\text{U}}(\theta_0)$.

Step-4 : Finally, solve for θ_0 from the equations $g_{\text{L}}(\theta_0) = \widehat{\theta}_{\text{ML}}$ and $g_{\text{U}}(\theta_0) = \widehat{\theta}_{\text{ML}}$. The two solutions of θ_0 thus obtained set the boundaries of the interval estimate of θ with minimum confidence level $(1 - \alpha)$.

[In fact, the curves $g_{\text{L}}^{-1}(\widehat{\theta}_{\text{ML}})$ and $g_{\text{U}}^{-1}(\widehat{\theta}_{\text{ML}})$ can act as $(1 - \alpha)$-level two sided confidence bounds for the parameter θ when $\widehat{\theta}_{\text{ML}}$ is the MLE of θ.] The above computer intensive approach of interval estimation comes in handy in several applications as we will see in later chapters.

Part-II
Exponential and Other Positively Skewed Distributions with Applications

Chapter 6

Exponential Distribution

6.1 Preliminaries

In Chapter 2, the exponential distribution has been mentioned briefly. A random variable X is said to have the exponential distribution with scale parameter β if its *pdf* is given by

$$f(x \mid \beta) = \frac{1}{\beta} e^{-x/\beta}, \quad x > 0 \text{ and } \beta > 0 \tag{6.1.1}$$

The mean of X or $E(X)$ is β, and hence if X represents the life span of a randomly selected object in hours, then β stands for the mean life of all such objects having the exponential (6.1.1) life time distribution. One can use the reparameterization $\lambda = 1/\beta$ and then the *pdf* (6.1.1) of X becomes

$$f(x \mid \lambda) = \lambda e^{-\lambda x}, \quad x > 0 \quad \text{and} \quad \lambda > 0 \tag{6.1.2}$$

Note that with the *pdf* (6.1.2), $E(X) = 1/\lambda$, and $\lambda = 1/E(X)$ is called the failure rate.

Henceforth an exponential distribution with mean β (i.e., the representation (6.1.1)) will be denoted as ' **Exp**(β)'. At the same token, the reparameterized version (6.1.2) will be denoted as 'Exp$(1/\lambda)$'.

Figure 6.1.1 plots $f(x \mid \beta)$ (6.1.1) for various values of β.

In queuing theory the exponential distribution is commonly used to model the probability distribution of a random variable that represents service time. For examples, the time is taken to complete service at a grocery checkout counter, at a doctor's office, at an airline check-in counter,

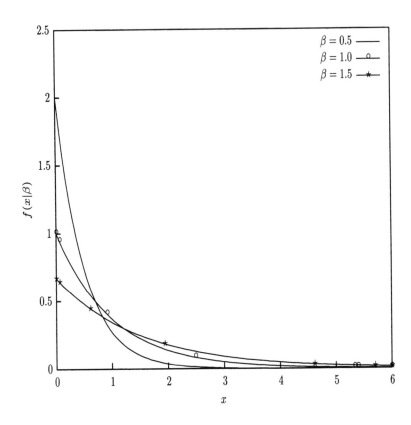

Figure 6.1.1: The Exponential *pdf*s.

at a bank teller, and so on. In reliability applications, the exponential distribution is frequently used to model the lifetimes of components which are subject to wear out. For examples, lifetime of electric bulbs, batteries, appliances, and transistors, etc., are modeled by the exponential distribution. This distribution has also been used successfully to model the distribution of the length of time between successive random events such as the time between two auto accidents on a particular stretch of a highway; the time between the arrivals of two customers at a ticket counter; and the time between breakdowns of a machine, etc.

From (6.1.1) or (6.1.2), it is seen that the distribution's support is the positive side of the real line, i.e., $\mathbb{R}^+$. Since 'time', in general, is a positive measure, an exponential random variable is often well-suited to represent quantitative characteristics measured in time units. Also, from the plots in Figure 6.1.1, it is evident that the distribution is heavily skewed to the right. Since very few service times for customers, life times of appliances, or failure times of machines exceed their expected values, the exponential distribution seems natural to model the probability distribution of these types of random variables.

One appealing feature of the exponential distribution is its mathematical simplicity. If a random variable X follows $\text{Exp}(\beta)$, then $Y = X/\beta$ follows $\text{Exp}(1)$. Therefore,

$$
\begin{aligned}
P(a \le X \le b) &= P\left(\frac{a}{\beta} \le Y \le \frac{b}{\beta}\right) \\
&= \int_{a/\beta}^{b/\beta} e^{-y} dy \\
&= e^{-a/\beta} - e^{-b/\beta}
\end{aligned}
\tag{6.1.3}
$$

Example 6.1.1: The service time at a drive-through window of a fast-food restaurant is assumed to be exponentially distributed with a mean of 2 minutes. What percentage of customers end up waiting more than 5 minutes?

Let the random variable X be the service time (the time from placing an order till delivery of the food) of a customer selected at random. Since X follows $\text{Exp}(2)$, the *pdf* of X is

$$
f(x) = \frac{1}{2}e^{-x/2}, \quad x \ge 0
$$

The probability that X is more than 5 minutes is

$$
\begin{aligned}
P(X > 5) &= 1 - P(0 \le X \le 5) \\
&= e^{-5/2} \quad \text{(from (6.1.3))} \\
&= 0.082
\end{aligned}
\tag{6.1.4}
$$

Thus 8.2% of the customers wait for more than 5 minutes.

Example 6.1.2: Suppose that the radiator of a particular type of imported cars has lifetime distributed as $\text{Exp}(\beta)$ with $\beta = 5$ years. If five of these cars are sold by a local dealer recently, what is the probability that at least two will have original radiators functioning at the end of 8 years?

Let p be the probability that a given radiator of a car mentioned above functions after 8 years. If X represents the life of a radiator, then X follows Exp(β) with $\beta = 5$ years. Hence,

$$p = P(X > 8) = \int_8^\infty \frac{1}{5} e^{-x/5} dx = e^{-8/5} \approx 0.2$$

The dealer sells five cars lately, and let Y be the number of cars (out of five) with the original radiator still functioning after eight years. It is seen that Y follows a binomial distribution, B(n, p), with $n = 5$ and $p = 0.2$. Therefore,

$$P(Y \geq 2) = 1 - P(Y \leq 1) = 1 - [P(Y = 0) + P(Y = 1)]$$

where

$$P(Y = y) = \binom{5}{y}(0.2)^y(0.8)^{5-y}, \quad y = 0, 1$$

Straightforward computation yields that

$$P(Y \geq 2) \approx 0.263$$

Therefore, 26.3% of the time at least two cars out of five will have original radiators functioning after 8 years.

In the following, we provide two real life data sets which tend to follow exponential distributions. These data sets are obtained from Sakai and Burris (1985).

Example 6.1.3: Differences in physical characteristics between males and females are well known for animals. similarly, sex-related differences in vegetative growth of dioecious plants have long been recognized and are of interest lately. For example, vegetative growth of females may be reduced because of nutrient drain of fruit production, or female plants may require larger size to attain the energy necessary for fruit production. Data from field studies have shown that females are associated with larger size in many short-lived species. In perennial species, females tend to exhibit reduced vigor, growth rate, and vegetative spread. Two varieties of 'trembling aspen' (Populus tremuloides) have been studied and males and females have been compared in terms of ramets (stems). Table 6.1.1 shows ramet size distributions (the frequency and relative frequency distributions) for 4289 females and 3498 males of Pellston Plain clones, and their sample probability histograms (or the relative frequency distributions) are displayed in Figure 6.1.2 and Figure 6.1.3.

| Ramet Size (in cm) (dbh) | Sex of the Plant | | | |
| | Female | | Male | |
	Freq.	Rel. Freq.	Freq.	Rel. Freq.
0.0 – 1.3	1450	0.338	1350	0.386
1.3 – 3.8	1363	0.318	1177	0.336
3.8 – 6.3	725	0.169	500	0.143
6.3 – 8.9	300	0.070	178	0.051
8.9 – 11.4	175	0.041	80	0.023
11.4 – 14.0	100	0.023	70	0.020
14.0 – 16.5	60	0.014	50	0.014
16.5 – 19.1	47	0.011	35	0.010
19.1 – 21.6	35	0.008	30	0.009
21.6 or more	34	0.008	28	0.008
Total	4289	1.000	3498	1.000

Table 6.1.1: Ramet Size Distributions.

A further generalization of the exponential distribution (6.1.1) is given by the two-parameter exponential distribution **Exp**(β, μ) with *pdf*

$$f(x \mid \beta, \mu) = \frac{1}{\beta} e^{-(x-\mu)/\beta}, \quad x > \mu \text{ and } \mu \in \mathbb{R}, \ \beta > 0 \qquad (6.1.5)$$

In the expression (6.1.5), the parameter μ is a location (or shift) parameter, and β is a scale parameter. If, for example, the life span of a particular brand of electric bulbs follows a two-parameter exponential distribution as mentioned earlier, then μ indicates the **minimum guarantee time**. Figure 6.1.4 shows plots of the two-parameter exponential *pdf* with $(\beta, \mu) = (1, 1)$ and $(2, 3)$.

In many real life applications, it is reasonable to assume $\mu = 0$. This may be due to some prior or expert knowledge, or after testing and accepting a null hypothesis $H_0 : \mu = 0$ against $H_1 : \mu \neq 0$. The special case thus obtained is the **one-parameter exponential** distribution **Exp**(β) discussed earlier which is essentially **Exp** $(\beta, 0)$.

If $\mu = 0$ and $\beta = 1$, then the two-parameter exponential distribution **Exp**$(1, 0)$ is called the standard exponential distribution.

Notational Clarification: From now on **Exp**(β, μ) indicates a two parameter exponential distribution with scale parameter β and location

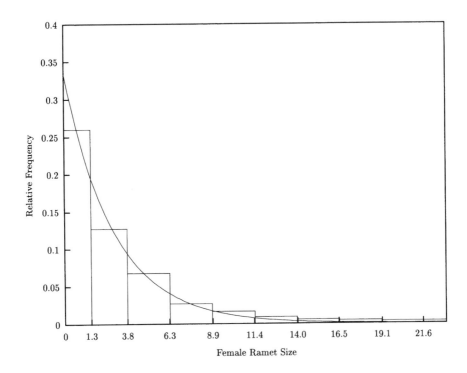

Figure 6.1.2: The relative frequency histogram and a matching exponential *pdf* of female ramet size with $\beta = 3.0$.

parameter μ. If not mentioned otherwise, **Exp**(β) indicates **Exp**($\beta, 0$), i.e., the one parameter exponential distribution with scale parameter β.

6.2   Characterization of Exponential Distribution

The easiest and the most widely accepted way to characterize the exponential distribution is through the **memory-less property** (see the equation (6.2.1) below). Suppose that the continuous random variable $X > 0$ represents the lifetime of a component. Also, let the future life time have the same probability distribution no matter how old the com-

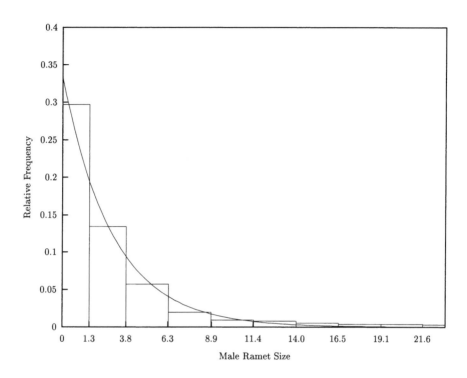

Figure 6.1.3: The relative frequency histogram of male ramet size and a matching exponential *pdf* with $\beta = 2.5$.

ponent is, i.e.,

$$P(X \leq x + c \mid X > c) = P(X \leq x) \tag{6.2.1}$$

for all $c > 0$ and all $x > 0$. If $f(x)$ and $F(x)$ denote the *pdf* and *cdf* of X respectively, then (6.2.1) yields

$$\frac{P(c < X \leq x + c)}{P(X > c)} = P(X \leq x)$$

$$\text{or,} \quad \frac{F(x + c) - F(c)}{1 - F(c)} = F(x) \quad \forall\, x > 0,\ c > 0 \tag{6.2.2}$$

Dividing both sides of (6.2.2) by x, and then taking the limit as x tending to 0 gives

$$\lim_{x \to 0} \frac{F(x + c) - F(c)}{x} \cdot \frac{1}{1 - F(c)} = \lim_{x \to 0} \frac{F(x)}{x}$$

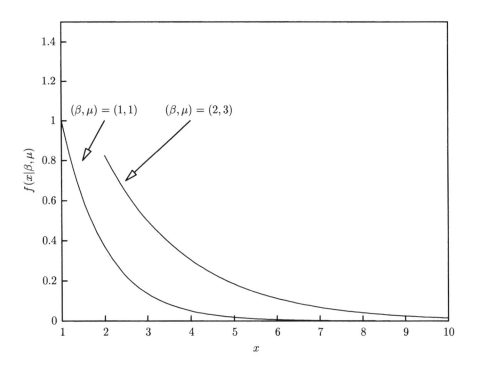

Figure 6.1.4: The *pdf*s of **Exp**(β, μ) with $(\beta, \mu) = (1, 1), (2, 3)$.

Using $F(0) = 0$, the above equation can be written as

$$\frac{F'(c)}{1 - F(c)} = F'(0)$$

Since $F'(0)$ is a constant, we can denote it as $F'(0) = a$. Now the above equation becomes

$$\frac{-F'(c)}{1 - F(c)} = -a \quad \forall c > 0 \tag{6.2.3}$$

Solving the above differential equation[1] (6.2.3), one gets

$$1 - F(c) = be^{-ac} \tag{6.2.4}$$

---

[1] For a function $g(x) = y$ and the differential equation $y'/y = k$ (constant), one can get the solution as $g(x) \propto exp(kx)$.

for some suitable constant b. Using the condition that $\lim_{c \to 0} F(c) = 0$, and replacing c by x, one can obtain the following result,

$$1 - F(x) = e^{-ax} \quad \text{or} \quad f(x) = ae^{-ax} \tag{6.2.5}$$

which is the same as $\text{Exp}(\beta, \mu)$ with $\mu = 0$ and $\beta = 1/a$.

Conversely, it can be seen easily that if X follows $\text{Exp}(\beta, 0)$, then the condition (6.2.1) is satisfied. Further, X follows $\text{Exp}(\beta, \mu)$ if and only if (6.2.1) is satisfied for all $c > \mu$ and $x > 0$.

Another characterization of the exponential distribution is given in terms of the order statistics.

Theorem 6.2.1 *Let $X_1, X_2, \ldots, X_n$ be continuous iid random variables and $X_{(1)} \leq X_{(2)} \leq \cdots \leq X_{(n)}$ be the order statistics, then if the conditional expectation $E(X_{(i+1)} - X_{(i)} | X_{(i)} = x)$ is independent of x for fixed i, $1 \leq i < n$, then X_i's have exponential distribution.*

Desu (1971) gave a different type of characterization of $\text{Exp}(\beta, 0)$, the key idea of which was to present a function of the order statistics having the same distribution as the one sampled. This is presented in the following theorem.

Theorem 6.2.2 *Let $X_1, X_2, \ldots, X_n$ be iid nonnegative random variables with a nondegenerate[2] cdf $F(\cdot)$ and $X_{(1)} = \min\left(X_1, X_2, \ldots, X_n\right)$. Then $nX_{(1)}$ follows the same cdf $F(\cdot)$ if and only if $F(x) = 1 - exp(-x/\beta)$, for $x \geq 0$, where β is a positive constant.*[3]

While Theorem 6.2.2 characterizes an exponential distribution in terms of the distribution of the smallest observed value, the following result, due to Ahsanullah (1977), characterizes it in terms of the distribution of the spacings of order statistics when the *iid* observations are thought to have a particular type of probability distributions.

Let F be the *cdf* of a nonnegative random variable and $\overline{F}(x) = 1 - F(x)$ for $x \geq 0$. The *cdf* F is called (i) **new better than used** (NBU) if

$$\overline{F}(x + y) \leq \overline{F}(x)\overline{F}(y) \quad \text{for } x, y \geq 0 \tag{6.2.6}$$

---

[2] That is, the range (or support) has more than one point.
[3] $F(x) = 1 - exp(-x/\beta)$ is the cdf of $\text{Exp}(\beta, 0)$ distribution.

and (ii) **new worse than used** (NWU) if

$$\overline{F}(x+y) \geq \overline{F}(x)\overline{F}(y) \quad \text{for } x, y \geq 0 \tag{6.2.7}$$

Remark 6.2.1 *If we think of a random variable X representing the life-time of some unit and has cdf F, then X being NBU (or NWU) means $P(X > x + y | X > x) \leq$ (or $\geq$) $P(X > y)$, i.e., the probability of addi-tional life of any used item is less (or greater) than that of a new item. In other words, NBU (or NWU) implies that survivability of an old item is lower (or higher) than that of a new one.*

Theorem 6.2.3 *Let $X_1, X_2, \ldots, X_n$ be iid random variables having cdf $F(x)$ that is strictly increasing on $[0, \infty)$. Then the following properies are equivalent:*

1. *$F(x) = 1 - exp(-x/\beta)$ for some $\beta > 0$ and $x > 0$*

2. *The variables $Y_i = (n-i)\Big(X_{(i+1)} - X_{(i)}\Big)$, $1 \leq i \leq n$, are identically distributed with cdf $F(x)$, and $F(\cdot)$ is either NBU or NWU*

Seshadri and Csörgö (1969) exploited further the idea of spacings of order statistics and characterized exponential distributions in terms of the distribution of spacings. Such characterizations eventually help us in testing the goodness of fit of an exponential distribution.

Theorem 6.2.4 *Let $X_1, X_2, \ldots, X_n$ be iid nonnegative random vari-ables with mean $\beta > 0$. Define $S = \sum_{i=1}^{n} X_i$ and $Z_j = \sum_{i=1}^{j} X_i/S$, $1 \leq j \leq n - 1$. Then the random variables Z_j's act like $(n - 1)$ ordered random variables from continuous Uniform over the interval $[0, 1]$ (i.e., CU[0, 1]) if and only if the random variables X_i's are $Exp(\beta, 0)$.*

Theorem 6.2.5 *Let $X_1, X_2, \ldots, X_n$ be iid nonnegative random vari-ables. Define $Y_i = (n+1-i)\Big(X_{(i)} - X_{(i-1)}\Big)$, $1 \leq i \leq n$, where $X_{(0)} = 0$. The random variables Y_i's follow $Exp(\beta, 0)$ if and only if X_i's are so.*

The extension of Theorem 6.2.4 to a more general case when X_i's are $Exp(\beta, \mu)$ is given in the next theorem.

Theorem 6.2.6 *Let $X_1, X_2, \ldots, X_n$ be iid random variables with mean β and $X_i > \mu$ for all $1 \leq i \leq n$. Define $Y_i = (n + i - 1)\Big(X_{(i)} - X_{(i-1)}\Big)$, $1 \leq i \leq n$, where $X_{(0)} = \mu$. Let $S = \sum_{i=1}^{n} Y_i$ and $Z_j^* = \sum_{i=1}^{j} Y_i/S$, $1 \leq j \leq n - 1$. Then the random variables Z_j^*'s act like $(n - 1)$ ordered CU[0,1] random variables if and only if X_i's are $Exp(\beta, \mu)$.*

6.3   Estimation of Parameter(s)

We first consider estimation of the scale parameter β in $\mathbf{Exp}(\beta)$ distribution.

Given *iid* observation $X_1, X_2, \ldots, X_n$ with common *pdf* (6.1.1) the likelihood function of the data is

$$L(\beta \mid X_1, X_2, \ldots, X_n) = \beta^{-n} exp\left\{ -\sum_{i=1}^{n} X_i/\beta \right\}, \quad \beta > 0 \qquad (6.3.1)$$

and hence the minimal sufficient statistic for β is $\sum_{i=1}^{n} X_i$ (or $\bar{X}$). The **maximum likelihood estimator** (MLE) as well as the method of moment estimator of β is

$$\widehat{\beta}_{\mathrm{ML}} = \sum_{i=1}^{n} X_i/n = \overline{X} \qquad (6.3.2)$$

It can be shown that

$$E(\widehat{\beta}_{\mathrm{ML}}) = 0, \quad Var(\widehat{\beta}_{\mathrm{ML}}) = \beta^2/n \qquad (6.3.3)$$

i.e., $\widehat{\beta}_{\mathrm{ML}}$ is an unbiased estimator of β for the model (6.1.1). Since $\widehat{\beta}_{\mathrm{ML}}$ is a function of the minimal sufficient statistic, it is also the unique minimum variance unbiased estimator (UMVUE) of β (see Theorem 3.4.4, Rao-Blackwell Theorem). Also, if $\eta = \eta(\beta)$ is a function of β, then the MLE of η is

$$\widehat{\eta}_{\mathrm{ML}} = \eta(\widehat{\beta}_{\mathrm{ML}}) \qquad (6.3.4)$$

However, $\widehat{\eta}_{\mathrm{ML}}$ is not necessarily an unbiased estimator of η.

For example, as a special case, let ξ_c be the $100c^{th}$ percentile point $(0 < c < 1)$ of the $\mathbf{Exp}(\beta)$-distribution; i.e., $c = F(\xi_c) = P(X \leq \xi_c) = 1 - exp(-\xi_c/\beta)$. Then $\xi_c = -\beta \, ln(1-c)$. Hence, the MLE of ξ_c is $\widehat{\xi}_{c(\mathrm{ML})} = -\widehat{\beta}_{\mathrm{ML}} \, ln(1-c)$.

Example 6.3.1: Consider the given ordered data set

$$\begin{array}{cccc} 48.96 & 54.67 & 100.75 & 120.68 \\ 148.34 & 228.43 & 293.45 & 415.82 \end{array}$$

which is to be modeled by a suitable exponential distribution. Estimate β by $\widehat{\beta}_{\mathrm{ML}} = \overline{X} = 177.3875$. Then get $\widehat{\xi}_c = -(176.3875) \, ln \, (1 - c)$, and this is plotted in Figure 6.3.1.

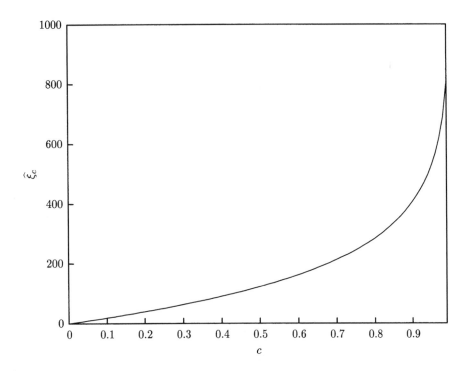

Figure 6.3.1: The graph of $\widehat{\xi}_c$ against $c \in (0, 1)$.

Note that the risk function of $\widehat{\beta}_{\mathrm{ML}}$ under the squared error loss (or the mean squared error) is same as the variance of $\widehat{\beta}_{\mathrm{ML}}$ (due to its unbiasedness). Therefore

$$Var(\widehat{\xi}_{c(\mathrm{ML})}) = (ln\,(1-c))^2 Var(\widehat{\beta}_{\mathrm{ML}}) = (ln\,(1-c))^2 \beta^2/n \qquad (6.3.5)$$

The following two diagrams plot $Var(\widehat{\xi}_{c(\mathrm{ML})})$ against β (with fixed $c = 0.90$) and c (with fixed $\beta = 1$), for $n = 8$.

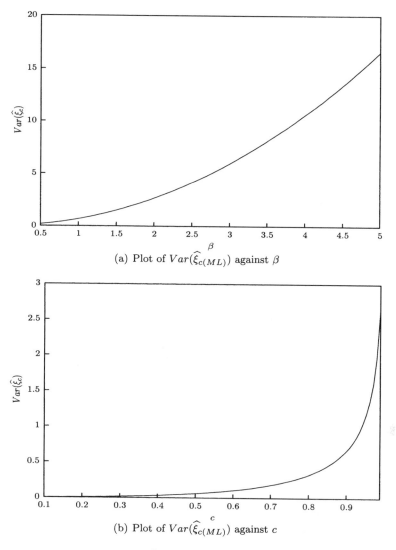

(a) Plot of $Var(\widehat{\xi}_{c(ML)})$ against β

(b) Plot of $Var(\widehat{\xi}_{c(ML)})$ against c

Figure 6.3.2: Plots of $Var(\widehat{\xi}_{c(ML)})$ for $n = 8$; (a) against β (with $c = 0.90$); (b) against c (with $\beta = 1$).

Now we consider a two-parameter exponential distribution with scale parameter β and location parameter μ.

Given *iid* observations $X_1, X_2, \ldots, X_n$ with *pdf* (6.1.5), the likelihood function of the data is

$$L(\beta, \mu \mid X_1, X_2, \ldots, X_n) = \beta^{-n} exp\left\{-\sum_{i=1}^{n}(X_i - \mu)/\beta\right\}$$

$$= \beta^{-n} exp\left\{-\sum_{i=1}^{n}(X_i - X_{(1)})/\beta\right.$$

$$\left. -n(X_{(1)} - \mu)/\beta\right\} \quad (6.3.6)$$

where $\beta > 0$ and $X_i \geq \mu$. Therefore, the minimal sufficient statistics for (β, μ) is $(\sum_{i=1}^{n}(X_i - X_{(1)}), X_{(1)})$. The MLEs of β and μ are

$$\widehat{\beta}_{ML} = \sum_{i=1}^{n}(X_i - X_{(1)})/n \quad \text{and} \quad \widehat{\mu}_{ML} = X_{(1)} \quad (6.3.7)$$

respectively. The statistics $\widehat{\beta}_{ML}$ and $\widehat{\mu}_{ML}$ are mutually independent, and

$$2n\widehat{\beta}_{ML}/\beta \sim \chi^2_{2(n-1)}, \quad \widehat{\mu}_{ML} \sim Exp\,(\beta/n, \mu) \quad (6.3.8)$$

Using the above properties (6.3.8), it is easy to see that

$$E(\widehat{\beta}_{ML}) = \beta(n-1)/n \quad \text{and} \quad E(\widehat{\mu}_{ML}) = \mu + \beta/n \quad (6.3.9)$$

Therefore, both $\widehat{\beta}_{ML}$ and $\widehat{\mu}_{ML}$ are biased estimators of their respective parameters under the model (6.1.5). Hence, the best unbiased estimators (UEs) of β and μ respectively are

$$\widehat{\beta}_U = n\widehat{\beta}_{ML}/(n-1) \quad \text{and} \quad \widehat{\mu}_U = \widehat{\mu}_{ML} - \widehat{\beta}_U/n \quad (6.3.10)$$

Under the model (6.1.5), one can find the first two population moments as

$$E(X) = \mu + \beta \quad \text{and} \quad E(X^2) = \beta^2 + (\mu + \beta)^2 \quad (6.3.11)$$

Hence, one can find the method of moments estimators (MMEs) of μ and β by equating the population moments (in (6.3.11)) with their sample counterparts as

$$\widehat{\beta}_{MM} = \left\{\sum_{i=1}^{n}(X_i - \overline{X})^2/n\right\}^{1/2} \quad \text{and} \quad \widehat{\mu}_{MM} = \overline{X} - \widehat{\beta}_{MM} \quad (6.3.12)$$

But since the MMEs are not functions of minimal sufficient statistics, they (or the unbiased estimators based on the MMEs) are worse than the

MLE (6.3.9) as well as the UE (6.3.10) in terms of variance. Therefore, one is better off by focusing only on the MLEs or UEs.

From (6.3.9) it is easy to see that

$$Bias(\widehat{\beta}_{\mathrm{ML}}) \quad = \quad E(\widehat{\beta}_{\mathrm{ML}} - \beta) = -\beta/n, \quad Var(\widehat{\beta}_{\mathrm{ML}}) = (n-1)\beta^2/n^2$$

and

$$Bias(\widehat{\mu}_{\mathrm{ML}}) \quad = \quad E(\widehat{\mu}_{\mathrm{ML}} - \mu) = \beta/n, \quad Var(\widehat{\mu}_{\mathrm{ML}}) = \beta^2/n^2 \quad (6.3.13)$$

From (6.3.13), we have the following MSE expressions of the MLEs:

$$
\begin{aligned}
MSE(\widehat{\beta}_{\mathrm{ML}}) &= (n-1)\beta^2/n^2 + \beta^2/n^2 = \beta^2/n \\
MSE(\widehat{\mu}_{\mathrm{ML}}) &= 2\beta^2/n^2
\end{aligned}
\quad (6.3.14)
$$

For the unbiased estimators in (6.3.10), we have

$$
\begin{aligned}
MSE(\widehat{\beta}_{\mathrm{U}}) &= Var(\widehat{\beta}_{\mathrm{U}}) = \beta^2/(n-1) \\
MSE(\widehat{\mu}_{\mathrm{U}}) &= Var(\widehat{\mu}_{\mathrm{U}}) = \beta^2/(n(n-1))
\end{aligned}
\quad (6.3.15)
$$

Thus, in term of MSE (i.e., risk under the squared error loss), for estimating β, the unbiased estimator $\widehat{\beta}_{\mathrm{U}}$ is comparable to the MLE $\widehat{\beta}_{\mathrm{ML}}$ except for small n; and for estimating μ, the unbiased estimator $\widehat{\mu}_{\mathrm{U}}$ gives almost 50% improvement over the MLE $\widehat{\mu}_{\mathrm{ML}}$ for large n. The following figures provide the MSE curves of the above estimators.

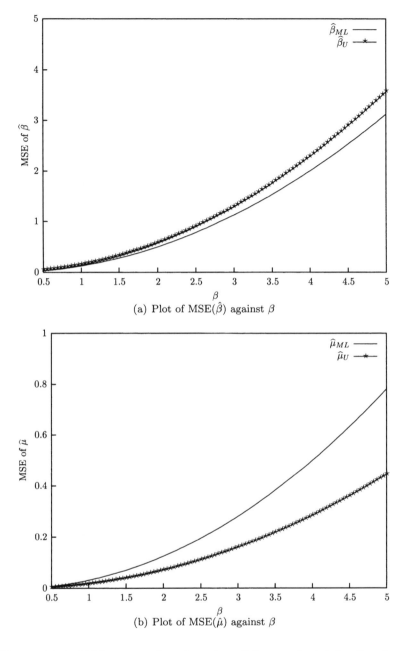

(a) Plot of MSE($\hat{\beta}$) against β

(b) Plot of MSE($\hat{\mu}$) against β

Figure 6.3.3: MSE curves plotted against β for $n = 8$: (a) for β estimation; (b) for μ estimation.

6.4   Goodness of Fit Tests for Exponential Distribution

In parametric inference where the probability distribution of the data is assumed to fall in a known family of distributions (called the probability model of the data), it is imperative that the practitioner needs to verify the validity of the model before going into further inferences. If the assumed model itself is doubtful in explaining the data, then such an assumption can lead to incorrect inferences.

In the case of exponential distribution (or model), one needs to test the validity of it when a data set is available. Given a data set, the easiest thing one can do is to draw a probability histogram of the data and if the histogram shows a pattern similar to those in Figure 6.1.1, then an exponential model may seem plausible. However, one should go through a rigorous checking of the model validity known as a **goodness of fit test** and this is described below. We first consider the $\mathbf{Exp}(\beta, 0)$ (or $\mathbf{Exp}(\beta)$) model, and then we generalize it to the $\mathbf{Exp}(\beta, \mu)$ model.

Suppose that a random sample $X_1, X_2, \ldots, X_n$ is to be tested to see if it could come from the $\mathrm{Exp}(\beta, 0)$ distribution. We consider testing the null hypothesis

$$\mathrm{H}_0 : X_i\text{'s are } iid \ \mathrm{Exp}(\beta, 0) \text{ random variables} \qquad (6.4.1)$$

against the alternative hypothesis

$$\mathrm{H}_1 : X_i\text{'s are not } iid \ \mathrm{Exp}(\beta, 0) \text{ random variables} \qquad (6.4.2)$$

Using the characterization in Theorem 6.2.4, testing H_0 in (6.4.1) is equivalent to test

$$\mathrm{H}_0^* : Z_i\text{'s } (1 \leq i \leq n-1) \text{ are ordered statistics from } \mathrm{CU}[0,1] \qquad (6.4.3)$$

where Z_i's are defined in Theorem 6.2.4. Therefore, the problem now becomes to test the goodness of fit of the ordered observations $Z_1, Z_2, \ldots, Z_n$ from $\mathrm{CU}[0, 1]$.

A well-known test statistic for testing H_0^* in (6.4.3) is

$$\mathrm{T}^* = -2 \sum_{i=1}^{n-1} ln Z_i \qquad (6.4.4)$$

which is due to Pearson (1933). If T^* is 'too small' or 'too large', then H_0^* is rejected. The probability distribution of T^* under H_0^* is $\chi^2_{2(n-1)}$ (i.e., a chi-square distribution with $2(n-1)$ degrees of freedom). Large values of T^* mean the points have moved toward 0 and small values of T^* imply the points have shifted toward 1. One can thus reject H_0^* (6.4.3), and hence H_0 (6.4.1) at level α if T^* is significant,[4] i.e., $T^* > \chi^2_{2(n-1),(\alpha/2)}$ or $T^* < \chi^2_{2(n-1),1-(\alpha/2)}$; and accept H_0 otherwise.

For testing a more general hypothesis

$$H_0 : X_i\text{'s are } iid \text{ Exp}(\beta, \mu) \text{ random variables} \qquad (6.4.5)$$

against a suitable H_1 (i.e., X_i's are not iid **Exp**(β, μ)), we can use the characterization provided by Theorem 6.2.6. In practical applications, however, note that both μ and β are unknown, and hence Y_1 (as defined in Theorem 6.2.6) can not be observed. Therefore, one should only use the spacings $Y_2, Y_3, \ldots, Y_n$ and then define

$$S^* = \sum_{i=2}^n Y_i \quad \text{and} \quad Z_j^{**} = \frac{1}{S^*} \sum_{i=2}^j Y_i, \text{ for } 2 \leq j \leq n-1 \qquad (6.4.6)$$

If the **Exp**(β, μ) model is correct for original random variables $X_1, X_2, \ldots, X_n$, then Z_j^{**}'s should act like $(n-2)$ ordered observations from CU$[0, 1]$. Similar to (6.4.4), a test statistic for testing H_0 in (6.4.5) is

$$T^{**} = -2 \sum_{i=2}^{n-2} \ln Z_i^{**} \qquad (6.4.7)$$

which, under H_0 in (6.4.5), follows $\chi^2_{2(n-2)}$ distribution. Therefore, reject H_0 if $T^{**} > \chi^2_{2(n-2),(\alpha/2)}$ or $T^{**} < \chi^2_{2(n-2),1-(\alpha/2)}$; and accept H_0 otherwise.

Example 6.4.1: Consider the given ordered data set

$$48.96 \quad 54.67 \quad 100.75 \quad 120.68$$
$$148.34 \quad 228.43 \quad 293.45 \quad 415.82$$

which is to be modeled by a suitable exponential distribution. Test the goodness of fit for exponential distributions.

First, we would like to see whether the given data are from **Exp**$(\beta, 0)$ by testing the hypotheses,

$$H_0 : X_i\text{'s are } iid \text{ **Exp**}(\beta, 0) \text{ random variables}$$

---

[4]Cutoff points are available from χ^2-distribution tables in the Appendix.

$$H_1 : X_i\text{'s are not } iid \textbf{ Exp}(\beta, 0) \text{ random variables}$$

As shown in (6.4.4), the test statistic for testing the above statements is

$$T^* = -2 \sum_{i=1}^{n-1} ln Z_i$$

where Z_i's are defined in Theorem 6.2.4. The following table provides necessary computational results which are needed for calculating T^*. It is noted that $S = 1411.10$.

i	X_i	$\sum_{k=1}^{i} X_k$	$Z_i = \sum_{k=1}^{i} X_k / S$	$ln Z_i$
1	48.96	48.96	0.034696	-3.36112
2	54.67	103.63	0.038743	-3.25081
3	100.75	204.38	0.071398	-2.63948
4	120.68	325.06	0.085522	-2.45898
5	148.34	473.40	0.105124	-2.25262
6	228.43	701.83	0.161881	-1.82090
7	293.45	995.28	0.207958	-1.57042
8	415.82	1411.10		

Table 6.4.1: Computational Results for Calculating T^*.

Adding the last column of Table 6.4.1, we can obtain the test statistic as

$$T^* = -2 \sum_{i=1}^{n-1} ln Z_i = (-2) \times (-17.3543) = 34.70865$$

At the significant level $\alpha = 0.05$, the Chi-square critical values with $2(n - 1) = 14$ degrees of freedom can be found from the χ^2-distribution table in the Appendix as $\chi^2_{14,(\alpha/2)} = 26.119$. Therefore, we reject the statement H_0 because $T^* = 34.70865 > \chi^2_{14,(\alpha/2)} = 26.119$. Hence, it is unlikely that the given data are from the exponential distribution $\textbf{Exp}(\beta, 0)$.

Now check whether the data could have come from a suitable $\textbf{Exp}(\beta, \mu)$ distribution. Adding the last column of Table 6.4.2, we can obtain the test statistic as

$$T^{**} = -2 \sum_{i=1}^{n-1} ln Z_i = (-2) \times (-11.51538) = 23.03076$$

i	Y_i	Z_j^{**}	$ln Z_j^{**}$
2	51.39	0.0105951	-4.54736
3	460.8	0.105599	-2.24811
4	219.23	0.150798	-1.89181
5	331.92	0.21923	-1.51762
6	1041.17	0.433889	-0.83497
7	910.28	0.621563	-0.47551
8	1835.55		

Table 6.4.2: Computational Results for Calculating T**.

At the significant level $\alpha = 0.05$, the Chi-square critical value with $2(n-2) = 12$ degrees of freedom can be found from the χ^2-distribution table in Appendix as $\chi^2_{12,(\alpha/2)} = 23.337$ and $\chi^2_{12,1-(\alpha/2)} = 4.404$. Therefore, we accept the statement H$_0$ because T$^{**} = 23.03076$ falls between the two cut-off points. Hence, it seems plausible that the given data are obtained from a two-parameter exponential distribution $\mathbf{Exp}(\beta, \mu)$.

Chapter 7

Gamma Distribution

7.1 Preliminaries

In Chapter 2, the gamma distribution and its basic properties were introduced. A random variable X is said to have a (two parameter) gamma distribution with scale parameter β and shape parameter α provided the *pdf* of X is given as

$$f(x\,|\,\alpha,\beta) = \frac{1}{\beta^\alpha \Gamma(\alpha)} x^{\alpha-1} e^{-x/\beta}, \quad 0 < x < \infty,\ \alpha > 0,\ \beta > 0 \quad (7.1.1)$$

The above *pdf* includes the following special cases:

(i) If $\alpha = k/2$, for any integer k, and $\beta = 2$, then we get the χ_k^2 (Chi-square with k degrees of freedom) *pdf* as

$$f(x\,|\,k) = \frac{1}{2^{k/2}\Gamma(k/2)} x^{k/2-1} e^{-x/2}, \quad 0 < x < \infty,\ k > 0 \quad (7.1.2)$$

(ii) If $\alpha = 1$, then we obtain the exponential distribution with scale parameter β as discussed in Chapter 6.

(iii) If $\beta = 1$, then the gamma *pdf* (7.1.1) is called the standard gamma distribution with shape parameter α as

$$f(x\,|\,\alpha) = \frac{1}{\Gamma(\alpha)} x^{\alpha-1} e^{-x}, \quad 0 < x < \infty,\ \alpha > 0 \quad (7.1.3)$$

From (7.1.3) one can have

$$\Gamma(\alpha) = \int_0^\infty x^{\alpha-1} e^{-x} dx \quad (7.1.4)$$

which, by integration by parts, yields

$$\Gamma(\alpha) = (\alpha - 1)\Gamma(\alpha - 1) = (\alpha - 1)(\alpha - 2)\Gamma(\alpha - 2) = \dots \quad (7.1.5)$$

As a result, if α is an integer, then

$$\Gamma(\alpha) = (\alpha - 1)! = (\alpha - 1)(\alpha - 2) \dots 3 \cdot 2 \cdot 1 \quad (7.1.6)$$

The quantity $\Gamma_c(\alpha)$ defined as

$$\Gamma_c(\alpha) = \int_0^c x^{\alpha-1} e^{-x} dx \quad (7.1.7)$$

is called an incomplete gamma function. Thus, if X follows a standard gamma distribution with shape parameter α, then the *cdf* of X is

$$P(X \leq x) = \frac{1}{\Gamma(\alpha)} \int_0^x u^{\alpha-1} e^{-u} du = \frac{\Gamma_x(\alpha)}{\Gamma(\alpha)} \quad (7.1.8)$$

A generalization of (7.1.1) known as a three-parameter gamma distribution is given as

$$f(x \mid \alpha, \beta, \mu) = \frac{1}{\beta^\alpha \Gamma(\alpha)} (x - \mu)^{\alpha-1} e^{-(x-\mu)/\beta}, \quad (7.1.9)$$

$$\mu < x < \infty, \ \alpha > 0, \ \beta > 0, \ \mu \in (-\infty, \infty)$$

where the extra parameter μ is called the shift or location parameter and α and β are the same as discussed in (7.1.1). This three-parameter gamma distribution is often called a **shifted gamma distribution**.

In most of the real life applications where gamma distribution is well suited one usually applies the regular two-parameter gamma distribution given in (7.1.1). If there is a compelling reason to use a three-parameter gamma distribution, then one should test a null hypothesis $H_0 : \mu = 0$ against $H_1 : \mu \neq 0$. If the null hypothesis is accepted, then the three-parameter gamma model can be reduced to the regular two-parameter one, otherwise the three-parameter gamma model is retained.

The following figure plots the *pdf* of a two-parameter gamma distribution (as given in (7.1.1)) for various values of α and β.

It can be seen from Figure 7.1.1 that, as α increases, the shape of the *pdf* becomes similar to the normal *pdf*. In fact, for the three-parameter gamma distribution (7.1.9), it can be verified that

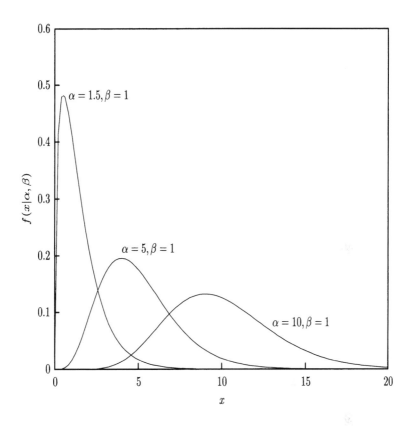

Figure 7.1.1: Two-parameter Gamma *pdf*s.

$$\lim_{\alpha \to \infty} P\left[\left(\frac{(X - \mu)}{\beta} - \alpha\right) \Big/ \sqrt{\alpha} \le u\right] = \Phi(u) \qquad (7.1.10)$$

and in particular if X follows a χ^2_k-distribution, then

$$\lim_{k \to \infty} P\left[\frac{(X - k)}{\sqrt{2k}} \le u\right] = \Phi(u)$$

where $\Phi(\cdot)$ is the standard normal *cdf*.

In real life problems, gamma distribution gives useful representation of many physical situations. It can be used to make realistic adjustments to exponential distribution in representing lifetime in 'life-testing' experiments. In the following, we provide some examples where gamma distribution seems to give a good fit. The shape parameter depends on such factors as the applied pressure and the composite wall thickness.

Example 7.1.1: Lifetimes of pressure vessels constructed of fiber/epoxy composite materials wrapped around metal liners may be modeled by a gamma distribution. Keating, Glaser, and Ketchum (1990) reported the failure times (in hours) of 20 similarly constructed vessels subjected to a certain constant pressure with the following observations:

274.0	1.7	871.0	1311.0	236.0
458.0	54.9	1787.0	0.75	776.0
28.5	20.8	363.0	1661.0	828.0
290.0	175.0	970.0	1278.0	126.0

Using the above data set, we make the following frequency distribution table with six classes over the range 0 to 1788 hours.

Lifetime of Vessels (hr)	Frequency	Relative Frequency
0.0 – 298.0	10	0.50 (50%)
298.0 – 596.0	2	0.10 (10%)
596.0 – 894.0	3	0.15 (15%)
894.0 – 1192.0	1	0.05 (5%)
1192.0 – 1490.0	2	0.10 (10%)
1490.0 – 1788.0	2	0.10 (10%)
Total	20	1.00 (100%)

Table 7.1.1: Frequency Table for Pressure Vessel Lifetime Data.

The following histogram (Figure 7.1.2) along with a matching gamma *pdf* are drawn with the relative frequency values in Table 7.1.1.

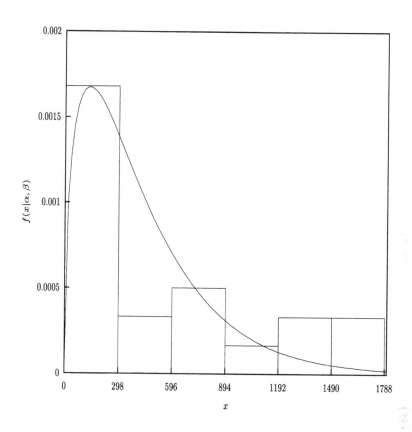

Figure 7.1.2: Histogram of the vessel lifetime data and a matching gamma
pdf with $\alpha = 1.45$ and $\beta = 300$.

Example 7.1.2: The following data on the times between successive
failures of air conditioning (A/C) equipment in Boeing 720 aircrafts have
been provided by Proschan (1963). The data in operating hours are as
follows:

90	10	60	186	61	49
14	24	56	20	79	84
44	59	29	118	25	156
310	76	26	44	23	62
130	208	70	101	208	

The following frequency table is made with four classes over the range from 9.0 to 311.0 hours. The endpoints are chosen conveniently to get class boundaries which do not coincide with the observations.

Operating Hours	Frequency	Relative Frequency
0.0 – 59.5	13	0.448 (44.8%)
59.5 – 119.5	10	0.345 (34.5%)
119.5 – 179.5	2	0.07 (7%)
179.5 – 239.5	3	0.10 (10.%)
239.5 – 311.0	1	0.03 (3%)
Total	29	$\approx$ 1.00 ($\approx$ 100%)

Table 7.1.2: Frequency Table for Operating Hours of A/C Equipment.

The following histogram (Figure 7.1.3) is drawn with the relative frequencies shown in Table 7.1.2.

Apart from the lifetime data, gamma distribution is widely used in seismology, hydrology, and economics. Such examples are given by McDonald and Jensen (1979), Bougeault (1982), Coe and Stern (1982), Singh and Singh (1985), among a host of other authors. The next example shows the use of gamma distribution in entomology.

Example 7.1.3: Matis, Rubink, and Makela (1992) used the gamma distribution to model the 'transit-time distribution' of Africanized honey bee spread through northern Guatemala and Mexico. Data were collected on the first capture times of the honey bee at various monitoring transects in northern Guatemala and on the eastern and western coasts of Mexico. The time intervals between consecutive sightings (in months) are reported where the distances between transects are fixed and the random variable is the time interval between consecutive observations. The transit time data (in months/ 100km) consists of 45 observations.

5.3	1.8	4.2	5.7	3.8	0.8	1.4	3.5	17.5
4.6	0.8	6.3	2.9	0.6	1.9	2.0	6.7	5.5
2.5	2.2	6.7	5.7	10.0	3.3	3.5	20.0	1.6
8.3	4.8	20.0	3.6	8.2	1.3	4.0	5.0	1.7
2.0	2.9	19.2	1.1	1.4	1.5	3.2	8.6	2.2

The following histogram (Figure 7.1.4) is drawn with the relative frequencies shown in Table 7.1.3.

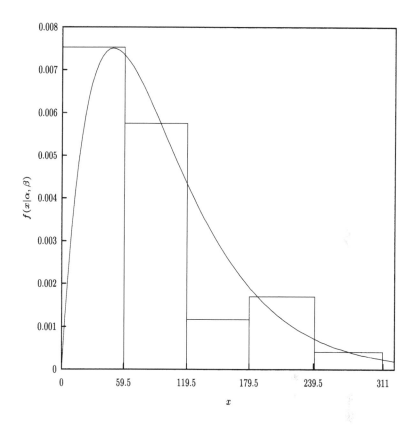

Figure 7.1.3: Histogram of the A/C equipment operating hours and a matching *pdf* with $\alpha = 2.0$ and $\beta = 49.0$.

Transit Time	Frequency	Relative Frequency
0.50 – 2.95	19	0.422 (42.2%)
2.95 – 5.40	12	0.267 (26.7%)
5.40 – 7.85	6	0.133 (13.3%)
7.85 – 10.30	4	0.089 (8.9%)
10.30 – 12.75	0	0.000 (0.0%)
12.75 – 15.20	0	0.000 (0.0%)
15.20 – 17.65	1	0.022 (2.2%)
17.65 – 20.10	3	0.067 (6.7%)
Total	45	$\approx$ 1.00 ($\approx$ 100%)

Table 7.1.3: Frequency Table for Transit Time of Africanized Honey Bee.

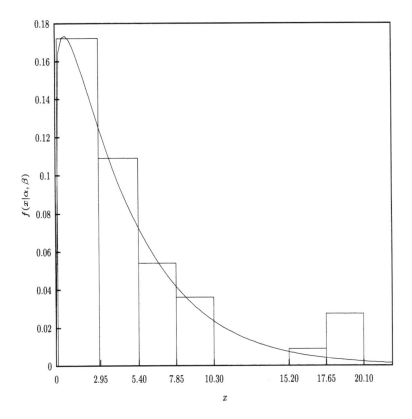

Figure 7.1.4: Histogram of Transit Time of Africanized Honey Bee and a matching *pdf* with $\alpha = 1.15$ and $\beta = 4.0$.

Notational Clarification: A two-parameter gamma distribution of the form (6.1.1) will be called $G(\alpha, \beta)$-distribution. And, a three-parameter gamma distribution of the form (7.1.9) will be called a $G(\alpha, \beta, \mu)$-distribution. Therefore, $G(\alpha, \beta)$ implies the $G(\alpha, \beta, 0)$-distribution, and vice-versa. Though α is used to denote the shape parameter, it is also used to indicate the significance level in a hypothesis testing problem and the distinction should be clear from the context.

7.2   Characterization of Gamma Distribution

The classical result on characterizing Gamma distribution was due to Lukacs (1955), part of which is widely used in statistical inference. This well-known result says that if X_1 and X_2 are two independent random variables with X_i having a Gamma(α_i, β) distribution, $i = 1, 2$, then $U = (X_1 + X_2)$ and $V = X_1/(X_1 + X_2)$ are independent. Moreover, U follows Gamma($\alpha_1 + \alpha_2, \beta$) distribution and V follows Beta(α_1, α_2) distribution. The proof of this result is straightforward. One needs to find the joint *pdf* of (U, V) using the method of transformation which factorizes into independent pieces with respective structures. In the following, Lukacs' (1955) original characterization result is presented.

Theorem 7.2.1 *Let X_1 and X_2 be independent nonnegative random variables with cdfs F_1 and F_2 (or pdfs f_1 and f_2) respectively. Then $U_1 = (X_1/X_2)$ and $U_2 = (X_1 + X_2)$ are independent if and only if F_1 and F_2 are gamma distributions (cdfs) with the same scale parameter.*

Khatri and Rao (1968) obtained the following characterization of the gamma distribution based on constancy of various regression functions.

Theorem 7.2.2 *(a) If $X_1, X_2, \dots, X_n$ ($n \geq 3$) are independent positive random variables and*

$$Y_1 = \sum_{i=1}^{n} b_{1i} X_i, \quad Y_j = \prod_{i=1}^{n} X_i^{b_{ji}}, \quad j = 2, 3, \dots, n, \quad b_{1i} \neq 0, \ 1 \leq i \leq n$$

with the $(n-1) \times n$ matrix $\left((b_{ji}) \right)$ ($j = 2, 3, \dots, n;\ i = 1, 2, \dots, n$) nonsingular, then the constancy of $E[Y_1 | Y_2, \dots, Y_n]$ ensures that X_i's must have a common gamma distribution.

(b) In the above conditions of (a), if $E[1/X_j] \neq 0$ ($j = 1, 2, \dots, n$) and $Z_1 = \sum_{i=1}^{n} b_{1i}/X_i$, $Z_j = \sum_{i=1}^{n} b_{ji} X_j$, $j = 2, 3, \dots, n$ with b_{ji}'s satisfying the same conditions as in (a), then the constancy of $E[Z_1 | Z_2, \dots, Z_n]$ ensures that each X_i has a gamma distribution, not necessarily the same for all i.

(c) Under the same conditions as in (a), if $E[X_1 \ln X_1] < \infty$, then the constancy of $E\left[\sum_{j=1}^{n} a_j X_j \mid \prod_{i=1}^{n} X_i^{b_i} \right]$ with $\sum_{j}^{n} a_j b_j = 0$, $|b_n| > \max_{1 \leq j \leq n-1}(|b_j|)$, and $a_j b_j/(a_n b_n) < 0$ for all $j = 1, 2, \dots, (n-1)$ ensures that X_1 has a gamma distribution.

(d) If $X_1, X_2, \ldots, X_n$ are iid positive random variables with $E[1/X_i] \neq 0$, $i = 1, 2, \ldots, n$, then constancy of

$$E\Big[\sum_{j=1}^{n} (a_j/X_j)\Big| \sum_{j=1}^{n} b_j X_j\Big]$$

with same conditions on a_j's and b_j's as in (c), ensures that the common distribution of X_i's is a gamma distribution.

Corollary 7.2.1 (i) In Theorem 7.2.2 (parts (a) and (b)), if one takes $b_{11} = \ldots = b_{1n} = 1$, $b_{j,j-1} = -1$, $b_{j,j} = 1$ and all other b_{ji}'s zero, then the condition in part (a) becomes the constancy of

$$E\Big[\sum_{j=1}^{n} X_j\Big| X_2/X_1, X_3/X_1, \ldots, X_n/X_1\Big]$$

and the condition of part (b) becomes the constancy of

$$E\Big[\sum_{j=1}^{n} 1/X_j\Big| X_2 - X_1, X_3 - X_1, \ldots, X_n - X_1\Big]$$

(ii) In Theorem 7.2.2 (parts (b) and (c)), if one takes $a_1 = a_2 = \ldots = a_n = 1$, $b_n = n - 1$, $b_1 = b_2 = \ldots = b_{n-1} = -1$, then the condition in part (c) reduces to the constancy of

$$E\Big[\sum_{j=1}^{n} X_j\Big| \prod_{j=1}^{n-1} (X_n/X_j)\Big]$$

and the condition in part (d) reduces to the constancy of

$$E\Big[\sum_{j=1}^{n} 1/X_j\Big| X_n - \overline{X}\Big]$$

where $\overline{X} = \sum_{j=1}^{n} X_j/n$.

Hall and Simons (1969) gave a characterization of the gamma distribution in terms of quadratic expression of two random variables as given below.

Theorem 7.2.3 Suppose X_1 and X_2 are independent random variables with finite second moments and characteristic functions $\psi_1(t)$ and $\psi_2(t)$, respectively. Then for some a_1, a_2,

$$E[X_i^2| X_1 + X_2] = a_i(X_1 + X_2)^2 \quad a.s., \quad i = 1, 2$$

if and only if for some real $\beta \neq 0$ and positive α_1, α_2,

$$\psi_1(t) = (1 - i\beta t)^{-\alpha_1} \quad and \quad \psi_2(t) = (1 - i\beta t)^{-\alpha_2}$$

moreover, $(1 - a_1 + a_2)\alpha_1 = (1 + a_1 - a_2)\alpha_2$, and $|a_1 - a_2| < 1$. In other words, assuming $X_1 + X_2 \neq 0$ a.s., if $X_1^2/(X_1 + X_2)^2$ and $X_2^2/(X_1 + X_2)^2$ each have constant regression on $(X_1 + X_2)$, then X_1 and X_2 or $-X_1$ and $-X_2$ have gamma distributions with identical scale parameter β.

7.3   Estimation of Parameters

We first consider estimation of parameters α and β in a two-parameter gamma distribution.

If $X_1, X_2, \ldots, X_n$ are independent random variables each having *pdf* (7.1.1), then the **maximum likelihood estimators** of α and β are $\widehat{\alpha}_{\mathrm{ML}}$ and $\widehat{\beta}_{\mathrm{ML}}$ respectively obtained by solving the system of equations:

$$\widehat{\alpha}_{\mathrm{ML}}\widehat{\beta}_{\mathrm{ML}} = \overline{X} \quad and \quad ln\widehat{\beta}_{\mathrm{ML}} + \psi(\widehat{\alpha}_{\mathrm{ML}}) = \frac{1}{n}\sum_{i=1}^{n} lnX_i \qquad (7.3.1)$$

where $\overline{X} = \sum_{i=1}^{n} X_i/n$ and $\psi(\alpha) = \frac{\partial}{\partial\alpha}ln\Gamma(\alpha)$, called the **di-gamma function**. Substituting $\widehat{\beta}_{\mathrm{ML}} = \overline{X}/\widehat{\alpha}_{\mathrm{ML}}$ in the second equation of (7.3.1), we obtain the following equation for $\widehat{\alpha}_{\mathrm{ML}}$

$$ln\widehat{\alpha}_{\mathrm{ML}} - \psi(\widehat{\alpha}_{\mathrm{ML}}) = ln\left[\frac{\text{arithmatic mean (AM) of } X_1, X_2, \ldots, X_n}{\text{geometric mean (GM) of } X_1, X_2, \ldots, X_n}\right]$$
$$(7.3.2)$$

Given the values of X_i, $i = 1, 2, \ldots, n$, one can now find the value of $\widehat{\alpha}_{\mathrm{ML}}$ by solving (7.3.2) which then in turn gives the value of $\widehat{\beta}_{\mathrm{ML}}$ as $\widehat{\beta}_{\mathrm{ML}} = \overline{X}/\widehat{\alpha}_{\mathrm{ML}}$.

From the above derivation it is obvious that $\widehat{\alpha}_{\mathrm{ML}}$ and $\widehat{\beta}_{\mathrm{ML}}$ do not have closed expressions, and as a result, it is impossible to study analytical sampling properties of these estimators for fixed sample sizes. However, some asymptotic properties (as $n \to \infty$) of $\sqrt{n}\widehat{\alpha}_{\mathrm{ML}}$ and $\sqrt{n}\widehat{\beta}_{\mathrm{ML}}$ are known to be as

$$\lim_{n\to\infty} E\left[\sqrt{n}\left(\widehat{\alpha}_{\mathrm{ML}} - \alpha\right)\right] = 0$$
$$\lim_{n\to\infty} E\left[\sqrt{n}\left(\widehat{\beta}_{\mathrm{ML}} - \beta\right)\right] = 0$$
$$\lim_{n\to\infty} Var\left(\sqrt{n}\widehat{\alpha}_{\mathrm{ML}}\right) = \frac{\alpha}{(\alpha\psi'(\alpha) - 1)} \qquad (7.3.3)$$

$$\lim_{n \to \infty} Var\left(\sqrt{n}\widehat{\beta}_{\mathrm{ML}}\right) = \frac{\beta^2 \psi'(\alpha)}{(\alpha \psi'(\alpha) - 1)}$$

$$\text{and} \quad \lim_{n \to \infty} Corr\left(\widehat{\alpha}_{\mathrm{ML}}, \widehat{\beta}_{\mathrm{ML}}\right) = -\frac{1}{\sqrt{\alpha \psi'(\alpha)}}$$

where $\psi(\alpha)$ is the di-gamma function mentioned above. In other words, asymptotically, both $\widehat{\alpha}_{\mathrm{ML}}$ and $\widehat{\beta}_{\mathrm{ML}}$ are unbiased and

$$Var\left(\widehat{\alpha}_{\mathrm{ML}}\right) = \frac{\alpha}{n(\alpha \psi'(\alpha) - 1)} + O\left(n^{-2}\right)$$

$$Var\left(\widehat{\beta}_{\mathrm{ML}}\right) = \frac{\beta^2 \psi'(\alpha)}{n(\alpha \psi'(\alpha) - 1)} + O\left(n^{-2}\right) \qquad (7.3.4)$$

Berman (1981) proved that for fixed sample size $n \geq 2$, the estimator $\widehat{\alpha}_{\mathrm{ML}}$ is always positively biased. Using the reparameterization $\lambda = 1/\beta$, it was also shown that $\widehat{\lambda}_{\mathrm{ML}} = 1/\widehat{\beta}_{\mathrm{ML}}$ is also positively biased, i.e., $E\left(\widehat{\lambda}_{\mathrm{ML}}\right) > \lambda$ for all $\alpha, \lambda > 0$. Shenton and Bowman (1972) have given the mean expressions of $\widehat{\alpha}_{\mathrm{ML}}$ and $\widehat{\lambda}_{\mathrm{ML}}$ as (assuming that $n > 3$)

$$E\left[\widehat{\alpha}_{\mathrm{ML}} - \alpha\right] = \frac{3\alpha}{(n-3)}\left\{1 - \frac{2}{9\alpha} + \frac{(n-1)}{27n\alpha^2} + \frac{7(n^2-1)}{162(n+3)n^2\alpha^3} + \cdots\right\}$$

$$E\left[\widehat{\lambda}_{\mathrm{ML}} - \lambda\right] = \frac{n\alpha\lambda}{(n-3)(n\alpha-1)}\left\{3 + \frac{1}{3\alpha} - \frac{3}{n\alpha} + \frac{(n-1)}{9n\alpha^2} + \cdots\right\}$$

$$\qquad (7.3.5)$$

From the expressions in (7.3.5), Anderson and Ray (1975) suggested the modified MLEs of α and λ as

$$\widehat{\alpha}_{\mathrm{MML}} = \frac{(n-3)}{n}\widehat{\alpha}_{\mathrm{ML}} + \frac{2}{3n} \quad \text{and}$$

$$\widehat{\lambda}_{\mathrm{MML}} = \widehat{\lambda}_{\mathrm{ML}}g\left(\widehat{\alpha}_{\mathrm{MML}}\right) \qquad (7.3.6)$$

$$\text{(i.e., } \quad \widehat{\beta}_{\mathrm{MML}} = \widehat{\beta}_{\mathrm{ML}}/g(\widehat{\alpha}_{\mathrm{MML}}))$$

where

$$g(\alpha) = \left\{1 + \frac{3n\alpha}{(n-3)(n\alpha-1)}\left[1 + \frac{1}{9\alpha} - \frac{1}{n\alpha} + \frac{(n-1)}{27n\alpha^2}\right]\right\}^{-1}$$

The above modified MLEs have smaller bias, and this reduction of bias is not gained at the cost of an increase in MSE (mean squared error). Extensive numerical study shows that $\widehat{\alpha}_{\mathrm{MML}}$ and $\widehat{\lambda}_{\mathrm{MML}}$ can provide substantial improvements over $\widehat{\alpha}_{\mathrm{ML}}$ and $\widehat{\lambda}_{\mathrm{ML}}$, respectively, in terms of bias and MSE for small sample sizes $(n \leq 20)$.

Using the parameters α and β (as in (7.1.1)), Shenton and Bowman (1969) have shown that

$$E\left[\left(\frac{\widehat{\alpha}_{\mathrm{ML}}}{\alpha}\right)^s\right] \approx \frac{n^s}{(n-3)(n-5)\cdots(n-2s-1)}\left(1 - \frac{s(s+1)}{3n\alpha}\right) \quad \text{and}$$

$$E\left[\frac{\widehat{\beta}_{\mathrm{ML}}}{\beta}\right] = 1 - \frac{1}{n} + O(\alpha^{-2}) \qquad (7.3.7)$$

It is seen that though $\widehat{\alpha}_{\mathrm{ML}}$ is positively biased, the estimator $\widehat{\beta}_{\mathrm{ML}}$ is negatively biased.

Since the MLEs (and hence the modified MLEs) of the gamma parameters α and β do not have closed expressions, a simple and convenient way of estimating parameters is through the method of moments. The method of moments estimators (MMEs) of α and β are obtained by equating the first two population moments with those sample analogues, i.e., $\widehat{\alpha}_{\mathrm{MM}}$ and $\widehat{\beta}_{\mathrm{MM}}$ are obtained by solving the following equations:

$$\widehat{\alpha}_{\mathrm{MM}}\widehat{\beta}_{\mathrm{MM}} = \overline{X} \quad \text{and} \quad \widehat{\alpha}_{\mathrm{MM}}\left(\widehat{\beta}_{\mathrm{MM}}\right)^2 = \frac{1}{n}\sum_{i=1}^n (X_i - \overline{X})^2$$

which produce

$$\widehat{\alpha}_{\mathrm{MM}} = \frac{n\left(\overline{X}\right)^2}{\sum_{i=1}^n (X_i - \overline{X})^2} \quad \text{and} \quad \widehat{\beta}_{\mathrm{MM}} = \frac{\sum_{i=1}^n (X_i - \overline{X})^2}{n\overline{X}} \qquad (7.3.8)$$

Though $\widehat{\alpha}_{\mathrm{MM}}$ and $\widehat{\beta}_{\mathrm{MM}}$ have simpler expressions, the MLEs (or the modified MLEs) perform consistently better than the MMEs in terms of bias as well as MSE even for small sample sizes.

We now consider estimation of parameters for a three-parameter gamma distribution.

If we have *iid* observations $X_1, X_2, \ldots, X_n$ from a three-parameter gamma distribution with parameters $\alpha, \beta,$ and μ (as shown in (7.1.9)), then the MLEs $\widehat{\alpha}_{\mathrm{ML}}, \widehat{\beta}_{\mathrm{ML}}$ and $\widehat{\mu}_{\mathrm{ML}}$ are obtained by solving the following system of equations:

$$\sum_{i=1}^n ln(X_i - \widehat{\mu}_{\mathrm{ML}}) - nln\widehat{\beta}_{\mathrm{ML}} - n\psi\left(\widehat{\alpha}_{\mathrm{ML}}\right) = 0$$

$$\sum_{i=1}^n \left(X_i - \widehat{\mu}_{\mathrm{ML}}\right) - n\widehat{\alpha}_{\mathrm{ML}}\widehat{\beta}_{\mathrm{ML}} = 0 \qquad (7.3.9)$$

$$\sum_{i=1}^n \left(X_i - \widehat{\mu}_{\mathrm{ML}}\right)^{-1} - n\left(\widehat{\beta}_{\mathrm{ML}}\left(\widehat{\alpha}_{\mathrm{ML}} - 1\right)\right)^{-1} = 0$$

From the third equation in (7.3.9), it is seen that if $\widehat{\alpha}_{\mathrm{ML}}$ is less than 1, then some X_i's must be less than $\widehat{\mu}_{\mathrm{ML}}$ which is contradictory since X_i's are all greater than μ and hence $\widehat{\mu}_{\mathrm{ML}}$ must be less than X_i's. Therefore, the system of equations in (7.3.9) has to be solved iteratively. One can use MME estimators of α, β, μ as starting (or initial) values; then use the first equation to determine a new value of $\widehat{\beta}_{\mathrm{ML}}$, given $\widehat{\alpha}_{\mathrm{ML}}$ and $\widehat{\mu}_{\mathrm{ML}}$; then the second equation for a new value of $\widehat{\mu}_{\mathrm{ML}}$, given $\widehat{\alpha}_{\mathrm{ML}}$ and $\widehat{\beta}_{\mathrm{ML}}$; and then the third equation for a new value of $\widehat{\alpha}_{\mathrm{ML}}$, given $\widehat{\beta}_{\mathrm{ML}}$ and $\widehat{\mu}_{\mathrm{ML}}$. Continue this process until the stabilized values of $\widehat{\alpha}_{\mathrm{ML}}, \widehat{\beta}_{\mathrm{ML}}$, and $\widehat{\mu}_{\mathrm{ML}}$ are achieved.

Example 7.3.1: Demonstrate the MLE computation by drawing a random sample of size $n = 10$ from a gamma distribution with $\alpha = 5, \beta = 1$, and $\mu = 0$, and then follow the iterative process until $\widehat{\alpha}_{\mathrm{ML}}, \widehat{\beta}_{\mathrm{ML}}$, and $\widehat{\mu}_{\mathrm{ML}}$ values (stabilized) are achieved.

First, a random sample of size $n = 10$ is generated from $G(5.0, 1.0, 0.0)$ distribution. The data set is:

1.8671   3.9899   5.4569   2.7314   3.7961   5.2041   7.4768   3.0366
3.7333   5.1969

Then calculate the MLEs using (7.3.9) to get $\widehat{\alpha}_{\mathrm{ML}} = 5.7375, \widehat{\beta}_{\mathrm{ML}} = 0.6586$, and $\widehat{\mu}_{\mathrm{ML}} = 0.4700$.

If we equate the first three population (i.e., $G(\alpha, \beta, \mu)$) moments with their sample counterparts, then we have the method of moments estimators $\widehat{\mu}_{\mathrm{MM}}, \widehat{\alpha}_{\mathrm{MM}}$, and $\widehat{\beta}_{\mathrm{MM}}$ obtained by solving the following three equations:

$$
\begin{aligned}
(i) \quad \mu + \alpha\beta &= \bar{X} \\
(ii) \quad \alpha\beta^2 &= S/n \\
(iii) \quad 2\beta(4\alpha^2 + 3\alpha + 1) &= T/S
\end{aligned}
\qquad (7.3.10)
$$

where $S = \sum_{i=1}^{n}(X_i - \bar{X})^2$ and $T = \sum_{i=1}^{n}(X_i - \bar{X})^3$. (The third equation above is obtained by equating the third sample moment with its population counterpart, and then using the second equation above.)

In the following, we provide an approach to extend the two-parameter **modified maximum likelihood estimation** for the three-parameter case.

After obtaining $\hat{\mu}_{\mathrm{ML}}$, $\hat{\alpha}_{\mathrm{ML}}$, and $\hat{\beta}_{\mathrm{ML}}$ (solving (7.3.9) as discussed earlier), get

$$X_i^* = X_i - \hat{\mu}_{\mathrm{ML}}, \ i = 1, 2, \ldots, n \qquad (7.3.11)$$

The new observations $X_1^*, \ldots, X_n^*$ are now modelled by a $G(\alpha, \beta)$ distribution (two parameter gamma model).

Obtain $\hat{\alpha}_{\mathrm{MML}}$ and $\hat{\beta}_{\mathrm{MML}}$ based on X_i^*'s as described in (7.3.6). Then, plug in $\hat{\alpha}_{\mathrm{MML}}$ and $\hat{\beta}_{\mathrm{MML}}$ in the third equation in (7.3.9) and solve for μ to get $\hat{\mu}_{\mathrm{MML}}$.

The above three methods (i.e., MM estimation, ML estimation, and modified ML estimation) are now used for estimating the parameters of a three-parametrer gamma distribution used to model the data set in Example 7.1.2. The estimates are given below.

$$\hat{\mu}_{\mathrm{MM}} = -7.89; \quad \hat{\alpha}_{\mathrm{MM}} = 1.73; \quad \text{and } \hat{\beta}_{\mathrm{MM}} = 52.96$$
$$\hat{\mu}_{\mathrm{ML}} = 2.76; \quad \hat{\alpha}_{\mathrm{ML}} = 1.85; \quad \text{and } \hat{\beta}_{\mathrm{ML}} = 42.73$$
$$\hat{\mu}_{\mathrm{MML}} = 0.63; \quad \hat{\alpha}_{\mathrm{MML}} = 1.38; \quad \text{and } \hat{\beta}_{\mathrm{MML}} = 59.86$$

In the following we provide the bias and MSE of the above three estimators through simulation.

Generate $X_1, \ldots, X_n$ from $G(\alpha, \beta, \mu)$ $M = 20,000$ times with various α values, $\beta = 1$ and $\mu = 0$. Estimate the bias and MSE of $\hat{\alpha}$, $\hat{\beta}$, $\hat{\mu}$ (following each of the methods discussed above).

$$
\begin{aligned}
Bias(\hat{\alpha}) &\approx \left(\sum_{j=1}^{M} \hat{\alpha}^{(j)}/M \right) - \alpha \\[2mm]
Bias(\hat{\beta}) &\approx \left(\sum_{j=1}^{M} \hat{\beta}^{(j)}/M \right) - \beta \qquad (7.3.12) \\[2mm]
Bias(\hat{\mu}) &\approx \left(\sum_{j=1}^{M} \hat{\mu}^{(j)}/M \right) - \mu
\end{aligned}
$$

and

$$
\begin{aligned}
MSE(\hat{\alpha}) &\approx \sum_{j=1}^{M} \left(\hat{\alpha}^{(j)} - \overline{\hat{\alpha}} \right)^2 / M \\[1mm]
MSE(\hat{\beta}) &\approx \sum_{j=1}^{M} \left(\hat{\beta}^{(j)} - \overline{\hat{\beta}} \right)^2 / M \qquad (7.3.13) \\[1mm]
MSE(\hat{\mu}) &\approx \sum_{j=1}^{M} \left(\hat{\mu}^{(j)} - \overline{\hat{\mu}} \right)^2 / M
\end{aligned}
$$

where $\hat{\alpha}^{(j)}$, $\hat{\beta}^{(j)}$, $\hat{\mu}^{(j)}$ are the estimated values of α, β, and μ based on the jth ($j = 1, \ldots, M$) replicated sample, and $\overline{\hat{\alpha}} = \sum_{j=1}^{M} \hat{\alpha}^{(j)}/M$,

$\overline{\beta} = \sum_{j=1}^{M} \hat{\beta}^{(j)}/M$, $\overline{\hat{\mu}} = \sum_{j=1}^{M} \hat{\mu}^{(j)}/M$. The biases and MSEs are plotted against α.

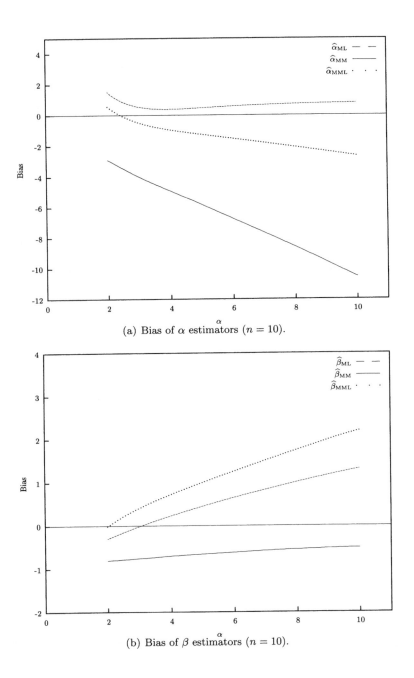

(a) Bias of α estimators ($n = 10$).

(b) Bias of β estimators ($n = 10$).

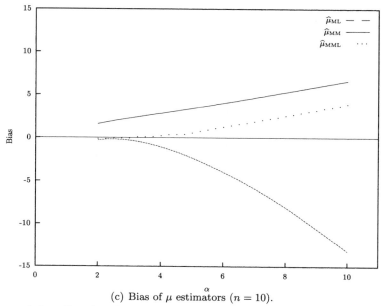

(c) Bias of μ estimators $(n = 10)$.

Figure 7.3.1: Graphs of bias of various estimators plotted against α when $n = 10$ and data come from $G(\alpha, 1, 0)$ distribution: (a) for α estimation, (b) for β estimation, (c) for μ estimation.

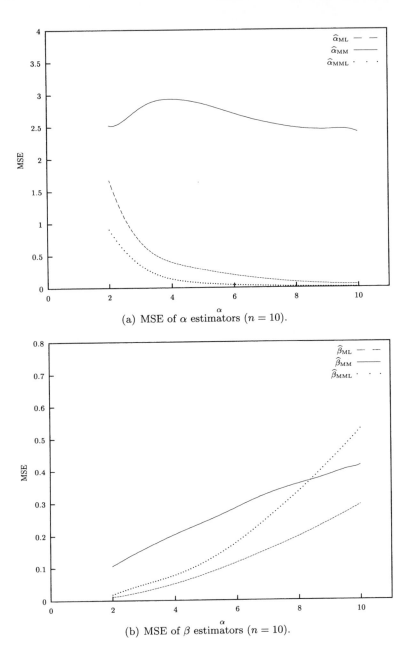

(a) MSE of α estimators ($n = 10$).

(b) MSE of β estimators ($n = 10$).

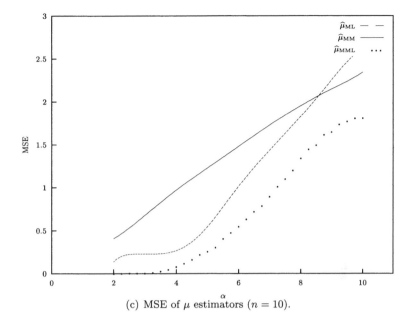

(c) MSE of μ estimators ($n = 10$).

Figure 7.3.2: Graphs of MSE of various estimators plotted against α when $n = 10$ and data come from $G(\alpha, 1, 0)$ distribution: (a) for α estimation, (b) for β estimation, (c) for μ estimation.

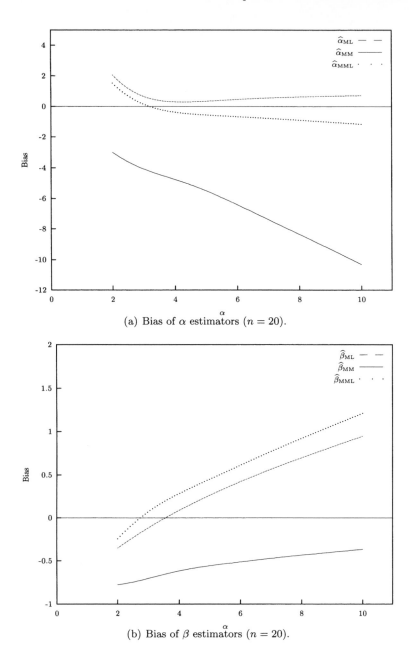

(a) Bias of α estimators ($n = 20$).

(b) Bias of β estimators ($n = 20$).

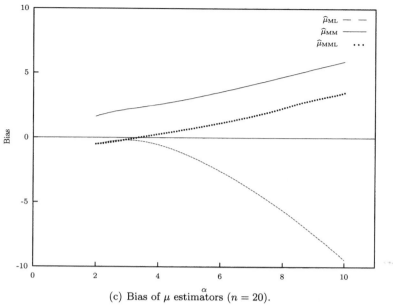

(c) Bias of μ estimators ($n = 20$).

Figure 7.3.3: Graphs of bias of various estimators plotted against α when $n = 20$ and data come from $G(\alpha, 1, 0)$ distribution: (a) for α estimation, (b) for β estimation, (c) for μ estimation.

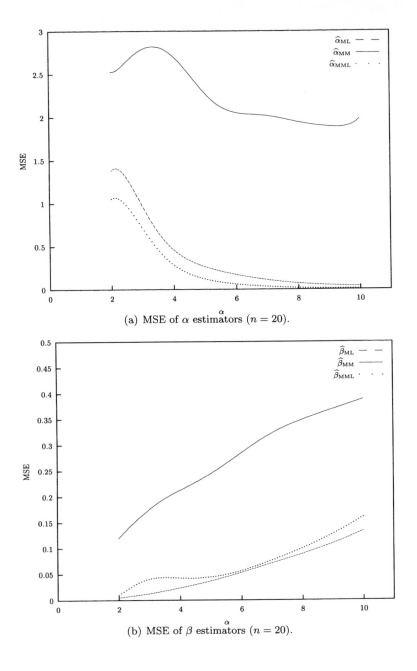

(a) MSE of α estimators ($n = 20$).

(b) MSE of β estimators ($n = 20$).

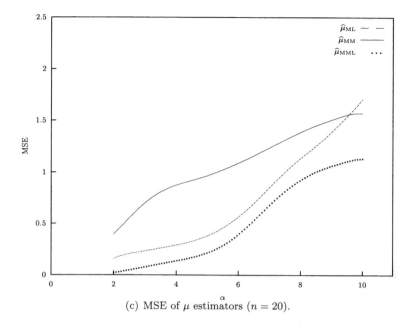

(c) MSE of μ estimators ($n = 20$).

Figure 7.3.4: Graphs of MSE of various estimators plotted against α when $n = 20$ and data come from $G(\alpha, 1, 0)$ distribution: (a) for α estimation, (b) for β estimation, (c) for μ estimation.

Remark 7.3.1 *From our extensive simulation study it appears that the modified maximum likelihood (MML) estimation has the best overall performance. Hence the MML estimation method is suggested for three parameter gamma distribution.*

7.4   Goodness of Fit Tests for Gamma Distribution

Dahiya and Gurland (1972) provided a general goodness of fit test for a large sample by making use of the central limit theorem. In the following, we provide a simplified version of the goodness of fit test applicable to a two-parameter gamma distribution (7.1.1).

To test whether a random sample $X_1, X_2, \ldots, X_n$ is following a $G(\alpha, \beta)$

distribution, we adopt the following notations:

$$k_1 = \frac{1}{n} \sum_{i=1}^{n} X_i = \overline{X}$$

$$k_2 = \frac{1}{n} \sum_{i=1}^{n} (X_i - \overline{X})^2$$

$$k_3 = \frac{1}{n} \sum_{i=1}^{n} (X_i - \overline{X})^3$$

$$h_1 = k_1 = \overline{X} \tag{7.4.1}$$

$$h_2 = \frac{k_2}{k_1} = \frac{\sum_{i=1}^{n} (X_i - \overline{X})^2}{n\overline{X}}$$

$$h_3 = \frac{k_3}{k_2} = \frac{\sum_{i=1}^{n} (X_i - \overline{X})^3}{\sum_{i=1}^{n} (X_i - \overline{X})^2}$$

Also, define

$$\boldsymbol{\theta}^* = (\theta_1^*, \theta_2^*)' = (\alpha\beta, \beta)' \quad \text{and} \tag{7.4.2}$$

$$\widehat{\boldsymbol{\theta}}^* = (k_1, k_2/k_1)' = \left(\overline{X}, \frac{1}{n} \sum_{i=1}^{n} (X_i - \overline{X})^2 / \overline{X}\right)'$$

Note that

$$\boldsymbol{\xi} = \boldsymbol{W}'\boldsymbol{\theta}^*, \quad \text{where}$$

$$\boldsymbol{W} = \begin{bmatrix} 1 & 0 & 0 \\ 0 & 1 & 2 \end{bmatrix} \tag{7.4.3}$$

Define the matrices J and G as

$$J = 3 \begin{pmatrix} 1 & 0 & 0 \\ -\frac{(2\alpha+1)}{\alpha} & \frac{1}{\alpha\beta} & 0 \\ (3\alpha+1) & -\frac{(3\alpha+2)}{\alpha\beta} & \frac{1}{\alpha\beta^2} \end{pmatrix} \tag{7.4.4}$$

and

$$G = ((g_{ij}))_{3\times 3} \tag{7.4.5}$$

where

$$g_{11} = \alpha\beta^2$$

$$g_{12} = 2\alpha(\alpha+1)\beta^3$$

$$g_{13} = 3\alpha(\alpha+1)(\alpha+2)\beta^4$$

$$g_{21} = 2\alpha(\alpha+1)\beta^3$$

$$g_{22} = 2\alpha(\alpha+1)(2\alpha+3)\beta^4$$

$$g_{23} = 6\alpha(\alpha+1)(\alpha+2)^2\beta^5$$

$$g_{31} = 3\alpha(\alpha+1)(\alpha+2)\beta^4$$

$$g_{32} = 6\alpha(\alpha+1)(\alpha+2)^2\beta^5$$

$$g_{33} = 3\alpha(\alpha+1)(\alpha+2)(3\alpha^2+15\alpha+20)\beta^6$$

The matrix $\boldsymbol{\Sigma}$ is defined as

$$\boldsymbol{\Sigma} = \boldsymbol{JGJ}' \qquad (7.4.6)$$

and then $\boldsymbol{\Sigma}$ is estimated by $\widehat{\boldsymbol{\Sigma}}$ where

$$\widehat{\boldsymbol{\Sigma}} = \left[\boldsymbol{\Sigma}\right]_{\boldsymbol{\theta}^*=\widehat{\boldsymbol{\theta}}^*} \qquad (7.4.7)$$

In other words, the matrix $\boldsymbol{\Sigma}$, whose elements are functions of α and β, is estimated by replacing α and β by $\widehat{\alpha}_{\mathrm{ML}}$ and $\widehat{\beta}_{\mathrm{ML}}$ respectively. Now define the test statistic $\widehat{\boldsymbol{Q}}$ as

$$\widehat{\boldsymbol{Q}} = n\boldsymbol{h}'\widehat{\boldsymbol{A}}\boldsymbol{h} \qquad (7.4.8)$$

where

$$\widehat{\boldsymbol{A}} = \widehat{\boldsymbol{\Sigma}}^{-1}\left(\boldsymbol{I}-\widehat{\boldsymbol{R}}\right)$$

$$\widehat{\boldsymbol{R}} = \boldsymbol{W}'\left(\boldsymbol{W}\widehat{\boldsymbol{\Sigma}}^{-1}\boldsymbol{W}'\right)^{-1}\boldsymbol{W}\widehat{\boldsymbol{\Sigma}}^{-1}$$

$$\text{and} \quad \boldsymbol{h} = (h_1,h_2,h_3)' \quad \text{(as in (7.4.1))}$$

Theorem 7.4.1 *The asymptotic null distribution*[1] *of* $\widehat{\boldsymbol{Q}}$ *is that of* χ_1^2.

Using the above result, for a large n, we reject gamma distribution as a suitable one for the data if $\widehat{\boldsymbol{Q}}$ exceeds the upper α-cutoff point of χ_1^2 distribution, i.e., reject gamma model if $\widehat{\boldsymbol{Q}} > \chi_{1,\alpha}^2$; and retain the gamma model otherwise. [Here α indicates the level of significance.]

As a demonstration of this goodness of fit test, we use the data set in Example 7.1.2 to justify the applicability of a Gamma distribution to model the data.

For simplicity, one can get $\widehat{\boldsymbol{\Sigma}}$ by directly using $\widehat{\boldsymbol{J}}$ and $\widehat{\boldsymbol{G}}$ obtained from $\boldsymbol{J}$ and $\boldsymbol{G}$ respectively, by replacing α and β by $\widehat{\alpha}_{\mathrm{ML}}$ and $\widehat{\beta}_{\mathrm{ML}}$ respectively.

---

[1]The distribution of $\widehat{\boldsymbol{Q}}$ when $X_1, X_2, \ldots, X_n$ are assumed to follow (7.1.1).

For the data set in Example 7.1.2, we have

$$\widehat{\alpha}_{ML} = 1.67 \quad \text{and} \quad \widehat{\beta}_{ML} = 49.98$$

$$\widehat{J} = \begin{bmatrix} 3 & 0 & 0 \\ -7.79641 & 0.0359425 & 0 \\ 18.03 & -0.251957 & 0.000719 \end{bmatrix}$$

$$\widehat{G} = \begin{bmatrix} 4171.66 & 1113390 & 306337000 \\ 1113390 & 352803000 & 112381000000 \\ 306337000 & 112381000000 & 40876100000000 \end{bmatrix}$$

Also,

$$\widehat{h} = \begin{bmatrix} 83.5172 \\ 57.9592 \\ 105.909 \end{bmatrix}$$

which gives

$$\widehat{Q} = 0.002692$$

Since $\chi^2_{1,\alpha}$ (with $\alpha = 0.05$) $= \chi^2_{1,0.05} = 3.841$ and $\widehat{Q} < \chi^2_{1,0.05}$, we accept the fact that a two-parameter gamma model is suitable for the data in Example 7.1.2.

Remark 7.4.1 *To verify the goodness of fit of a three parameter gamma model, first obtain the $\widehat{\mu}_{MML}$ (as given in the previous section). Transform the original data $X_1, X_2, \ldots, X_n$ (iid $G(\alpha, \beta, \mu)$) into $X_1^*, X_2^*, \ldots, X_n^*$ where $X_i^* = X_i - \widehat{\mu}_{MML}$. Carry out the goodness of fit test of a two parameter gamma model on the transformed data $X_1^*, X_2^*, \ldots, X_n^*$. If a $G(\alpha, \beta)$ is found suitable for the transformed data $X_1^*, X_2^*, \ldots, X_n^*$, then it indicates that a $G(\alpha, \beta, \mu)$ model would be reasonable for the original data $X_1, X_2, \ldots, X_n$.*

Remark 7.4.2 *It is always desirable to fit a simpler model (probability distribution) to the data so that the inferences become easier. In the case of a two-parameter gamma model one may wish to test $H_0 : \alpha = 1$ against $H_1 : \alpha \neq 1$. Note that $\alpha = 1$ gives the **Exp**(β) distribution. We apply the computational approach (discussed in Section 5.5) in testing the above H_0 for the data set in Example 7.1.2 as shown below. It should be noted that the same approach can be followed for testing $H_0 : \alpha = \alpha_0$ for any $\alpha_0(> 0)$ other than 1.*

Testing $H_0 : \alpha = 1$ vs. $H_1 : \alpha \neq 1$ for the data set in Example 7.1.2.

Step-1 : Obtain $\widehat{\alpha}_{\mathrm{ML}} = 1.67$ and $\widehat{\beta}_{\mathrm{ML}} = 49.98$ (using (7.3.1)).

Step-2 : (a) Set $\alpha = 1$ (the H_0 value of α). Then find the restricted MLE of β (with H_0 restriction on α) as $\widehat{\beta}_{\mathrm{RML}} = 83.51724$. (Notice that when $\alpha = 1$, the $G(\alpha, \beta)$ model reduces to the **Exp**(β) model for which the MLE of β is $\bar{X}$.)
(b) Now generate $(X_1, \ldots, X_{29})$ *iid* from $G(1, 83.51724)$, i.e., **Exp**(83.51724). For this data recalculate $\widehat{\beta}_{\mathrm{ML}}$ and $\widehat{\alpha}_{\mathrm{ML}}$ following (7.3.1). Repeat this $M = 10,000$ times and retain only the $\widehat{\alpha}_{\mathrm{ML}}$ values, denoted as $\widehat{\alpha}_{01}, \widehat{\alpha}_{02}, \ldots, \widehat{\alpha}_{0M}$.

Step-3 : Order the above and get $\widehat{\alpha}_{0(1)} \leq \ldots \leq \widehat{\alpha}_{0(M)}$; and using the 5% significance level get the cutoff points $\widehat{\alpha}_{\mathrm{L}} = \widehat{\alpha}_{0(250)} = 0.6874$ and $\widehat{\alpha}_{\mathrm{U}} = \widehat{\alpha}_{0(9750)} = 1.7287$.

Step-4 : Since $\widehat{\alpha}_{\mathrm{ML}} = 1.67$ falls between the two cut-off points given above, we accept the $H_0 : \alpha = 1$ at 5% level, i.e., the same data set can be modelled by a one-parameter exponential (**Exp**(β)) distribution.

Remark 7.4.3 *One can invert the acceptance region of the hypothesis testing procedure discussed above to find a confidence interval for α as shown below. A similar approach can be followed to find a confidence interval for β also.*

Finding a 95% confidence interval for α for the data set in Example 7.1.2.

Step-1 : Consider $\alpha_0 = 0.5, 1.0, 1.5, \ldots, 5.0$.

Step-2 : For each α_0 above, pretend to test $H_0 : \alpha = \alpha_0$ vs. $H_1 : \alpha \neq \alpha_0$. Follow the steps after Remark 7.4.2 and obtain the cutoff points $\widehat{\alpha}_{\mathrm{L}}$ and $\widehat{\alpha}_{\mathrm{U}}$. Note that $\widehat{\alpha}_{\mathrm{L}}$ and $\widehat{\alpha}_{\mathrm{U}}$ are very much dependent on α_0 (the value of α used in H_0).

Step-3 : Plot $\widehat{\alpha}_{\mathrm{L}}$ and $\widehat{\alpha}_{\mathrm{U}}$ simultaneously against α_0.

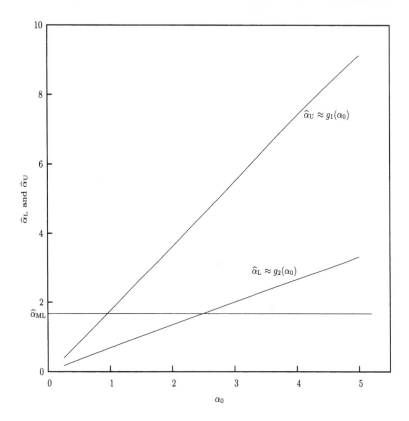

Figure 7.4.1: Plots of $\widehat{\alpha}_{\mathrm{L}}$ and $\widehat{\alpha}_{\mathrm{U}}$ against α_0 for the data set in Example 7.1.2.

If necessary, one can approximate both $\widehat{\alpha}_{\mathrm{L}}$ and $\widehat{\alpha}_{\mathrm{U}}$ by suitable non-linear regression functions of α_0 as $\widehat{\alpha}_{\mathrm{L}} \approx g_1(\alpha_0)$ and $\widehat{\alpha}_{\mathrm{U}} \approx g_2(\alpha_0)$.

Step-4 : Draw the horizontal line at $\widehat{\alpha}_{\mathrm{U}} = \widehat{\alpha}_{\mathrm{L}} = \widehat{\alpha}_{\mathrm{ML}} = 1.67$. The values of $\widehat{\alpha}_{\mathrm{L}}$ and $\widehat{\alpha}_{\mathrm{U}}$ intersected by the horizontal line provide a 95% confidence interval for α. If one uses g_1 and g_2 as defined in Step-3, then the bounds of the confidence interval are $g_1^{-1}(\widehat{\alpha}_{\mathrm{ML}}) \approx 0.88$ and $g_2^{-1}(\widehat{\alpha}_{\mathrm{ML}}) \approx 2.56$.

Chapter 8

Weibull Distribution

8.1 Preliminaries

In Chapter 2, we have seen the basic properties of a Weibull distribution. A random variable X is said to have a (two-parameter) Weibull distribution, or Weibull(α, β), with parameters α and β provided the *pdf* of X is given as

$$f(x \mid \alpha, \beta) = \frac{\alpha}{\beta^\alpha} x^{\alpha-1} e^{-(x/\beta)^\alpha}, \quad 0 < x < \infty, \ \alpha > 0, \ \beta > 0 \qquad (8.1.1)$$

When $\beta = 1$, the *pdf* (8.1.1) is reduced to

$$f(x \mid \alpha, 1) = \alpha x^{\alpha-1} e^{-x^\alpha}, \quad x > 0, \ \alpha > 0 \qquad (8.1.2)$$

and this is known as the Standard Weibull distribution.

The distribution is named after Waloddi Weibull, a Swedish physicist, who used it to represent the probability distribution of the breaking strength of materials. The use of the Weibull distribution in reliability and quality control work has been advocated by Kao (1958, 1959).

Note that if X follows (8.1.1), and one makes a transformation of

$$Y = \left(\frac{X}{\beta} \right)^\alpha \qquad (8.1.3)$$

then the random variable Y follows a standard exponential distribution, i.e., Y has *pdf* e^{-y}, $y > 0$. Conversely, if Y (in (8.1.3)) follows a standard exponential distribution, then $X = \beta Y^{1/\alpha}$ has a Weibull(α, β) distribution.

The following diagram provides graphs of Weibull(α, β) *pdfs* for various values of α and β.

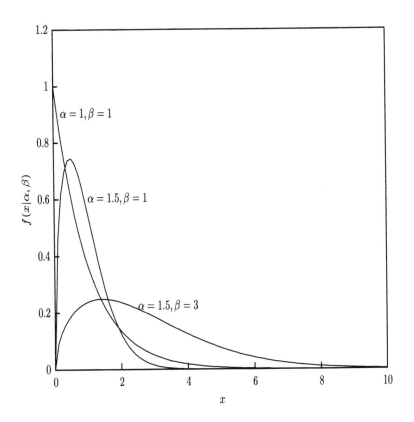

Figure 8.1.1: Two-parameter Weibull *pdf*s.

From the above figure, it is clear that the Weibull(α, β) family includes *pdfs* from highly positively skewed to almost symmetric ones. When $\alpha = 1$, the distribution becomes exponential distribution with scale parameter β. Differentiating (8.1.1), one can get

$$f'(x \mid \alpha, \beta) = \frac{\alpha}{\beta^{\alpha}} x^{\alpha-2} e^{-(x/\beta)^{\alpha}} \left[(\alpha - 1) - \frac{\alpha}{\beta^{\alpha}} x^{\alpha} \right] \qquad (8.1.4)$$

It is seen that for $0 < \alpha \le 1$, the mode of the *pdf* is at zero, and $f(x \mid \alpha, \beta)$ is monotonically decreasing in $x > 0$. For $\alpha > 1$, the *pdf* has a unique mode at $x_0 = \beta \big((\alpha - 1)/\alpha \big)^{1/\alpha}$ which tends to β as $\alpha \to \infty$.

The cumulative distribution function based on the *pdf* (8.1.1) is

$$F(x) = P(X \le x) = 1 - \exp\left\{-\left(\frac{x}{\beta}\right)^{\alpha}\right\} \qquad (8.1.5)$$

and the hazard rate, $r(x)$, the instantaneous failure rate (if the random variable X denotes the lifetime) is

$$r(x) = \frac{f(x \,|\, \alpha, \beta)}{1 - F(x)} = \frac{\alpha}{\beta^{\alpha}} x^{\alpha - 1} \qquad (8.1.6)$$

This hazard rate can also be obtained by differentiating the hazard function $R(x)$ defined as

$$R(x) = -ln(1 - F(x)) \qquad (8.1.7)$$

Note that, in the special case, where $\alpha = 1$, one obtains the exponential distribution where the hazard rate is a constant.

A further generalization of the model in (8.1.1) is possible by introducing a shift or location parameter which gives rise to a three-parameter Weibull distribution with *pdf*

$$f(x \,|\, \alpha, \beta, \mu) = \frac{\alpha}{\beta^{\alpha}} (x - \mu)^{\alpha - 1} e^{-((x-\mu)/\beta)^{\alpha}},$$

$$x > \mu, \ -\infty < \mu < \infty, \ \alpha > 0, \ \beta > 0 \quad (8.1.8)$$

and its cumulative distribution function (*cdf*)

$$F(x) = 1 - e^{-((x-\mu)/\beta)^{\alpha}}, \quad x > \mu \qquad (8.1.9)$$

Note that for any value of $\beta > 0$,

$$F(\beta + \mu) = 1 - e^{-1} \approx 0.6321 = 63.21\% \qquad (8.1.10)$$

It means that $(\beta + \mu)$ is approximately the 63rd percentile point of the Weibull distribution given in (8.1.8). The above three-parameter Weibull distribution will be denoted by Weibull(α, β, μ) or simply by W(α, β, μ) distribution.

If X follows a W(α, β, μ) distribution, then $Y' = \{(X - \mu)/\beta\}^{\alpha}$ follows a standard exponential distribution (similar to the arguments for (8.1.3)).

In the following, we present some data sets where the Weibull distribution seems to provide a good fit.

Example 8.1.1: Lieblein and Zelen (1956) gave the results of the test of endurance, in millions of revolutions (m.o.r.), of 23 ball bearings with the following ordered observations:

17.88	28.92	33.00	41.52	42.12
45.60	48.48	51.84	51.96	54.12
55.56	67.80	68.63	68.64	68.88
84.12	93.12	98.64	105.12	105.84
127.92	128.04	173.40		

The following table gives the frequency distribution of the above observations.

Endurance (in m.o.r.)	Frequency	Relative Frequency
17.5 – 48.7	7	0.304 (30.4%)
48.7 – 79.9	8	0.348 (34.8%)
79.9 – 111.1	5	0.217 (21.7%)
111.1 – 142.3	2	0.087 (8.7%)
142.3 – 173.5	1	0.043 (4.3%)
Total	23	1.000 (100%)

Table 8.1.1: Frequency Table of the Test of Endurance Data.

The histogram in Figure 8.1.2 is drawn with the above relative frequencies.

Example 8.1.2: The following data set is taken from Cox and Oakes (1984) and consists of ten values of the number of cycles to failure when springs are subjected to various stress levels. For the following data , the stress level is 950 N/mm^2, and the values, given in units of 1000 cycles, are:

$$225 \quad 171 \quad 198 \quad 189 \quad 189$$
$$135 \quad 162 \quad 135 \quad 117 \quad 162$$

Since the above data set is small (with $n = 10$), we can make the frequency table with three classes only as given in Table 8.1.2.

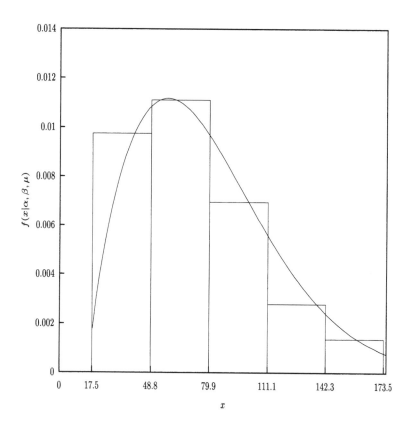

Figure 8.1.2: Histogram of the Test of Endurance Data and a matching $W(\alpha, \beta, \mu)$ *pdf* with $\alpha = 1.7$, $\beta = 70.0$, $\mu = 16.0$.

Cycles to Failure (in '000 of Cycles)	Frequency	Relative Frequency
117 – 153	3	0.3 (30%)
153 – 189	5	0.5 (50%)
189 – 225	2	0.2 (20%)
Total	10	1.0 (100%)

Table 8.1.2: Frequency Table for Springs' Failure Data.

The histogram in Figure 8.1.3 is drawn with the frequency distribution in Table 8.1.2 which is then approximated by a three-parameter Weibull *pdf* with $\alpha = 2.4, \beta = 78.0$, and $\mu = 99.0$.

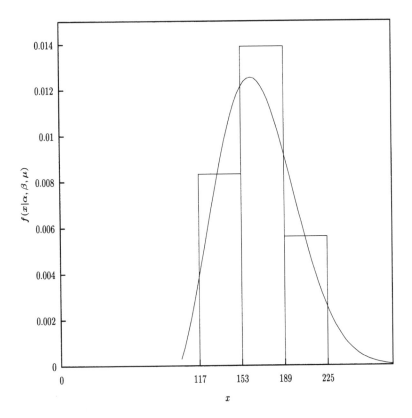

Figure 8.1.3:   Histogram of Springs' Failure Data and a matching $W(\alpha, \beta, \mu)$ *pdf* with $\alpha = 2.4$, $\beta = 78.0$, $\mu = 99.0$.

Example 8.1.3: Suppose that the service life, in years, of a particular type of camera battery is a random variable whose probability distribution can be modeled by a two-parameter $W(\alpha, \beta)$ distribution with $\alpha = 0.5$ and $\beta = 2.0$.

(a) What is the probability that such a battery life will exceed 2 years?

(b) How long such a battery is expected to last?

Let the random variable X represent the service life (in years) of a camera battery. It is given that X follows a $W(\alpha, \beta)$-distribution.

(a) P(battery life will exceed 2 years) $= P(X > 2)$

$$= exp\left\{-(2/\beta)^{\alpha}\right\} \quad \text{(by (8.1.5))}$$

$$= \quad exp\left\{-(2/2)^{0.5}\right\} \quad \text{(since } \alpha = 0.5 \text{ and } \beta = 2.0\text{)}$$
$$= \quad 36.79\% \approx 37\%$$

(b) Expected lifetime of a battery $= E[X]$

$$= \quad \beta\,\Gamma\big(1 + 1/\alpha\big)$$
$$= \quad (2.0)\,\Gamma\big(1 + 1/0.5\big)$$
$$= \quad 2\,\Gamma(3) = 4 \text{ years}$$

Apart from modeling lifetimes of components, Weibull distribution has applications in many other real life problems ranging from wind power generation to rainfall intensity. For such applications, one can see Carlin and Haslett (1982), Barros and Estevan (1983), Tuller and Brett (1984), Wilks (1989), and Johnson and Haskell (1983).

Notational Clarification: A two-parameter Weibull distribution of the form (8.1.1) will be called a $W(\alpha, \beta)$-distribution, whereas a three-parameter Weibull distribution of the form (8.1.8) will be called a $W(\alpha, \beta, \mu)$-distribution. Hence, $W(\alpha, \beta, 0)$-distribution is nothing but the $W(\alpha, \beta)$-distribution and vice-versa. Again, as mentioned in the previous chapter, here α is primarily used to denote the shape parameter, and it should be clear from the context if it is used to indicate the level of significance in testing a hypothesis.

8.2   Characterization of Weibull Distribution

Characterizations of the Weibull distribution are best done through the hazard function and failure rate distribution as shown by Roy and Mukherjee (1986).

Recall that the hazard function R of a *cdf* F of a nonnegative random variable is defined as

$$R(x) = -ln(1 - F(x)), \quad x > 0 \qquad (8.2.1)$$

Note that by differentiating $R(x)$ *w.r.t.* x, one gets the failure rate, i.e., $R'(x) = f(x)/\{1 - F(x)\} = r(x)$. If $F(x) = 1 - \exp\{-(x/\beta)^{\alpha}\}$, $x > 0$; i.e., the *cdf* of a Weibull distribution, then $R(x) = (x/\beta)^{\alpha}$. The following two results characterize a two-parameter Weibull distribution through R.

Theorem 8.2.1 *The hazard function R of a random variable X satisfies the condition:*

$$R(xy) \cdot R(1) = R(x) \cdot R(y) \quad \text{for all } x > 0 \text{ and } y > 0 \qquad (8.2.2)$$

if and only if X follows a $W(\alpha, \beta)$ distribution.

Another variant of the above result is given in the next theorem.

Theorem 8.2.2 *A nonnegative random variable X follows a $W(\alpha, \beta)$ distribution, if and only if its hazard function R satisfies the following conditions:*

(i) *$R(xy) \cdot R(1) = R(x) \cdot R(y)$, with $R(1) > 0$ for all $x > 0$ and at least one value of $y > 0$, $y \neq 1$; and*

(ii) *$xR'(x)/R(x)$ is nondecreasing in $x > 0$.*

One may be interested in characterizing a Weibull distribution through the hazard rate (or the failure rate) instead of the hazard function (see (8.1.6)).

If X is a random variable with *cdf* F, then it is well-known that $F(X)$ follows continuous uniform $(CU[0,1])$ distribution. It is easy to see that the probability distribution of $R(X)$, where R being the hazard function, is completely known as $P\big(R(X) \leq t\big) = 1 - \exp(-t)$. Therefore, the probability distribution of $R(X)$ cannot characterize that of X. However, the failure rate of X may characterize the probability distribution of X. For instance, X follows an exponential distribution if and only if the failure rate of X is a constant (i.e., $R'(X)$ has a degenerate distribution). This exponential characterization can be extended for the Weibull distribution as follows.

Theorem 8.2.3 *Let X be a random variable with failure rate $r(\cdot)$ such that*

(i) *r is strictly increasing with $r(0) = 0$; and*

(ii) *$r(X)$ follows $W(\alpha', \beta')$ with $\alpha' > 1$ if and only if X follows $W(\alpha, \beta)$ where $\alpha > 1$, $(1/\alpha) + (1/\alpha') = 1$, and $\beta' = (\alpha/\beta)^{1/(\alpha-1)}$*

Dubey (1966) has given a different type of characterization of the Weibull distribution in terms of the smallest order statistics.

Theorem 8.2.4 *Let $X_1, X_2, \ldots, X_n$ be independent and identically distributed random variables. Then $\min_{1 \le i \le n}(X_i)$ has a Weibull distribution if and only if the common distribution of the X_i's is a Weibull distribution. Moreover, if X_i's have pdf (8.1.8), then $\min_{1 \le i \le n}(X_i)$ has the same pdf with β replaced by $\beta_* = \beta n^{-1/\alpha}$.*

Further characterizations of the Weibull distribution in terms of entropy maximization and Fisher-information minimization are given by Roy and Mukherjee (1986), but these are omitted here because they involve generalized gamma distribution which is beyond the scope of this handbook.

8.3   Estimation of Parameters

First we consider estimation of parameters α and β of a two-parameter Weibull(α, β) distribution. Given independent and identically distributed observations $X_1, X_2, \ldots, X_n$ having the common *pdf* (8.1.1), the maximum likelihood estimators $\widehat{\alpha}_{\mathrm{ML}}$ and $\widehat{\beta}_{\mathrm{ML}}$ of α and β, respectively, satisfy the equations

$$\widehat{\beta}_{\mathrm{ML}} = \left(\frac{1}{n} \sum_{i=1}^{n} X_i^{\widehat{\alpha}_{\mathrm{ML}}} \right)^{1/\widehat{\alpha}_{\mathrm{ML}}}$$

$$\widehat{\alpha}_{\mathrm{ML}} = \left[\left(\sum_{i=1}^{n} X_i^{\widehat{\alpha}_{\mathrm{ML}}} \ln X_i \right) \Big/ \left(\sum_{i=1}^{n} X_i^{\widehat{\alpha}_{\mathrm{ML}}} \right) - \left(\frac{1}{n} \sum_{i=1}^{n} \ln X_i \right) \right]^{-1}$$

(8.3.1)

If there is an unknown shift parameter μ involved, i.e., if the random sample comes from a three-parameter Weibull distribution with *pdf* (8.1.8), then apart from α and β, μ has to be estimated from the data. In this case, in (8.3.1) one needs to replace X_i by $(X_i - \widehat{\mu}_{\mathrm{ML}})$, and there is an additional third equation

$$\left(\widehat{\alpha}_{\mathrm{ML}} - 1 \right) \sum_{i=1}^{n} \left(X_i - \widehat{\mu}_{\mathrm{ML}} \right)^{-1} = \frac{\widehat{\alpha}_{\mathrm{ML}}}{\left(\widehat{\beta}_{\mathrm{ML}} \right)^{\widehat{\alpha}_{\mathrm{ML}}}} \sum_{i=1}^{n} \left(X_i - \widehat{\mu}_{\mathrm{ML}} \right)^{\widehat{\alpha}_{\mathrm{ML}} - 1}$$ (8.3.2)

If the value of $\widehat{\mu}_{\mathrm{ML}}$ (after simplifying (8.3.2) and (8.3.1) with X_i replaced by $(X_i - \widehat{\mu}_{\mathrm{ML}})$) is smaller than $\min_{1 \le i \le n}(X_i)$, then it is the MLE of μ.

Otherwise, the MLE of μ is taken as $\widehat{\mu}_{\mathrm{ML}} = \min_{1 \leq i \leq n}(X_i) = X_{(1)}$, and one needs to find $\widehat{\alpha}_{\mathrm{ML}}$ and $\widehat{\beta}_{\mathrm{ML}}$ by solving (8.3.2) and (8.3.1) with X_i replaced by $(X_i - X_{(1)})$.

The aforementioned MLEs are known to have the desirable properties of consistency and asymptotic normality under fairly general conditions. In particular, if $\alpha > 2$, the usual regularity conditions are satisfied for $\widehat{\alpha}_{\mathrm{ML}}$ and $\widehat{\beta}_{\mathrm{ML}}$ and some weaker ones for $\widehat{\mu}_{\mathrm{ML}}$. Weibull parameter MLEs are 'regular' (i.e., have usual asymptotic distribution) if and only if at least one of the following conditions holds: (a) $\alpha > 2$, (b) the shift parameter μ is known, and (c) the sample is censored from below. (See Harter (1971) for more on this topic.) When $0 < \alpha \leq 1$, the estimate $\widehat{\mu} = X_{(1)}$ is a super-efficient estimator of μ (efficient when $\alpha = 1$), but no true MLE for the other two parameters exist. When $1 < \alpha < 2$, MLEs exist but their asymptotic variance-covariance matrix is meaningless because the determinant of the information matrix becomes negative, thereby yielding negative values for asymptotic variances of the MLEs. When $\alpha = 2$ (Rayleigh distribution), the determinant of the information matrix vanishes.

Dubey (1965) has shown that for a two-parameter Weibull(α, β) distribution, assuming that MLEs of α and β are regular and the sample size n is sufficiently large,

$$\lim_{n \to \infty} Var\left(\sqrt{n}\widehat{\beta}_{\mathrm{ML}}\right) = \left[1 + \frac{(\psi(2))^2}{\psi'(1)}\right]\left(\frac{\beta}{\alpha}\right)^2 = 1.087 \left(\frac{\beta}{\alpha}\right)^2$$

$$\lim_{n \to \infty} Var\left(\sqrt{n}\widehat{\alpha}_{\mathrm{ML}}\right) = \frac{\alpha^2}{\psi'(1)} = 0.608\alpha^2 \qquad (8.3.3)$$

$$\lim_{n \to \infty} Corr\left(\widehat{\alpha}_{\mathrm{ML}}, \widehat{\beta}_{\mathrm{ML}}\right) = \frac{\psi(2)}{\sqrt{\psi'(1) + (\psi(2))^2}} = 0.313$$

where $\psi(u) = \partial/\partial u\big(\ln \Gamma(u)\big)$ is the di-gamma function.

For the two-parameter case, the method of moments estimators (MMEs) of α and β are obtained by solving the equations:

$$\overline{X} = \beta \Gamma\left(1 + \frac{1}{\alpha}\right) \quad \text{and} \quad \frac{1}{n}\sum_{i=1}^{n} X_i^2 = \beta^2 \Gamma\left(1 + \frac{2}{\alpha}\right) \qquad (8.3.4)$$

The values of α and β thus obtained are denoted by $\widehat{\alpha}_{\mathrm{MM}}$ and $\widehat{\beta}_{\mathrm{MM}}$, respectively. Another way of getting MMEs is to compute $\widehat{\alpha}_{\mathrm{MM}}$ by solving the following equation first

$$\frac{\left(\Gamma\left(1+1/\widehat{\alpha}_{\text{MM}}\right)\right)^2}{\Gamma\left(1+2/\widehat{\alpha}_{\text{MM}}\right)} = \frac{n\overline{X}^2}{\sum_{i=1}^{n} X_i^2} \tag{8.3.5}$$

and then get $\widehat{\beta}_{\text{MM}}$ by

$$\widehat{\beta}_{\text{MM}} = \frac{\overline{X}}{\Gamma\left(1+1/\widehat{\alpha}_{\text{MM}}\right)} \tag{8.3.6}$$

In the presence of a shift parameter μ (i.e., for a three-parameter Weibull model), the parameter MMEs are found by solving the following three equations for α, β, and μ, and denoted by $\widehat{\alpha}_{\text{MM}}$, $\widehat{\beta}_{\text{MM}}$, and $\widehat{\mu}_{\text{MM}}$, respectively:

$$\overline{X} = \mu + \beta\Gamma\left(1+\frac{1}{\alpha}\right)$$

$$\frac{1}{n}\sum_{i=1}^{n} X_i^2 = \mu^2 + 2\mu\beta\Gamma\left(1+\frac{1}{\alpha}\right) + \beta^2\Gamma\left(1+\frac{2}{\alpha}\right) \tag{8.3.7}$$

$$\frac{1}{n}\sum_{i=1}^{n} X_i^3 = \mu^3 + 3\mu\beta^2\Gamma\left(1+\frac{2}{\alpha}\right) + 3\mu^2\beta\,\Gamma\left(1+\frac{1}{\alpha}\right) + \beta^3\Gamma\left(1+\frac{3}{\alpha}\right)$$

subject to the restrictions that $\widehat{\mu}_{\text{MM}} < X_{(1)}$, $\widehat{\alpha}_{\text{MM}} > 0$ and $\widehat{\beta}_{\text{MM}} > 0$.

Although maximum likelihood estimators (MLEs) are more computationally tedious to obtain than the other convenient estimators, MLEs perform better than the other estimators as shown by Thoman, Bain, and Antle (1969).

Maximum likelihood estimators for Weibull parameters with censored and grouped data sets can also be obtained. Zanakis and Kyparisis (1986) have discussed the computational aspects of obtaining MLEs when one has only:

(i) $X_{(1)} \leq X_{(2)} \leq \cdots \leq X_{(m)}$ $(1 \leq m \leq n)$, i.e., first m, out of n, ordered observations (i.e., right censoring)

(ii) $X_{(r+1)} \leq X_{(r+2)} \leq \cdots \leq X_{(m)}$ $(1 \leq r \leq m \leq n)$, i.e., all but lower r and upper $(n-m)$ ordered observations (i.e., left and right censored)

(iii) Left and right censored grouped data, i.e., the censored observations $X_{(r+1)} \leq X_{(r+2)} \leq \cdots \leq X_{(m)}$ are not observed individually, but are instead divided into k groups with the group boundaries given by $d_0 = X_{(r+1)} < d_1 < \ldots < d_k = X_{(m)}$

The proposed numerical methods tend to be either fast but not very reliable, or sufficiently reliable but too slow. This is expected in view of the intrinsic complexity involved in maximum likelihood estimation which is essentially a nonlinear optimization problem.

Example 8.3.1: To demonstrate how the above estimation techniques could be followed, we use the data set presented in Example 8.1.1 (by Lieblein and Zelen (1956)).

First we assume a two-parameter Weibull model for the data (i.e., Weibull(α, β), μ is assumed to be 0). Using (8.3.1) and (8.3.5) to (8.3.6), our MLEs and MMEs are

$$\widehat{\alpha}_{\mathrm{ML}} = 2.10204, \quad \widehat{\beta}_{\mathrm{ML}} = 81.8779$$

and   (8.3.8)

$$\widehat{\alpha}_{\mathrm{MM}} = 2.01568, \quad \widehat{\beta}_{\mathrm{MM}} = 81.5069$$

Now let us assume a three-parameter Weibull model (i.e., Weibull(α, β, μ)) and using the system of equations discussed earlier, the MLEs and MMEs are

$$\widehat{\alpha}_{\mathrm{ML}} = 1.5943, \quad \widehat{\beta}_{\mathrm{ML}} = 63.8793, \quad \widehat{\mu}_{\mathrm{ML}} = 14.8761$$

and   (8.3.9)

$$\widehat{\alpha}_{\mathrm{MM}} = 2.0027, \quad \widehat{\beta}_{\mathrm{MM}} = 81.0261, \quad \widehat{\mu}_{\mathrm{MM}} = 0.4181$$

Remark 8.3.1 *Maximum likelihood estimation of Weibull parameters is always a bit tricky. It is clear that if $\alpha < 1$ is allowed, then the likelihood function becomes unbounded. To avoid this problem, one usually restricts α to be greater than or equal to 1. With $\alpha > 1$, the failure rate of a three-parameter Weibull distribution is*

$$r(x) = \frac{\alpha(x - \mu)^{\alpha - 1}}{\beta^{\alpha}}, \quad x > \mu$$

which is an increasing function. Increasing failure rates are associated with common materials that wear out with usage. On the other hand, the case $\alpha < 1$, which implies a decreasing failure rate, represents work-hardened materials.

Theorem 8.3.1 (*Johnson and Haskell* (*1983*)) *Assume that* $(\alpha, \beta, \mu) \in \Theta$ *where* Θ *is given by*

$$\Theta = \left\{ 1 < \alpha < \overline{\alpha} < \infty, 0 < \underline{\beta} < \beta < \overline{\beta} < \infty, -\infty < \underline{\mu} < \mu < \overline{\mu} < \infty \right\}$$

for suitable constants $\overline{\alpha}$, $\underline{\beta}$, $\overline{\beta}$, $\underline{\mu}$, *and* $\overline{\mu}$. *Then* $(\widehat{\alpha}_{ML}, \widehat{\beta}_{ML}, \widehat{\mu}_{ML})$ *is a consistent estimator of* (α, β, μ); *in other words, as* $n \to \infty$, $(\widehat{\alpha}_{ML}, \widehat{\beta}_{ML}, \widehat{\mu}_{ML})$ *converges to* (α, β, μ) *in probability.*

In many applied problems, as shown by Johnson and Haskell (1983), one is interested in a lower tolerance bound for a Weibull distribution. Let ξ_c denote the population $100c^{th}$ percentile point $(0 < c < 1)$; i.e.,

$$c = F(\xi_c) = 1 - \exp\left\{ -\left(\frac{\xi_c - \mu}{\beta} \right)^{\alpha} \right\}$$

and ξ_c can be obtained as

$$\xi_c = \mu + \beta \left\{ -ln(1 - c) \right\}^{1/\alpha} \tag{8.3.10}$$

Therefore, the MLE of ξ_c is given by

$$\widehat{\xi}_{c(ML)} = \widehat{\mu}_{ML} + \widehat{\beta}_{ML} \left\{ -ln(1 - c) \right\}^{1/\widehat{\alpha}_{ML}} \tag{8.3.11}$$

Due to the complexities associated with maximum likelihood estimation as discussed above, several modifications have been suggested to avoid computational problems. In the following, we provide one such technique based upon the *profile likelihood* (PL) suggested by Lockhart and Stephens (1994).

For a given random sample $X_1, X_2, \dots, X_n$ following a three-parameter Weibull distribution (8.1.8), the likelihood function is

$$L(\alpha, \beta, \mu) = \prod_{i=1}^{n} \left[\frac{\alpha}{\beta} \left(\frac{X_i - \mu}{\beta} \right)^{\alpha - 1} \exp\left\{ -\left(\frac{X_i - \mu}{\beta} \right)^{\alpha} \right\} \right] \tag{8.3.12}$$

The profile likelihood $L^*(\mu_0)$ is defined for given μ_0 as

$$L^*(\mu_0) = \sup_{\alpha, \beta} L(\alpha, \beta, \mu = \mu_0) \tag{8.3.13}$$

A plot of $ln\, L^*(\mu_0)$ against μ_0 usually takes one of the following three forms.

Case (i) There is a local minimum for μ_0 close to $X_{(1)}$ which gives a saddlepoint for the likelihood, and a local maximum for μ_0 further from $X_{(1)}$, giving the desired maximum likelihood solution for μ.

Case (ii) There is a local minimum for μ_0, but no local maximum occurs.

Case (iii) The likelihood (or the log-likelihood) steadily decreases as μ_0 moves away from $X_{(1)}$ toward $-\infty$.

By eliminating β from the equations in (8.3.1) (with X_i replaced by $(X_i - \mu)$) and (8.3.2), the two equations in μ and α can be written as

$$\frac{1}{\alpha} - \frac{\left\{\sum_{i=1}^n (X_i - \mu)^\alpha \, ln(X_i - \mu)\right\}}{\left\{\sum_{i=1}^n (X_i - \mu)^\alpha\right\}} + \frac{1}{n}\sum_{i=1}^n ln(X_i - \mu) = 0 \quad (8.3.14)$$

and

$$\left(\frac{\alpha - 1}{\alpha}\right)\sum_{i=1}^n (X_i - \mu)^{-1} - n\frac{\left\{\sum_{i=1}^n (X_i - \mu)^{\alpha-1}\right\}}{\left\{\sum_{i=1}^n (X_i - \mu)^\alpha\right\}} = 0 \quad (8.3.15)$$

The above two equations produce $\hat{\alpha}_{\text{PL}}$ and $\hat{\mu}_{\text{PL}}$ which are then used to get $\hat{\beta}_{\text{PL}}$ by the equation

$$\hat{\beta}_{\text{PL}} = \left\{\frac{1}{n}\sum_{i=1}^n (X_i - \hat{\mu}_{\text{PL}})^{\hat{\alpha}_{\text{PL}}}\right\}^{1/\hat{\alpha}_{\text{PL}}} \quad (8.3.16)$$

To get the solutions $\hat{\alpha}_{\text{PL}}$ and $\hat{\mu}_{\text{PL}}$, we now fix $\mu = \mu_0$ and plot the solutions α^* of Equation (8.3.14) and α_* of Equation (8.3.15) against μ_0. It is seen that as $\mu_0 \to X_{(1)}$ from below, $\alpha^* \to 0$ and $\alpha_* \to 1$. If the graphs of α^* and α_* cross, then there is either the minimum or the maximum in case (i), or the minimum in case (ii); and if the graphs do not cross then it is case (iii) (which is observed to happen rarely).

Given a data set, it will be helpful to decide whether the situation is case (iii) without trying to solve the likelihood equations. For case (iii), as $\mu_0 \to -\infty$, the plots of α^* and α_* have parallel asymptotes. Let the limiting difference between α_* and α^* as

$$\Delta = \lim_{\mu_0 \to -\infty} (\alpha_* - \alpha^*) \quad (8.3.17)$$

To find the value of Δ, define $\overline{X} = \sum_{i=1}^{n} X_i/n$, $S_0 = \sum_{i=1}^{n} X_i^2/n$ and $T_r = \sum_{i=1}^{n} (X_i)^r exp\left(-\gamma X_i \right)$, $r = 0, 1, 2$; where the quantity γ is the solution of $\gamma(\overline{X} - T_1/T_0) = 1$. The value of γ can easily be found by iteration, starting with $\gamma = -1$ in the expressions of T_0 and T_1. Define $D = \overline{X}T_0 + \gamma(T_2 - \overline{X}T_1)$; γ is the limiting slope of the lines, and Δ is given by

$$\Delta = \left\{ \overline{X} \, T_0 - \gamma(S_0 T_0 - T_2)/2 \right\}/D \qquad (8.3.18)$$

A negative value of Δ means that we have case (iii), in which case a three parameter Weibull model with μ unknown looks questionable.

For case (i) and case (ii) it is important to establish whether or not a saddlepoint exists; and if it does then we have case (i). The saddlepoint is usually very close to $X_{(1)}$. Thus it can be detected by starting with $\mu_o = X_{(1)} - \epsilon$, where ϵ is sufficiently small so that $\alpha_* - \alpha^*$ is positive, and then decreasing μ_0 in very small steps until the saddlepoint is passed (i.e., $(\alpha_* - \alpha^*)$ becomes negative). The steps in μ_0 can then be increased until once again $(\alpha_* - \alpha^*) = 0$, and at that point the values of μ_0 and α become the PL solutions $\widehat{\mu}_{PL}$ and $\widehat{\alpha}_{PL}$ respectively. If no saddlepoint exists then we are in case (ii), for which $(\alpha_* - \alpha^*)$ will decrease to a minimum positive value and then start to increase again. In this case the MLE of μ is $X_{(1)}$ which is biased. The bias in the estimate $\widehat{\mu}_{PL} = X_{(1)}$ is approximately $\beta n^{-1/\alpha}$. Since α and β are unknown, they must be estimated and then used to reduce the bias in the estimate $\widehat{\mu}_{PL} = X_{(1)}$. The whole process of estimating α, β, and μ now goes through an iteration process. The likelihood equation $\partial(lnL)/\partial\alpha = 0$ is needed, i.e.,

$$\frac{n}{\alpha} + \sum_{i=1}^{n} ln(X_i - \mu) - \beta^{-\alpha} \sum_{i=1}^{n} (X_i - \mu)^\alpha \, ln(X_i - \mu) - $$

$$\left\{ n - \beta^{-\alpha} \sum_{i=1}^{n} (X_i - \mu)^\alpha \right\} ln\beta = 0 \qquad (8.3.19)$$

Suppose that $\widehat{\mu}_{(l)}$, $\widehat{\beta}_{(l)}$, and $\widehat{\alpha}_{(l)}$ are the estimates at the lth iteration. Then the improved estimates $\widehat{\mu}_{(l+1)}$, $\widehat{\beta}_{(l+1)}$, $\widehat{\alpha}_{(l+1)}$ are obtained as follows.

Step 1 : $\widehat{\mu}_{(l+1)} = X_{(1)} - \widehat{\beta}_{(l)} n^{-1/\widehat{\alpha}_{(l)}}$

Step 2 : Solve the Equation (8.3.19) for $\widehat{\alpha}_{(l+1)}$ using $\mu = \widehat{\mu}_{(l+1)}$ and $\beta = \widehat{\beta}_{(l)}$

Step 3 : Use Equation (8.3.16) to get $\widehat{\beta}_{(l+1)}$ using $\widehat{\mu}_{\text{PL}} = \widehat{\mu}_{(l+1)}$ and $\widehat{\alpha}_{\text{PL}} = \widehat{\alpha}_{(l+1)}$

Iteration of these three steps continues until the required accuracy for $\widehat{\alpha}_{\text{PL}}$ is obtained. Initial estimates $\widehat{\alpha}_0$ and $\widehat{\beta}_0$ may be found by setting $\widehat{\mu}_0 = X_{(1)}$ and continuing with Steps (2) and (3) given above, but using only the $(n-1)$ sample values $X_{(2)}, X_{(3)}, \dots, X_{(n)}$ (since otherwise the term $ln(X_{(1)} - \mu)$ with $\mu = \widehat{\mu}_{(0)} = X_{(1)}$ in (8.3.19) causes problem). When the iteration stops, we get the estimates $\widehat{\mu}_{\text{PL}}$, $\widehat{\beta}_{\text{PL}}$, and $\widehat{\alpha}_{\text{PL}}$ for case (ii).

Example 8.3.2: Here we need to estimate the parameters of a three-parameter Weibull distribution for the data sets in Example 8.1.1 and Example 8.1.2. The parameter estimates are given below.

Example 8.1.1			
Method	$\widehat{\mu}$	$\widehat{\alpha}$	$\widehat{\beta}$
Usual MLE	14.876	1.5943	63.8793
Lockhart and Stephens' PLE	9.1717	1.5759	63.6851

Example 8.1.2			
Method	$\widehat{\mu}$	$\widehat{\alpha}$	$\widehat{\beta}$
Usual MLE	99.0109	2.3755	78.2417
Lockhart and Stephens' PLE	100.161	1.5648	73.3424

8.4   Goodness of Fit Tests for Weibull Distribution

In this section we discuss the goodness of fit tests for two-parameter and three-parameter Weibull distributions as suggested by Coles (1989) and Lockhart and Stephens (1994).

First we discuss the goodness of fit test for a two parameter Weibull distribution with *pdf* (8.1.1) (i.e., with *cdf* (8.1.5)). Coles' (1989) goodness-of-fit-test is based on Michael's (1983) stabilized probability plot and Blom's (1958) procedure of parameter estimation.

To determine whether an ordered sample $Y_{(1)} \leq \ldots \leq Y_{(n)}$ has been drawn from a population with *cdf* $G_Y((y-\delta)/\eta)$, of location-scale family, the plotting coordinates $(d_i, e_i), 1 \leq i \leq n$, of the stabilized probability plot are calculated as

$$d_i = (2/\pi)Sin^{-1}\left\{\sqrt{(i-0.5)/n}\right\} \quad \text{and}$$

$$e_i = (2/\pi)Sin^{-1}\left\{\sqrt{G_Y((Y_{(i)}-\delta)/n)}\right\} \tag{8.4.1}$$

Note that, if the ordered sample $Y_{(1)} \leq \ldots \leq Y_{(n)}$ really comes from the distribution with *cdf* $G_Y((y-\delta)/\eta)$ then:

Case (i) The coordinates $(d_i, e_i), 1 \leq i \leq n$, are expected to be close to the line joining the points $(0,0)$ and $(1,1)$

Case (ii) As shown by Michael (1983), the plotted points (d_i, e_i) have approximately equal variances due to the asymptotic distribution of e_i's

The two-parameter Weibull distribution with *cdf* (8.1.5) is not a location-scale type, however, a log transformation yields a variable with a location-scale type *cdf*. In other words, if $X_1, \ldots, X_n$ are *iid* $W(\alpha, \beta)$, if and only if $Y_1 = lnX_1, \ldots, Y_n = lnX_n$ are *iid* with *cdf*

$$G_Y(y) = G_Y((y-\delta)/\eta) = 1 - exp\left[-exp\{(y-\delta)/\eta\}\right] \tag{8.4.2}$$

where $\delta = ln\beta$ and $\eta = 1/\alpha$. Thus, instead of checking whether an ordered sample $X_{(1)} \leq \ldots \leq X_{(n)}$ has been drawn from a $W(\alpha, \beta)$ distribution with unknown parameters, one can test whether $Y_{(1)} \leq \ldots \leq Y_{(n)}$ has been drawn from the distribution with *cdf* (8.4.2).

The first test statistic to test the claim that the *cdf* (8.4.2) is the true distribution is

$$\Delta^* = \max_{1 \leq i \leq n} \left|d_i - \widehat{e}_i\right|, \text{ where} \tag{8.4.3}$$

$$\widehat{e}_i = e_i \quad \text{with } \delta \text{ and } \eta \text{ replaced by } \widehat{\delta} \text{ and } \widehat{\eta} \tag{8.4.4}$$

The estimates $\widehat{\delta}$ and $\widehat{\eta}$ in (8.4.4) are due to Blom (1958) and obtained through the steps as given below.

Step 1 : Calculate

$$u_i = -\left(\frac{n-i+1}{n+1}\right)ln\left(\frac{n-i+1}{n+1}\right);$$
$$1 \leq i \leq n-1, \; 0 = u_o = u_n$$

and

$$\nu_i = u_i \frac{n!}{(i-1)!(n-i)!} \int_0^1 ln(-ln(1-t))t^{i-1}(1-t)^{n-i} \, dt$$
$$1 \leq i \leq n$$

Step 2 : Calculate

$$c_{1i} = u_i - u_{i+1} \quad \text{and} \quad c_{2i} = \nu_i - \nu_{i+1}, \; 0 \leq i \leq n-1$$

Step 3 : Invert the matrix

$$C = \begin{bmatrix} \sum_{i=1}^{n} c_{1i}^2 & \sum_{i=1}^{n} c_{1i}c_{2i} \\ \sum_{i=1}^{n} c_{1i}c_{2i} & \sum_{i=1}^{n} c_{2i}^2 \end{bmatrix}$$

and get

$$C^{-1} = \begin{bmatrix} c^{11} & c^{12} \\ c^{21} & c^{22} \end{bmatrix}$$

Step 4 : Finally, the unbiased nearly best linear estimates of δ and η are obtained as

$$\widehat{\delta} = \sum_{i=1}^{n} d_{1i}Y_{(i)} \quad \text{and} \quad \widehat{\eta} = \sum_{i=1}^{n} d_{2i}Y_{(i)}$$

where

$$d_{ri} = u_i\left\{c^{r1}(c_{1i} - c_{1(i-1)}) + c^{r2}(c_{2i} - c_{2(i-1)})\right\}, \quad r = 1, 2$$

Coles (1989) provided simulated critical values for Δ^* (in (8.4.3)) based on 6,000 replications for selected values of n and significance level. Note that large values of Δ^* indicate lack-of-fit. In the following we have provided refined simulated critical values based on 10,000 replications.

The second test statistic to test the claim that the *cdf* (8.4.2) is the true distribution is

$$\Delta^{**} = \text{correlation coefficient of } (d_i, \widehat{e}_i), \; 1 \leq i \leq n \quad\quad (8.4.5)$$

n	Level of significance		
	0.01	0.05	0.10
5	0.2048	0.2047	0.1956
10	0.1889	0.1543	0.1429
15	0.1446	0.1309	0.1217
20	0.1440	0.1275	0.1159
25	0.1431	0.1190	0.1087
30	0.1319	0.1127	0.1023
35	0.1250	0.1060	0.0973
40	0.1183	0.1018	0.0932
45	0.1150	0.0978	0.0891
50	0.1099	0.0942	0.0862
75	0.0966	0.0808	0.0710
100	0.0861	0.0719	0.0660
300	0.0554	0.0460	0.0420
500	0.0438	0.0371	0.0340

Table 8.4.1: Simulated Critical Values for Δ^* with 10,000 Replications.

Since stablized probability plots from distributions other than (8.4.2) tend to be nonlinear, the statistic Δ^{**} is a reasonable measure of goodness of fit. Hence small values of Δ^{**}, as opposed to large values of Δ^*, are significant, i.e., if Δ^{**} value is smaller than the one listed in Table 8.4.2, then it indicates a lack-of-fit.

Instructions to calculate simulated critical values of the test statistics Δ^* and Δ^{**}.

Step-1 : (a) Fix the value of n, say, $n = 5$.

(b) Calculate u_i, v_i, C^{-1} etc. i.e., Steps 1 through 3 after (8.4.4) (since they do not depend on the observations).

(c) Generate observations $Y_1^{(1)}, \ldots, Y_n^{(1)}$ from (8.4.2) with $\delta = 0$ and $\eta = 1$; or generate observations $X_1^{(1)}, \ldots, X_n^{(1)}$ from (8.1.1) with $\alpha = \beta = 1$ and then transform to $Y_1^{(1)} = lnX_1^{(1)}, \ldots$ $Y_n^{(1)} = lnX_n^{(1)}$. The superscript $^{(1)}$ indicates the first replication.

n	Level of significance		
	0.01	0.05	0.10
5	0.8428	0.8965	0.9179
10	0.9050	0.9329	0.9458
15	0.9330	0.9526	0.9610
20	0.9480	0.9636	0.9697
25	0.9584	0.9704	0.9757
30	0.9654	0.9754	0.9795
35	0.9705	0.9786	0.9824
40	0.9738	0.9811	0.9844
45	0.9771	0.9834	0.9860
50	0.9790	0.9850	0.9873
75	0.9864	0.9900	0.9916
100	0.9898	0.9924	0.9937
300	0.9966	0.9975	0.9979
500	0.9978	0.9985	0.9987

Table 8.4.2: Simulated Critical Values for Δ^{**} with 10,000 Replications.

(d) Calculate $\widehat{\delta}^{(1)} = \sum_{i=1}^{n} d_{1i} Y_{(i)}^{(1)}$, $\widehat{\eta}^{(1)} = \sum_{i=1}^{n} d_{2i} Y_{(i)}^{(1)}$.

(e) Calculate d_i and $\widehat{e}_i^{(1)} = \left(\frac{2}{\pi}\right) Sin^{-1} \left\{ \sqrt{G_Y((Y_{(i)}^{(1)} - \widehat{\delta}^{(1)})/\widehat{\eta}^{(1)})} \right\}$
where $G_Y((y - \delta)/\eta)$ expression is given in (8.4.2). $1 \le i \le n$.

(f) Calculate

$$\Delta^{*(1)} = \max_{1 \le i \le n} \left| d_i - \widehat{e}_i^{(1)} \right|$$

$$\Delta^{**(1)} = \text{correlation coefficient of } (d_i, \widehat{e}_i^{(1)}), \quad 1 \le i \le n$$

These are the test statistic values from the first replication.

Step-2 : Repeat the above steps (1)(c) to (1)(f) for a large number of replications, say, M. Thus we get

$$\Delta^{*(1)}, \Delta^{*(2)}, \ldots, \Delta^{*(M)}$$

and

$$\Delta^{**(1)}, \Delta^{**(2)}, \ldots, \Delta^{**(M)}$$

Step-3 : Order the values $\Delta^{*(1)}, \Delta^{*(2)}, \ldots, \Delta^{*(M)}$ (smallest to largest) and with $M = 10,000$ get Δ^* value with rank $(10,000)(1 - \alpha)$. This is the α-level critical value for Δ^*. If $\alpha = 0.05$, then get the value of Δ^* (from the ordered replicated values) that has rank 9500. Similarly, for Δ^{**}, get the value with rank $(10,000)(\alpha)$, i.e., 500.

Example 8.4.1: To demonstrate the usage of the goodness of fit test procedures for a two-parameter Weibull distribution, we consider the data set (with $n = 23$) presented in Example 8.1.1. We obtain

$$\Delta^* = 0.08562 \quad \text{and} \quad \Delta^{**} = 0.9860$$

Note that the critical values for Δ^* (Table 8.4.1) are monotonically decreasing with n. On the other hand, the critical values for Δ^{**} (Table 8.4.2) are monotonically increasing with n. For $n = 23$, $\Delta^* = 0.08562 < 0.1190 = $ critical value for $(n = 25) < $ critical value for $(n = 23)$. Thus Weibull seems to be a plausible model. The same conclusion is arrived at using the statistic Δ^{**} with similar argument.

In the following we now discuss about the goodness of fit test for a three-parameter Weibull distribution based on the empirical distribution function.

Let $X_{(1)} \leq \ldots \leq X_{(n)}$ be the order statistics of the *iid* observation $X_1, \ldots, X_n$ which is thought to be coming from the distribution with *pdf* (8.1.8). The following steps are to be followed.

Step 1 : First estimate the unknown parameters α, β, μ by $\widehat{\alpha}, \widehat{\beta}, \widehat{\mu}$ as described in Section 8.3 based on the profile likelihood approach of Lockhart and Stephens (1994).

Step 2 : Make the transformation

$$Z_{(i)} = F(X_{(i)} | \widehat{\alpha}, \widehat{\beta}, \widehat{\mu}) = 1 - exp\left[-\left\{(X_{(i)} - \widehat{\mu})/\widehat{\beta}\right\}^{\widehat{\alpha}}\right], i = 1, 2, \ldots, n$$

Step 3 : The three test statistics based on the empirical distribution are calculated as

$$\Delta_1 = \sum_{i=1}^{n}\left\{Z_{(i)} - \frac{(2i-1)}{2n}\right\}^2 + \frac{1}{12n}$$

$$\Delta_2 = \Delta_1 - \eta(\overline{Z} - 0.5)^2, \quad \overline{Z} = \sum_{i=1}^{n} Z_{(i)}/n$$

$$\Delta_3 = -n - \frac{1}{n}\sum_{i=1}^{n}(2i-1)\Big[lnZ_{(i)} + ln\big\{1 - Z_{(n-i+1)}\big\}\Big]$$

Step 4 : Let $\widehat{\eta} = 1/\widehat{\alpha}$. The following table gives the critical values for the above three test statistics at different significance levels. The null hypothesis that the data $X_1, \dots, X_n$ comes from the model (8.1.8) is rejected if the test statistic used is greater than the critical value for a specific level. When $0 < \widehat{\eta} < 0.5$, the critical point is obtained from the table corresponding to $\widehat{\eta}$. But for $\widehat{\eta} \geq 0.5$, the last line labelled as $\widehat{\eta} = 0.5$ should be used. Also note that the table does not use the sample size n. For a value of $\widehat{\eta}$, $0 < \widehat{\eta} < 0.5$, which is not listed in the table, linear interpolation can be done to get approximate critical values.

$\widehat{\eta}$	Level of significance								
	0.01			0.05			0.10		
	Δ_1	Δ_2	Δ_3	Δ_1	Δ_2	Δ_3	Δ_1	Δ_2	Δ_3
0.00	0.144	0.143	0.836	0.103	0.102	0.617	0.085	0.084	0.522
0.05	0.145	0.144	0.845	0.104	0.103	0.623	0.086	0.085	0.527
0.10	0.147	0.146	0.856	0.105	0.104	0.631	0.087	0.086	0.534
0.15	0.149	0.148	0.869	0.106	0.105	0.640	0.088	0.087	0.541
0.20	0.152	0.150	0.885	0.108	0.107	0.650	0.089	0.088	0.549
0.25	0.154	0.152	0.902	0.110	0.108	0.662	0.091	0.089	0.529
0.30	0.157	0.154	0.923	0.112	0.110	0.676	0.093	0.091	0.570
0.35	0.161	0.157	0.947	0.114	0.111	0.692	0.094	0.092	0.583
0.40	0.165	0.159	0.974	0.117	0.113	0.711	0.097	0.094	0.598
0.45	0.170	0.162	1.006	0.120	0.115	0.732	0.099	0.095	0.615
0.50	0.175	0.166	1.043	0.124	0.118	0.757	0.102	0.097	0.636

Table 8.4.3: Critical Values for the Test Statistics $\Delta_1, \Delta_2, \Delta_3$.

Example 8.4.2: To demonstrate the usage of the above goodness of fit test procedures for a three-parameter Weibull distribution we consider the data set (with $n = 10$) presented in Example 8.1.2. For the data set we obtain

$$\Delta_1 = 0.0842, \quad \Delta_2 = 0.0827, \quad \text{and} \quad \Delta_3 = 0.4664$$

With $\widehat{\eta} = 0.6391 > 0.5$ we should compare the above Δ-values with the critical values corresponding to $\widehat{\eta} = 0.5$ in Table 8.4.3.

Note that the values of Δ_1, Δ_2, and Δ_3 are smaller than the respective critical values available from the Table 8.4.3. Hence a three-parameter Weibull model seems plausible for the given data set.

Remark 8.4.1 *As discussed in Remark 7.4.2, here also we seek to find out whether a simpler Weibull model can be appropriate for a given data set. Once a two-parameter $W(\alpha, \beta)$ is found suitable, one can further test whether $\alpha = 1$ or not. Notice that $\alpha = 1$ reduces $W(\alpha, \beta)$ to $\mathbf{Exp}(\beta)$ which is much easier to work with. In the following we demonstrate the usage of our computational approach to test $H_0 : \alpha = 1$ vs. $H_1 : \alpha \neq 1$. Again, it should be noted that the approach can be used for testing $H_0 : \alpha = \alpha_0$ for any $\alpha_0(> 0)$ other than 1. Also, the same can be done for testing a H_0 (against a suitable H_1) on β.*

Testing $H_0 : \alpha = 1$ vs. $H_1 : \alpha \neq 1$ for the data set in Example 8.1.1.

Step-1 : Obtain $\widehat{\alpha}_{\mathrm{ML}} = 2.10204$ and $\widehat{\beta}_{\mathrm{ML}} = 81.8779$ (using (8.3.1)).

Step-2 : (a) Set $\alpha = 1$ (the H_0 value of α). Then find the restricted MLE of β (with the H_0 restriction on α) as $\widehat{\beta}_{\mathrm{RML}} = 72.22391$. (Notice that when $\alpha = 1$, the $W(\alpha, \beta)$ model reduces to the $\mathbf{Exp}(\beta)$ model for which the MLE of β is $\overline{X}$.)

(b) Now generate $(X_1, X_2, \ldots, X_{23})$ *iid* from $W(1, 72.22391)$, i.e., $\mathbf{Exp}(72.22391)$. For this data recalculate $\widehat{\beta}_{\mathrm{ML}}$ and $\widehat{\alpha}_{\mathrm{ML}}$ following (8.3.1). Repeat this $M = 10,000$ times and retain only the $\widehat{\alpha}_{\mathrm{ML}}$ values, denoted as $\widehat{\alpha}_{01}, \widehat{\alpha}_{02}, \ldots, \widehat{\alpha}_{0M}$.

Step-3 : Order the above and get $\widehat{\alpha}_{0(1)}, \widehat{\alpha}_{0(2)}, \ldots, \widehat{\alpha}_{0(M)}$; and using the 5% significance level get the cutoff points $\widehat{\alpha}_{\mathrm{L}} = \widehat{\alpha}_{0(250)} = 0.7628$ and $\widehat{\alpha}_{\mathrm{U}} = \widehat{\alpha}_{0(9750)} = 1.4847$.

Step-4 : Since $\widehat{\alpha}_{\mathrm{ML}}$ exceeds the above upper bound, we reject $H_0 : \alpha = 1$ at 5% level, and accept $H_1 : \alpha \neq 1$, i.e., the Weibull model cannot be reduced to an exponential one.

Remark 8.4.2 *Similar to what we did in the earlier chapter we can now invert the acceptance region for testing a hypothesis on α to find a confidence interval for α. A similar approach can be followed to test a hypothesis on β, and then invert the acceptance region to find a suitable confidence interval for β.*

Finding a 95% confidence interval for α for the data set in Example 8.1.1.

Step-1 : Consider $\alpha_0 = 1.0, 1.5, 2.0, \ldots, 5.0$.

Step-2 : For each α_0 above, pretend to test $H_0 : \alpha = \alpha_0$ vs. $H_1 : \alpha \neq \alpha_0$. Follow the steps after Remark 8.4.1 and obtain the cutoff points $\widehat{\alpha}_L$ and $\widehat{\alpha}_U$ which are functions of α_0 (the value of α under H_0).

Step-3 : Plot $\widehat{\alpha}_L$ and $\widehat{\alpha}_U$ simultaneously against α_0. If necessary, approximate both $\widehat{\alpha}_L$ and $\widehat{\alpha}_U$ as functions of α_0 by nonlinear regression functions as $\widehat{\alpha}_L \approx g_1(\alpha_0)$ and $\widehat{\alpha}_U \approx g_2(\alpha_0)$. See Figure 8.4.1 below.

Step-4 : The curves of $\widehat{\alpha}_L$ and $\widehat{\alpha}_U$ (as functions of α_0) intersected by the horizontal line $\widehat{\alpha}_U = \widehat{\alpha}_L = \widehat{\alpha}_{ML}$ provide a 95% confidence interval for α. In other words, a 95% two sided confidence interval for α is $(g_1^{-1}(\widehat{\alpha}_{ML}), g_2^{-1}(\widehat{\alpha}_{ML})) \approx (1.37, 2.86)$.

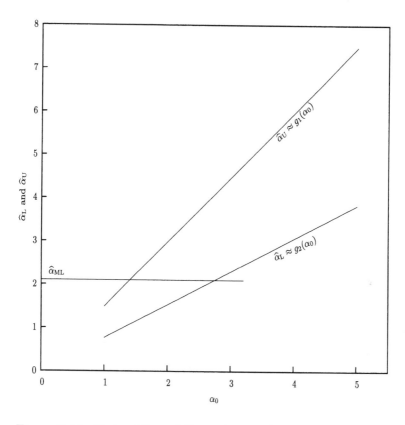

Figure 8.4.1: Plots of $\widehat{\alpha}_{\mathrm{L}}$ and $\widehat{\alpha}_{\mathrm{U}}$ against α_0 for the data set in Example 8.1.1.

Chapter 9

Extreme Value Distributions

9.1 Preliminaries

A random variable X is said to have a Type-I Extreme Value (EV) distribution with parameters μ and β provided the *pdf* of X is given as

$$f_1(x \mid \mu, \beta) \quad = \quad \frac{1}{\beta} e^{-(x-\mu)/\beta} \, exp\left\{-e^{-(x-\mu)/\beta}\right\}, \; x \in \mathbb{R}, \; \beta > 0,$$

$$\mu \in \mathbb{R} \qquad (9.1.1)$$

The above positively skewed *pdf* with range over the whole real line has the *cdf*

$$F_1(x) = P(X \le x) = exp\left\{-e^{-(x-\mu)/\beta}\right\} \qquad (9.1.2)$$

and henceforth the distribution will be called as EV-I(β, μ) distribution. When $\beta = 1$ and $\mu = 0$, the above *pdf* (9.1.1) reduces to

$$f_1(x \mid 0, 1) = e^{-x} exp\left\{-e^{-x}\right\} \qquad (9.1.3)$$

and this is known as EV-I$(1, 0)$ or standard Type I Extreme Value Distribution.

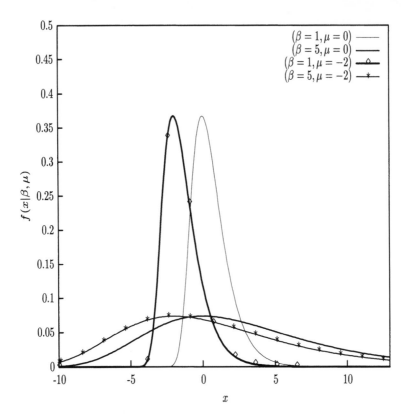

Figure 9.1.1: Type-I EV distribution *pdf*s for various β and μ.

The above Type-I Extreme Value Distribution is also known as **double exponential** distribution due to the structure of the distribution function (9.1.2) (presence of two 'e' s).

There are two other types of extreme value distributions widely used by the researchers. These are Type-II and Type-III extreme value distributions with three parameters given as:

(a) Type-II extreme value distribution (EV-II (α, β, μ)) with *pdf*

$$f_2(x\,|\,\alpha, \beta, \mu) \;\; = \;\; \left(\frac{\alpha}{\beta}\right)\left(\frac{x-\mu}{\beta}\right)^{-(\alpha+1)} exp\left\{-\left(\frac{x-\mu}{\beta}\right)^{-\alpha}\right\},$$

$$x \geq \mu, \;\; \beta > 0, \;\; \mu \in \mathbb{R}, \;\; \alpha > 0 \qquad (9.1.4)$$

(b) Type-III Extreme Value Distribution (EV-III(α, β, μ)) with *pdf*

$$f_3(x\,|\,\alpha, \beta, \mu) \;=\; \left(\frac{\alpha}{\beta}\right)\left(\frac{\mu - x}{\beta}\right)^{(\alpha-1)} exp\left\{-\left(\frac{\mu - x}{\beta}\right)^{\alpha}\right\},$$
$$x \leq \mu, \; \beta > 0, \; \mu \in \mathbb{R}, \; \alpha > 0 \qquad (9.1.5)$$

The *cdf*s corresponding to above Type-II and Type-III *pdf*s are respectively

$$F_2(x) = P(X \leq x) = \begin{cases} 0 & \text{if } x < \mu \\ exp\left\{-(\frac{x-\mu}{\beta})^{-\alpha}\right\} & \text{if } x \geq \mu \end{cases} \qquad (9.1.6)$$

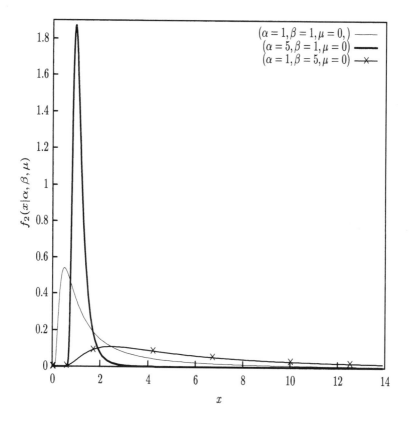

Figure 9.1.2: Type-II EV distribution *pdf*s for various α, β, and μ.

and

$$F_3(x) = P(X \leq x) = \begin{cases} exp\left\{-(\frac{\mu-x}{\beta})^{\alpha}\right\} & \text{if } x \leq \mu \\ 1 & \text{if } x > \mu \end{cases} \qquad (9.1.7)$$

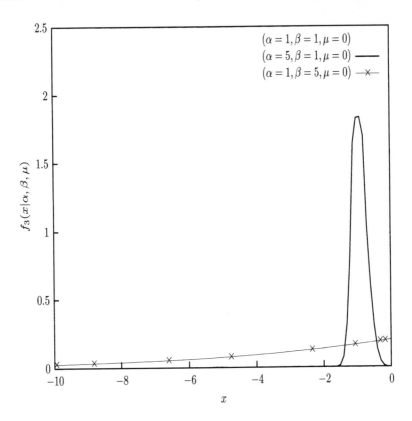

Figure 9.1.3: Type-III EV distribution *pdf*s for various α , β, and μ.

The standard Type-II Extreme Value Distribution is obtained when $\alpha = 1$, $\beta = 1$, and $\mu = 0$ (i.e., EV-II $(1, 1, 0)$) with *pdf*

$$f_2(x \mid 1, 1, 0) = x^{-2} e^{-1/x} , \quad x > 0 \tag{9.1.8}$$

Similarly, the standard Type-III Extreme Value Distribution (i.e., EV-III $(1, 1, 0)$) has *pdf*

$$f_3(x \mid 1, 1, 0) = e^x , \quad x \leq 0 \tag{9.1.9}$$

The name 'extreme value' is attached to the above mentioned distributions because they can be obtained, under certain conditions, as limiting distributions (as $n \to \infty$) of the **largest value** of n independent random variables each having the same continuous distribution.

The Type-I limiting distribution of the **smallest value** of n *iid* random variables has the *cdf* $G_1(x)$ given as

$$G_1(x) = 1 - exp\left\{-e^{(x-\mu)/\beta}\right\} \tag{9.1.10}$$

with *pdf*

$$g_1(x \mid \mu, \beta) \;=\; \frac{1}{\beta}e^{(x-\mu)/\beta}exp\left\{-e^{(x-\mu)/\beta}\right\},$$
$$x \in \mathbb{R}, \ \beta > 0, \ \mu \in \mathbb{R} \tag{9.1.11}$$

Note that, since $\min\{X_1^*, \ldots, X_n^*\} = -\max\{-X_1^*, \ldots, -X_n^*\}$, the above *pdf* (9.1.11) can be obtained from (9.1.1) by replacing X by $(-X)$ and μ by $(-\mu)$. In a similar way one can easily derive Type-II and Type-III limiting distributions of the smallest value of *iid* observations from (9.1.4) and (9.1.5).

If $X_1^*, X_2^*, \ldots, X_n^*$ are independent random variables with common *pdf* $f^*(x)$, then the cumulative distribution function (*cdf*) of $X_{(n)}^* = \max\{X_1^*, \ldots, X_n^*\}$ is $F_{(n)}^*(x) = [F^*(x)]^n$, where $F^*(x)$ is the common *cdf* of X_i^*'s, $1 \le i \le n$. It is obvious that

$$\lim_{n\to\infty} F_{(n)}^*(x) = \begin{cases} 1 & \text{if } F^*(x) = 1 \\ \\ 0 & \text{if } F^*(x) < 1 \end{cases}$$

which is of very little interest. The limiting distribution is thus found by considering a sequence $\left\{(X_{(n)}^* - \mu_n^*)/\sigma_n^*\right\}$ where the constants μ_n^* and σ_n^* may depend on n only (μ_n^* and σ_n^* are not necessary measures of location and scale of the distribution of $X_{(n)}^*$). The EV-I distribution is found by taking $\sigma_n^* = 1$, and the other two types are derived by taking $\sigma_n^* \ne 1$.

In a fundamental paper Gnedenko (1943) established certain correspondences between the parent distribution $F^*(x)$ (from which the observations $X_1^*, \ldots, X_n^*$ come) and the type of EV distribution to which the limiting distribution belongs. It was shown that the conditions relate essentially to the behavior of $F^*(x)$ for large (or small) values of x if the limiting distribution of the largest (or smallest) value of $\{X_1^*, \ldots, X_n^*\}$ is to be considered. It is quite possible that $X_{(n)}^* = \max\{X_1^*, \ldots, X_n^*\}$ and $X_{(1)}^* = \min\{X_1^*, \ldots, X_n^*\}$ may have a different limiting distributions.

Gnedenko's (1943) conditions, which are necessary and sufficient, do characterize the extreme value distributions described above. The conditions are given in the following:

(i) For Type-I EV distribution : Let ξ_c^* be the $(100c)^{th}$ percentile point of the distribution $F^*(\cdot)$, i.e., $F^*(\xi_c^*) = c$ $(0 < c < 1)$. Then the condition is:

$$\lim_{n\to\infty} n\left[1 - F^*\left\{\xi_{1-1/n}^* + x(\xi_{1-1/(ne)}^* - \xi_{1-1/n}^*)\right\}\right] = e^{-x} \quad (9.1.12)$$

(ii) For Type-II EV distribution:

$$\lim_{x\to\infty} \frac{1 - F^*(x)}{1 - F^*(bx)} = b^\alpha \quad \text{for any } b > 0,\ \alpha > 0 \qquad (9.1.13)$$

(iii) For Type-III EV distribution: There exists $d \in \mathbb{R}$ such that

$$\lim_{x\to\infty} \frac{1 - F^*(bx + d)}{1 - F^*(x)} = b^\alpha \quad \text{for any } b > 0,\ \alpha > 0 \qquad (9.1.14)$$

where $F^*(d) = 1$, $F^*(x) < 1$ for $x < d$

The extreme value distributions are used widely in many applied problems ranging from industrial engineering to clinical studies. Fuller (1914) first considered the concept of 'extreme value' in connection with flood control. Later Gumbel (1941, 1944, 1945, 1949), while studying the return of flood flows, gave the theory of EV distribution a solid foundation. In hydrological studies one is concerned about the height of a dam or a river-bank wall which is exceeded by annual maximum hourly or daily or weekly water level with a (small) probability p, which can vary from site to site depending upon the potential amount of damage from flooding. While one looks at maximum value of water level recorded over a fixed period for flood control, the opposite, i.e., minimum value of water level recorded over a fixed period is used for building ports or docks to maintain a certain level of navigation. For applications of EV distributions in sea-level studies see Tawn (1992) and many other references therein.

Gumbel (1937a, 1937b) used extreme value distributions for modelling human life times radioactive emissions. Other areas where EV distributions have found applications include meteorological data (Thom (1954), Jenkinson (1955)), corrosion studies (Aziz (1955), Eldredge (1957)), environmental studies (DasGupta and Bhaumik (1995)), and analyzing the results of horse races (Henery (1984)). Also, in industrial engineering problems the job characteristics like ovality, eccentricity, etc. follow an Extreme Value Distribution (see e.g., DasGupta, Ghosh, and Rao (1981),

DasGupta (1993)).

Example 9.1.1. DasGupta and Bhaumik (1995) reproduced a data set (Table 9.1.1) on ozone concentration in stratosphere which protects the earth from ultraviolet (UV) radiation emitted by the sun. The data set was originally reported by Pyle (1985). It has been observed that the ozone level is most stable at midnight and varies greatly during the day time. Also, the ozone level and its fluctuation are dependent on the altitude. The following table gives minimum of four observations (at an interval of 15 minutes) per hour recorded in terms of percentage deviation from midnight values for a diurnal cycle for two different altitudes (40 and 48 km from the earth's surface).

Hour	Altitude 40 km	48 km	Hour	Altitude 40 km	48 km
0	0.00	0.00	13	2.64	-5.21
1	0.1	0.1	14	3.43	-5.86
2	0.20	0.15	15	4.14	-6.00
3	0.15	0.15	16	4.43	-5.71
4	0.20	0.20	17	4.14	-4.14
5	0.25	0.25	18	1.00	-1.43
6	0.30	0.30	19	0.00	-0.14
7	-1.00	-0.21	20	-0.25	-0.25
8	-1.14	0.86	21	-0.20	-0.20
9	-0.71	-0.57	22	-0.15	-0.15
10	0.21	-2.71	23	-0.10	-0.10
11	1.43	-4.14	24	-0.05	-0.05

Table 9.1.1: Ozone Level Values (in percentage deviation from midnight = 0 hour) at Two Altitudes.

The following two frequency tables with five subgroups each have been created for the ozone level data for altitudes 40 and 48 km respectively.

| Intervals | Freq- | Relative | Intervals | Freq- | Relative |
	uency	frequency		uency	frequency
(-1.135, -0.021)	10	41.67%	(-5.995, -4.623)	4	16.67%
(-0.021, 1.093)	8	33.33%	(-4.623, -3.251)	2	8.33%
(1.093, 2.207)	1	4.17%	(-3.251, -1.879)	1	4.17%
(2.207, 3.321)	1	4.17%	(-1.879, -0.507)	2	8.33%
(3.321, 4.435)	4	16.67%	(-0.507, 0.865)	15	62.50%
Totals	24	100%	Totals	24	100%

(a) Altitude = 40 km   (b) Altitude = 48 km

Table 9.1.2: Frequency Tables for the Ozone Level Data.

The following two histograms are drawn with the above relative frequencies.

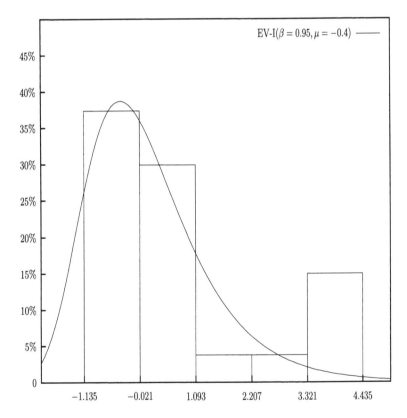

Figure 9.1.4: (a) Histogram for altitude 40 km and a matching EV-I *pdf*.

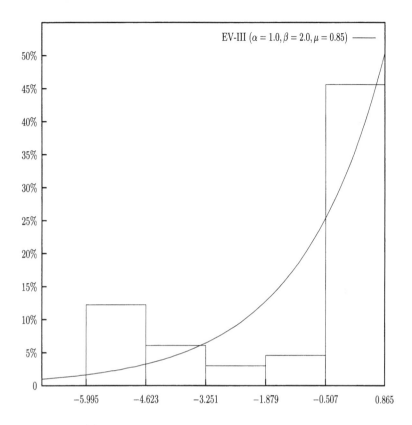

Figure 9.1.4: (b) Histogram for altitude 48 km and a matching EV-III *pdf*.

Example 9.1.2. The following data set (Table 9.1.3) was reported by Press (1950) where frequency distribution of maximum gust-velocity measurements per traverse was obtained and extreme value theory was applied to model this observed frequency distribution.

Note that in the table gust-velocities are all even values. For a better representation of the data and to draw a relative frequency histogram, each gust-velocity x can be thought of representing the interval $(x-1, x+1)$. This is given in the following Table 9.1.4, and the corresponding histogram is given in Figure 9.1.5.

An interesting relationship among the three EV distributions discussed earlier is that by simple '*ln*' (natural logarithm) transformations we can transform Type-II and Type-III EV distributions to Type-I EV

Gust-velocity (pfs)	Frequency	Gust-velocity (pfs)	Frequency
2	4	22	17
4	11	24	18
6	27	26	8
8	48	28	7
10	62	30	6
12	58	32	3
14	55	34	1
16	60	36	2
18	61	38	1
20	36		

Table 9.1.3: Frequency Distribution of 485 Maximum Gust-Velocity Measurements.

Interval	Relative frequency	Interval	Relative frequency
1.0–3.0	0.82%	21.0–23.0	3.51%
3.0–5.0	2.26%	23.0–25.0	3.71%
5.0–7.0	5.57%	25.0–27.0	1.65%
7.0–9.0	9.90%	27.0–29.0	1.44%
9.0–11.0	12.78%	29.0–31.0	1.24%
11.0–13.0	11.96%	31.0–33.0	0.62%
13.0–15.0	11.34%	33.0–35.0	0.21%
15.0–17.0	12.37%	35.0–37.0	0.41%
17.0–19.0	12.58%	37.0–39.0	0.21%
19.0–21.0	7.42%		

Table 9.1.4: Relative Frequency Distribution of Gust-Velocity Measurements.

distribution as shown below.

If X follows EV-II (α, β, μ) (i.e., has *pdf* (9.1.4)) then $Y = ln(X - \mu)$ follows EV-I (β_*, μ_*) (with *pdf* (9.1.1)) where $\beta_* = 1/\alpha$ and $\mu_* = \ln \beta$. Note that the range of the new random variable Y is the whole real line.

Similarly, if X follows EV-III (α, β, μ) (i.e., has *pdf* (9.1.5)) then $Y = -ln(\mu - X)$ follows EV-I (β_*, μ_*) (with *pdf* (9.1.1)) where $\beta_* = 1/\alpha$

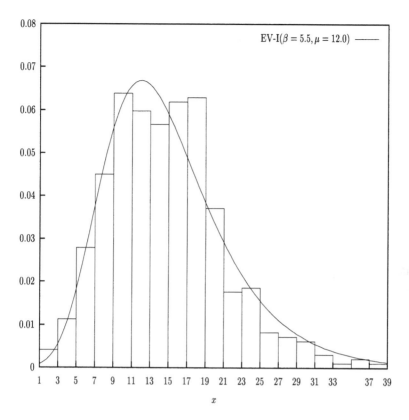

Figure 9.1.5: Relative frequency histogram of gust-velocity data and a matching EV-I *pdf*.

and $\mu_* = -ln\beta$. Again, note that the new random variable Y has the whole real line as its range.

Converse of the above two properties are also true. If X follows EV-I (β, μ) (as given in (9.1.1)) then:

(a) $Y_1 = \mu + exp(X)$ follows EV-II$(\alpha_*, \beta_*, \mu_*)$ where $\alpha_* = 1/\beta$, $\beta_* = exp(\mu)$ and $\mu_* = \mu$

(b) $Y_2 = \mu - exp(-X)$ follows EV-III$(\alpha_{**}, \beta_{**}, \mu_{**})$, where $\alpha_{**} = 1/\beta$, $\beta_{**} = exp(-\mu)$ and $\mu_{**} = \mu$

Also, note that if X follows Type-I EV distribution (with *pdf* (9.1.1)) then $U = exp\{-(X - \mu)/\beta\}$ has the standard exponential distribution, i.e., the *pdf* of U is e^{-u}, $u \geq 0$. Using this property of EV-I (β, μ), the moment generating function (*mgf*) of X can be found as

$$E(e^{tX}) = e^{t\mu}\Gamma(1 - \beta t), \text{ for } \beta|t| < 1 \qquad (9.1.15)$$

From (9.1.15) one can find the mean of X as

$$E(X) = \mu - \beta\Gamma'(1), \text{ where } \Gamma'(1) = \frac{\partial}{\partial x}\Gamma(x)|_{x=1} \qquad (9.1.16)$$

and the variance of X as

$$Var(X) = \beta^2\left[\Gamma''(1) - (\Gamma'(1))^2\right], \text{ where } \Gamma''(1) = \frac{\partial^2}{\partial x^2}\Gamma(x)|_{x=1} \quad (9.1.17)$$

The EV-I (β, μ) is unimodal with mode at μ. From the distribution function (9.1.2) the $100\gamma^{th}$ $(0 < \gamma < 1)$ percentile point ξ_γ can be found by solving the equation

$$exp\left\{-e^{-(\xi_\gamma - \mu)/\beta}\right\} = \gamma$$

which yields

$$\xi_\gamma = \mu - \beta ln(-ln\gamma) \qquad (9.1.18)$$

In particular, the median (i.e., $\xi_{0.5}$) is obtained by using $\gamma = 0.5$, i.e., $\xi_{0.5} \approx \mu + (0.366)\beta$.

The following table gives the percentiles for standard Type-I EV distribution (i.e., EV-I $(1, 0)$).

γ	ξ_γ	γ	ξ_γ	γ	ξ_γ
0.0001	-1.957	0.100	-1.100	0.950	1.866
0.005	-1.750	0.250	-0.705	0.975	2.416
0.010	-1.640	0.500	-0.164	0.990	3.137
0.025	-1.468	0.750	0.521	0.995	3.679
0.050	-1.306	0.900	1.305	0.999	4.936

Table 9.1.5: Percentiles of the Standard Type-I Extreme Value Distribution.

9.2   Characterizations of Extreme Value Distributions

In the previous section it has been discussed how Type-II and Type-III EV distributions are related to the Type-I EV distribution (through the simple 'ln' transformation). Therefore, in this section we talk about characterization of Type-I EV distribution only. However, note that, Gnedenko's (1943) necessary and sufficient conditions do serve as most general characterization of the three types of EV distributions.

It is seen that a random variable X has a Type-I EV distribution, if and only if $exp\{-(X - \mu)/\beta\}$ has the standard exponential distribution. Therefore, the characterization results for exponential distribution can also be used for Type-I EV distribution, by applying them to $exp\{-(X - \mu)/\beta\}$.

Dubey (1966) characterized the Type-I EV distribution by the distribution of the smallest order statistics from the same distribution.

Theorem 9.2.1 *Let* $X_1, X_2, \ldots, X_n$ *be iid observations and* $X_{(1)} = \min$ $\{X_1, X_2, \ldots, X_n\}$. *Then,* $X_{(1)}$ *follows a Type-I EVD distribution with pdf (9.1.11) if and only if* X_i's *are so.*

Ballerini (1987) provided a characterization of the Type-I EV distribution based on record analysis as given below.

Theorem 9.2.2 *Suppose* $\{X_i\}_{i \geq 0}$ *be a sequence of iid nondegenerate continuous random variables with cdf F. Assume that* $EX_0^+ < \infty$ *(where* $X_0^+ = X_0 I(X_0 > 0)$ *). Define* $Y_i = (X_i + ic)$ *where* $c > 0$ *and* $i \geq 0$. *Let* $Y_{(n)} = \max\{Y_1, \ldots, Y_n\}$, *and* $1_n = I[Y_n > Y_{(n-1)}] = I[\text{record occurs at time } n]$. *If for every* $c > 0$, $Y_{(n)}$ *is independent of* 1_n *for each n, then the cdf F has the form (9.1.2) for some* μ *real and* $\beta > 0$.

Among the other characterizations of the Type-I EV distribution, prominent are Leadbetter, Lindgren, and Rootzén (1983) where it has been shown that EV-I (β, μ) is the only 'max-stable' distribution with the entire real line as its support. Also see Sethuraman (1965) for characterization of all three types of EV distributions in terms of 'complete confounding' of random variables.

Before we go to the next section, we would like to observe the close relationship between the Weibull distribution and the Type-I Extreme Value Distribution.

Result 9.2.1 *Suppose* X *follows a two parameter Weibull distribution* $(W(\alpha, \beta))$ *with pdf (8.1.1). Then (i)* $Y = lnX$ *has the EV-I* (β_*, μ_*) *of smallest values with pdf (9.1.11) where the location parameter* $\mu_* = ln\beta$ *and scale parameter* $\beta_* = 1/\alpha$; *(ii)* $Y = -lnX$ *has the EV-I* (β_*, μ_*) *of*

*largest values with pdf (9.1.1) where the location parameter $\mu_* = -\ln\beta$
and scale parameter $\beta_* = 1/\alpha$.*

*Conversely, if X follows EV-I (β, μ) with pdf (9.1.1) (or (9.1.11)),
then $Y = exp(X)$ (or $exp(-X)$) follows a two-parameter Weibull dis-
tribution $(W(\alpha_*, \beta_*))$ with pdf (8.1.1) where $\alpha_* = 1/\beta_*$ and $\beta_* = exp(\mu)$
(or $exp(-\mu)$).*

Remark 9.2.1 *A similar relationship between the Weibull distribution
and the other two types of EV distributions can be established easily
through the relationships that exist between Type-I EV distribution and
the Type-II and Type-III EV distributions as discussed earlier.*

Another interesting result is the relationship between the Type-I EV
distribution and the logistic distribution (see Section 2.2) as stated below.

Result 9.2.2 *Let X_1 and X_2 be two independent random variables having the Type-I EV distributions with common scale parameter β but different (possibly) location parameters μ_1 and μ_2 respectively (i.e., X_i follows
EV-I (β, μ_i), $i = 1, 2$). Then $Y = (X_1 - X_2)$ follows a two-parameter
logistic distribution with location parameter $\mu = (\mu_1 - \mu_2)$ and scale parameter β.*

9.3   Estimation of Parameters

First we consider estimation of parameters for the Type-I EV distribution
only. Then we consider the other two types of EV distribution.

Given *iid* observations $X_1, X_2, \ldots, X_n$ having the common *pdf* (9.1.1),
the maximum likelihood estimators $\widehat{\beta}_{\mathrm{ML}}$ and $\widehat{\mu}_{\mathrm{ML}}$ of β and μ respectively
satisfy the equations:

$$\sum_{i=1}^{n} exp\{-(X_i - \widehat{\mu}_{\mathrm{ML}})/\widehat{\beta}_{\mathrm{ML}}\} = n \qquad (9.3.1)$$

$$(\overline{X} - \widehat{\mu}_{\mathrm{ML}}) - \frac{1}{n}\sum_{i=1}^{n}(X_i - \widehat{\mu}_{\mathrm{ML}})exp\{-(X_i - \widehat{\mu}_{\mathrm{ML}})/\widehat{\beta}_{\mathrm{ML}}\} = \widehat{\beta}_{\mathrm{ML}} \qquad (9.3.2)$$

where $\overline{X} = \sum_{i=1}^{n} X_i/n$. Further simplification of (9.3.1) gives $\widehat{\mu}_{\mathrm{ML}}$ in
terms of $\widehat{\beta}_{\mathrm{ML}}$ as

$$\widehat{\mu}_{\mathrm{ML}} = -\widehat{\beta}_{\mathrm{ML}}ln\left\{\frac{1}{n}\sum_{i=1}^{n} exp(-X_i/\widehat{\beta}_{\mathrm{ML}})\right\} \qquad (9.3.3)$$

which after plugging in (9.3.2) gives $\widehat{\beta}_{\mathrm{ML}}$ in terms of the following equation:

$$\widehat{\beta}_{\mathrm{ML}} = \overline{X} - \{\sum_{i=1}^{n} X_i exp(-X_i/\widehat{\beta}_{\mathrm{ML}})\}/\{\sum_{i=1}^{n} exp(-X_i/\widehat{\beta}_{\mathrm{ML}})\} \qquad (9.3.4)$$

The above MLEs, which are asymptotically unbiased, are obtained through numerical methods and have the following asymptotically properties (Downton (1966)).

$$\lim_{n\to\infty} Var(\sqrt{n}\widehat{\mu}_{\mathrm{ML}}) = \{1 + 6(1-\gamma)^2/\pi^2\}\beta^2 \approx (1.1087)\beta^2$$

$$\lim_{n\to\infty} Var(\sqrt{n}\widehat{\beta}_{\mathrm{ML}}) = 6\beta^2/\pi^2 \approx (0.6079)\beta^2$$

$$\lim_{n\to\infty} Corr(\widehat{\mu}_{\mathrm{ML}}, \widehat{\beta}_{\mathrm{ML}}) = (\gamma-1)\sqrt{\frac{6}{n^2}} \Big/ \sqrt{1 + 6(\gamma-1)^2/\pi^2} \approx 0.313$$

$$(9.3.5)$$

where $\gamma = 0.57721566\ldots$ is Euler's constant. However, for fixed n, the above MLEs are biased.

In many practical problems one does not have a complete data set due to censoring. For instance, out of a random sample of size n the smallest r_1 and the highest r_2 sample values are not available, i.e., we observe only the ordered values $X_{(r_1+1)} \leq X_{(r_1+2)} \leq \cdots \leq X_{(n-r_2)}$. Harter and Moore (1968) provided the method of deriving the MLEs of μ and β based on the doubly censored observations $X_{(r_1+1)} \leq \cdots \leq X_{(n-r_2)}$ as given below.

Define $Z_i = (X_{(i)} - \mu)/\beta$, $F(Z_i) = \frac{1}{2} \pm \frac{1}{2} \mp exp\{-exp(\pm Z_i)\}$, $i = (r_1+1), \ldots, (n-r_2)$. Also, $f(Z_i) = exp\{\pm Z_i - exp(\pm Z_i)\}$, and the upper (lower) signs being those for the distribution of smallest (largest) values (with *cdf* (9.1.10) ((9.1.2)) as mentioned in Section 9.1). The MLEs of μ and β based on the doubly censored (DC) observations, denoted by $\widehat{\mu}_{\mathrm{ML}}^{\mathrm{DC}}$ and $\widehat{\beta}_{\mathrm{ML}}^{\mathrm{DC}}$, are obtained by solving the following two equations for μ and β:

$$\mp \frac{(n-r_1-r_2)}{\beta} \pm \frac{1}{\beta} \sum_{i=r_1+1}^{n-r_2} exp(\pm Z_i) - r_1 f(Z_{r_1+1})/[\beta F(Z_{r_1+1})]$$

$$+ r_2 f(Z_{n-r_2})/[\beta\{1 - F(Z_{n-r_2})\}] = 0 \qquad (9.3.6)$$

and

$$\mp \frac{1}{\beta} \sum_{i=r_1+1}^{n-r_2} Z_i \pm \frac{1}{\beta} \sum_{i=r_1+1}^{n-r_2} Z_i exp(\pm Z_i) - \frac{(n-r_1-r_2)}{\beta} - r_1 Z_{r_1+1}$$

$$f(Z_{r_1+1})/[\beta F(Z_{r_1+1})] + r_2 Z_{n-r_2} f(Z_{n-r_2})/[\beta\{1 - F(Z_{n-r_2})\}] = 0$$
$$(9.3.7)$$

To find the asymptotic dispersion matrix of $(\widehat{\beta}_{ML}^{DC}, \widehat{\mu}_{ML}^{DC})$ we define the following terms:

$$\Gamma_\omega(y) = \int_0^\omega u^{y-1} e^{-u} du$$

$$\Gamma'_\omega(a) = \frac{d}{dy}[\Gamma_\omega(y)]|_{y=a}$$

$$\Gamma''_\omega(a) = \frac{d^2}{dy^2}[\Gamma_\omega(y)]|_{y=a}$$

Let $q_1 = r_1/n$, $q_2 = r_2/n$, and $p = 1 - (q_1 + q_2) = 1 - (r_1 + r_2)/n$. Keeping q_1 and q_2 fixed, as $n \to \infty$, $Z_{r_1+1} \to \widehat{z}_1$ and $Z_{n-r_2} \to \widehat{z}_2$ such that

$$F(\widehat{z}_1) = \int_{-\infty}^{\widehat{z}_1} f(t)\, dt = q_1$$

$$1 - F(\widehat{z}_2) = \int_{\widehat{z}_2}^{\infty} f(t)\, dt = q_2$$

For the distribution of the smallest values, the information matrix is $I = ((I_{ij}))$, where

$$
\begin{aligned}
I_{11} = {}& \frac{n}{\beta^2}\Big[-p - 2\left\{\Gamma'_{-lnq_2}(1) - \Gamma'_{-ln(1-q_1)}(1)\right\} + \Gamma''_{-\ln q_2}(2) - \Gamma''_{-ln(1-q_1)}(2) \\
& +2\left\{\Gamma'_{-lnq_2}(2) - \Gamma'_{-ln(1-q_1)}(2)\right\} - q_2 lnq_2\, ln(-lnq_2)\{2 + ln(-lnq_2)\} \\
& +(1-q_1)ln(1-q_1)ln\{-ln(1-q_1)\}[2 + ln\{-ln(1-q_1)\} \\
& +ln(1-q_1)ln\{-ln(1-q_1)\}/q_1\,]\Big]
\end{aligned}
$$

$$I_{22} = \frac{n}{\beta^2}[p + q_2 ln\, q_2 - (1-q_1)ln(1-q_1)]$$

$$
\begin{aligned}
I_{21} = I_{12} = {}& \frac{n}{\beta^2}\Big[\Gamma'_{-lnq_2}(2) - \Gamma'_{-ln(1-q_1)}(2) - q_2\, lnq_2\, ln(-lnq_2) \\
& +(1-q_1)ln(1-q_1)ln\{-ln(1-q_1)\} \\
& +(\frac{1}{q_1} - 1)\{ln(1-q_1)\}^2 ln\{-ln(1-q_1)\}\Big]
\end{aligned}
$$
$$(9.3.8)$$

The asymptotic dispersion matrix of $(\widehat{\beta}_{ML}^{DC}, \widehat{\mu}_{ML}^{DC})$ is I^{-1}.

If one considers the distribution of the largest values instead of the distribution of smallest values, then it can be shown that by interchanging q_1 and q_2, the values of I_{11} and I_{22} remain the same, and I^{12} (the $(1,2)$-element of the matrix I^{-1}) retains the same absolute value but changes the sign.

In the case of single (or one-sided) censoring, where we observe only the r smallest observations $X_{(1)} \leq X_{(2)} \leq \ldots \leq X_{(r)}$ from a sample of size n, Schüpbach and Hüsler (1983) proposed simple looking unbiased estimators of μ and β for the model (9.1.11) given as

$$\widehat{\beta}_{\text{SH}} = \left\{ -\sum_{i=1}^{l} X_{(i)} + \left(\frac{l}{r-l}\right) \sum_{i=l+1}^{r} X_{(i)} \right\} / (n k_{r,n})$$

$$\widehat{\mu}_{\text{SH}} = \bar{X}_{(r)} + d_{r,n} \widehat{\beta}_{\text{SH}} \tag{9.3.9}$$

where $\bar{X}_{(r)} = \sum_{i=1}^{r} X_{(i)}/r$, $r = [pn] < n$, $l = [qn] < r$, $0 < p < 1$, $0 < q < 1$. Also, l is chosen such that the variance of $\widehat{\beta}_{\text{SH}}$ is minimized, and $k_{r,n}$ and $d_{r,n}$ are suitable constants. It has been observed that for $0 < p < 0.8$, the optimal choice of l is $(r-1)$. For $0.8 \leq p < 1.0$, Schüpbach and Hüsler (1983) provided the values of optimal l and $k_{r,n}$ as well as $d_{r,n}$ as given below in Table 9.3.1.

n	r	l	$k_{r,n}$	$d_{r,n}$	n	r	l	$k_{r,n}$	$d_{r,n}$
5	4	3	0.800	0.894		14	12	1.27	0.800
10	9	7	1.067	0.751		13	11	0.982	0.907
	8	7	0.962	0.918	17	16	13	1.240	0.687
11	10	8	1.120	0.737		15	13	1.158	0.789
	9	8	1.016	0.890		14	12	1.016	0.889
12	11	9	1.166	0.726	18	17	14	1.268	0.681
	10	9	1.065	0.866		16	14	1.187	0.778
13	12	10	1.207	0.716		15	13	1.048	0.873
	11	9	1.013	0.846	19	18	15	1.294	0.676
14	13	11	1.244	0.707		17	15	1.214	0.769
	12	10	1.055	0.829		16	14	1.078	0.859
15	14	12	1.277	0.699	20	19	16	1.318	0.672
	13	11	1.093	0.814		18	16	1.239	0.760
	12	11	1.019	0.927		17	15	1.105	0.846
16	15	13	1.307	0.693		16	14	0.9905	0.932

Table 9.3.1: Values of l, $k_{r,n}$, and $d_{r,n}$ for Selected n and r ($0.8 \leq r/n < 1.0$).

In the case of complete sample $X_{(1)} \leq X_{(2)} \leq \ldots \leq X_{(n)}$, the estimators in (9.3.9) remain same except that

$l = [0.84n] = $ largest integer not exceeding $0.84n$

$d_{r,n}$ is replaced by $\gamma = 0.57721566$ (Euler's constant)

$k_{r,n}$ is replaced by k_n; given in the following Table 9.3.2 (Engelhardt and Bain (1977))

n	k_n	n	k_n	n	k_n
2	0.6931	22	1.4609	42	1.5208
3	0.9808	23	1.4797	43	1.5303
4	1.1507	24	1.4975	44	1.4891
5	1.2674	25	1.5142	45	1.4984
6	1.3545	26	1.4479	46	1.5075
7	1.1828	27	1.4642	47	1.5163
8	1.2547	28	1.4796	48	1.5248
9	1.3141	29	1.4943	49	1.5331
10	1.3644	30	1.5083	50	1.5411
11	1.4079	31	1.5216	51	1.5046
12	1.4461	32	1.4665	52	1.5126
13	1.3332	33	1.4795	53	1.5204
14	1.3686	34	1.4920	54	1.5279
15	1.4004	35	1.5040	55	1.5352
16	1.4293	36	1.5156	56	1.5424
17	1.4556	37	1.5266	57	1.5096
18	1.4799	38	1.4795	58	1.5167
19	1.3960	39	1.4904	59	1.5236
20	1.4192	40	1.5009	60	1.5304
21	1.4408	41	1.5110	∞	1.5692

Table 9.3.2: Values of k_n for $2 \leq n < \infty$.

Another simple way of estimating the Type-I *EV* Distribution parameters is to match the log-log (natural) transformation of the empirical distribution function with that of the *cdf* and then use a simple regression method.

Note that the *cdf* in (9.1.2) (distribution of largest values) yields

$$-ln(-lnF_1(x)) = (x - \mu)/\beta$$

$$\text{i.e.,} \quad x = \mu + \beta\{-ln(-lnF_1(x))\} \qquad (9.3.10)$$

On the other hand, the empirical distribution function based on the doubly censored observations $X_{(r_1+1)}, \ldots, X_{(n-r_2)}$ is $\widehat{F}_1(x)$ where

$$\widehat{F}_1(x) = \{\# \, X'_{(i)}s \le x\}/(n - r_1 - r_2) \qquad (9.3.11)$$

For $i = (r_1 + 1), (r_1 + 2), \ldots, (n - r_2)$, we have from (9.3.11),

$$\widehat{F}_1(X_{(i)}) = (i - r_1)/(n - r_1 - r_2) \qquad (9.3.12)$$

Now in (9.3.10) replace $F_1(x)$ by $\widehat{F}_1(\cdot)$ for $x = X_{(r_1+1)}, \ldots, X_{(n-r_2)}$, yielding

$$X_{(i)} = \mu + \beta\{-ln(-ln((i - r_1)/(n - r_1 - r_2)))\}$$
$$\text{i.e.,} \quad X_{(i)} = \mu + \beta\{-ln(ln(n - r_1 - r_2) - ln(i - r_1))\} \quad (9.3.13)$$

Use a simple linear regression approach (where $X_{(i)}$'s play the role of dependent variable values and $\{-ln(ln(n-r_1-r_2)-ln(i-r_1))\}$'s are the independent variable values) to estimate μ and β, and call the resultant estimators $\widehat{\mu}_{\mathrm{R}}$ and $\widehat{\beta}_{\mathrm{R}}$ respectively.

Identical steps can be followed for the distribution of smallest values (i.e., *cdf* (9.1.10)).

Also note that when $r_1 = 0$ and $r_2 = n$, we have the full sample (uncensored).

In the following we use the data set in Example 9.1.2 to estimate the parameters of a Type-I EV distribution.

Method	$\widehat{\beta}$	$\widehat{\mu}$
MLE	5.2446	11.7128
Schűpbach & Hűster	5.21717	11.6112
Regression	4.81539	11.8520

Table 9.3.3: Estimated Parameters with Various Methods.

Since this data set is **large** (with $n = 485$), we apply the bootstrap method to estimate the bias and *MSE* of the above three types of estimates. Number of replications used here is 10,000.

Method	Estimators	Averages	Bias	MSE
MLE	$\widehat{\mu}_{\mathrm{ML}}$	11.7164	0.0035898	0.0652417
	$\widehat{\beta}_{\mathrm{ML}}$	5.2350	-0.0095709	0.0288717
Schűpbach & Hűster	$\widehat{\mu}_{\mathrm{SH}}$	11.6135	0.0023443	0.0590269
	$\widehat{\beta}_{\mathrm{SH}}$	5.2125	-0.0046892	0.0326424
Regression	$\widehat{\mu}_R$	11.8623	0.0102647	0.0678648
	$\widehat{\beta}_R$	4.7981	-0.0172493	0.0339656

Table 9.3.4: Estimated Mean, Bias, and MSE of the Parameter Estimates Using Bootstrap.

9.4 Goodness of Fit Tests for Extreme Value Distributions

In this section we discuss the goodness of fit tests for extreme value distributions of Type-I (i.e., EV-I(β, μ)) only.

Shapiro and Brain (1987) proposed a goodness of fit test for the Type-I limiting distribution of the **smallest value** of *iid* random variables, i.e., for the distribution with *pdf* (9.1.11).

To test that a random sample $X_1, X_2, \ldots, X_n$ is following the EV-I(β, μ) with *pdf* (9.1.11), we first order the observations $X_{(1)} \leq X_{(2)} \leq \ldots \leq X_{(n)}$ and then define the following quantities:

$$\omega_i = ln(\frac{n+1}{n-i+1}), \ i = 1, 2, \ldots, (n-1)$$

$$\omega_n = n - \sum_{i=1}^{n-1} \omega_i$$

$$\omega_{n+i} = \omega_i(1 + ln\omega_i) - 1, \ i = 1, 2, \ldots, (n-1) \qquad (9.4.1)$$

$$\omega_{2n} = 0.4228n - \sum_{i=1}^{n-1} \omega_{n+i}$$

$$A_1 = \sum_{i=1}^{n} \omega_i X_{(i)} \ \text{ and } \ A_2 = \sum_{i=1}^{n} \omega_{n+i} X_{(i)}$$

The test statistic to test goodness of fit is

$$W_n = \frac{(0.6079 A_2 - 0.2570 A_1)^2}{n \sum_{i=1}^{n}(X_i - \bar{X})^2} \qquad (9.4.2)$$

which provides a two-tail test of the null hypothesis (i.e., the data follows (9.1.11)). The distribution of W_n under the null hypothesis is not analytically tractable. However, the above authors carried out a comprehensive simulation work based on which percentile points of the null distribution of W_n were expressed as a function of the sample size. For $0 < \alpha < 1$, let $W_{n,\alpha}$ be the $(100\alpha)^{th}$ percentile point, i.e.,

$$\alpha = P(W \le W_{n,\alpha} \mid \text{null hypothesis is true}) \qquad (9.4.3)$$

Shapiro and Brain (1987) showed that

$$W_{n,\alpha} \approx \alpha_0 + \alpha_1(ln(n)) + \alpha_2(ln(n))^2 \qquad (9.4.4)$$

The values of α, α_0, α_1 and α_2 are provided in the following.

$0 < \alpha < 1$	α_0	α_1	α_2
0.005	0.10102	0.04249	0.005882
0.025	0.11787	0.08550	-0.002048
0.050	0.13200	0.10792	-0.006487
0.950	1.46218	-0.21111	0.012914
0.975	1.64869	-0.26411	0.016840
0.995	1.91146	-0.31361	0.017669

Table 9.4.1: Values of α, α_0, α_1, and α_2.

Using the formula (9.4.4) and the Table 9.4.1, the percentile points of W_n under the null hypothesis can be found as given below.

n	α					
	0.005	0.025	0.050	0.950	0.975	0.995
5	0.18464	0.25017	0.28889	1.15586	1.26724	1.45249
10	0.23004	0.30388	0.34610	1.04455	1.12984	1.28303
15	0.25922	0.33439	0.37668	0.98519	1.05696	1.19176
20	0.28110	0.35563	0.39708	0.94565	1.00862	1.13054
25	0.29873	0.37186	0.41217	0.91645	0.97304	1.08506
30	0.31358	0.38498	0.42402	0.89354	0.94521	1.04921
35	0.32644	0.39597	0.43369	0.87485	0.92255	1.01981
40	0.33780	0.40540	0.44183	0.85915	0.90358	0.99503

Table 9.4.2: Values of $W_{n,\alpha}$ for Selected Values of n and α.

To test that a random sample $X_1, X_2, \ldots, X_n$ is following the EV-I(β, μ) with *pdf* (9.1.1) (i.e., the limiting distribution of the largest value of *iid* observations), one needs to work with $X_1^*, X_2^*, \ldots, X_n^*$ where $X_i^* = -X_i$. Note that X_i^* follows (9.1.11) if and only if X_i follows (9.1.1). Hence, one works with X_i^*'s, defines the W_n statistic based on $X_{(1)}^* \leq X_{(2)}^* \leq \ldots \leq X_{(n)}^*$, and make use of $W_{n,\alpha}$ as discussed above.

At level α, one accepts the null hypothesis if $W_{n,(\alpha/2)} \leq W_n \leq W_{n,1-(\alpha/2)}$; and rejects otherwise.

As an application, we consider the data set (of size $n = 24$) given in the Example 9.1.1. The values for the altitude of 40 km are used to see if the *pdf* (9.1.11) is appropriate for the data. Our ordered observations are (row wise):

$$
\begin{array}{cccccc}
-1.14, & -1.00, & -0.71, & -0.25, & -0.20, & -0.15, \\
-0.10, & -0.05, & 0.00, & 0.00, & 0.10, & 0.15, \\
0.20, & 0.20, & 0.21, & 0.25, & 0.30, & 1.00, \\
1.43, & 2.64, & 3.43, & 4.14, & 4.14, & 4.43.
\end{array}
$$

i	ω_i	i	ω_i	i	ω_i
1	0.0408	9	0.4463	17	1.1394
2	0.0834	10	0.5108	18	1.2730
3	0.1278	11	0.5798	19	1.4271
4	0.1744	12	0.6539	20	1.6094
5	0.2231	13	0.7340	21	1.8326
6	0.2744	14	0.8210	22	2.1203
7	0.3285	15	0.9163	23	2.5257
8	0.3857	16	1.0217		

Table 9.4.3: The Coefficients W_i's for $n = 24$.

Since $n = 24$, we first calculate the W_i values, $1 \leq i \leq (n - 1) = 23$. We now have $A_1 = 48.1970$, $A_2 = 69.6708$, $W_n = 0.585897$. With $\alpha = 0.1$, $W_{n,\,(\alpha/2)} = 0.409457$ and $W_{n,\,1-(\alpha/2)} = 0.921693$. Since $W_{n,(\alpha/2)} < W_n < W_{n,1-(\alpha/2)}$, we accept the fact that the above data set can be modelled by EV-I(β, μ).

Chapter 10

System Reliability

10.1 Preliminaries

Complex machines, henceforth referred to as 'systems', are made of smaller devices, called components. The life span of the whole system depends on the life span of the individual components which are often independent of each other. Depending on the nature of the system (i.e., how the components are connected with each other), the system works if some or all of the components work. On the other hand, the system fails as soon as some or all of the components fail. In this chapter we shall deal mainly with reliability of two types of systems — *Series System* and *Parallel System.*

One can handle a system reliability in two ways as described below.

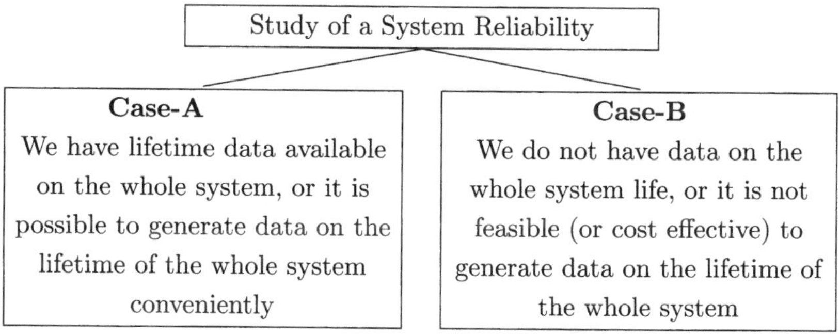

```
          ┌──────────────────────────────────┐
          │   Study of a System Reliability   │
          └──────────────────────────────────┘
```

Case-A	Case-B
We have lifetime data available on the whole system, or it is possible to generate data on the lifetime of the whole system conveniently	We do not have data on the whole system life, or it is not feasible (or cost effective) to generate data on the lifetime of the whole system

Figure 10.1.1:   Two possible ways of handling a system reliability.

Case-A: It is possible to have lifetime data on the whole system.

In this case usually the data tend to follow a positively skewed distribution, and hence can be modelled by one of the distributions discussed in the earlier chapters (Chapter 6 to Chapter 9).

As an example of this case, consider the problem of studying the lifetime of a particular brand of fax machines. Though a fax machine has numerous smaller components in it, it is possible to build several fax machines at a relatively low cost and to study how long they can survive. The data thus generated can then be modelled by a suitable positively skewed distribution.

Case-B: It is difficult to have lifetime data on the whole system.

In many real life cases it is not possible to generate lifetime data on the whole system at an acceptable cost. But it is possible to test individual components and to generate lifetime data on these individual components. If one knows how the individual components are connected (or placed) to make the whole system work, then individual component data can be used (along with the structural knowledge of the whole system) to study the lifetime of the whole system.

As an example of this case, consider the problem of studying the lifetime of a communication satellite. Various components like battery charger, solar panel, and data transmitter, etc. are integral parts of the satellite, and they must operate to make the whole satellite functional. Instead of launching a satellite and then observing its lifetime to collect a single costly observation, it makes sense to test the component prototypes in a simulated environment at a much lower cost, and then use that information to draw inferences about the satellite's lifetime.

For notational and other convenience one can unify the above two cases (A and B) theoretically under Case-B. When lifetime data is available on the whole system then just pretend that the whole system consists of a single component (the system itself), and then we can draw inferences just as we do under the above Case-B. On the other hand, under Case-B, think of a system with k components and we have lifetime data on the individual components. When $k = 1$, we are under the above Case-A. Therefore, in the rest of the chapter, we consider systems with k components, assuming that the lifetime data are available on the individual components.

10.2   Single Component Systems

Let the lifetime of a component be represented by a nonnegative random variable X with *pdf* $f(x)$ and *cdf* $F(x)$. The reliability of the component at time t is the probability that the component does not fail on or before time t, i.e.,

$$\bar{F}(t) = P(X > t) = \int_t^\infty f(x)\, dx = 1 - P(X \le t) = 1 - F(t) \quad (10.2.1)$$

Often, we assume a known structure for $f(x)$, the *pdf* of lifetime, which may involve one or more unknown parameters. In such a case one can estimate the unknown parameter, and hence the component reliability from a sample of observations on X. For example, one can test n prototypes of the component and observe their life spans to obtain the data $X_1, X_2, \ldots, X_n$, where X_i represents the life span of the ith prototype. In the following we discuss inferences on reliability using some of the distributions discussed in earlier chapters.

Exponential Distribution

Example 10.2.1. A random sample of size 8 has been obtained on the lifetime (in hours) of a component with the following observations:

$$
\begin{array}{cccc}
48.96 & 54.67 & 100.75 & 120.68 \\
148.34 & 228.43 & 293.45 & 415.82
\end{array}
$$

Assuming an exponential model estimate the component reliability.

To address the above problem we will use both the one-parameter exponential model (i.e., **Exp**(β) model (6.1.1)) and the two-parameter exponential model (i.e., **Exp**(β, μ) model (6.1.5)).

Case I: One Parameter Exponential Model

Step 1 : Goodness of fit test (see Section 6.4):

(i) $S = \sum_{i=1}^n X_i = 1411.10$

(ii) $Z_j = \sum_{i=1}^j X_i / S, \quad 1 \le j \le (n-1)$, values are given respectively as (in increasing order)

$$
\begin{array}{cccc}
0.034696 & 0.073439 & 0.144837 & 0.230359 \\
0.335483 & 0.497364 & 0.705322 &
\end{array}
$$

(iii) $T^* = -2\sum_{i=1}^{n-1}\ln Z_i = 23.024823$

(iv) $\chi^2_{2(n-1),(\alpha/2)}$ and $\chi^2_{2(n-1),1-(\alpha/2)}$ cut-off points with significance level $\alpha = 0.05$ are respectively 26.12 and 5.63. Since T^* falls in between these two values, at 5% level, the model **Exp**(β) seems plausible for the above data set.

Step 2 : Parameter estimation:

(i) The MLE (also the UMVUE) of the model parameter β is $\widehat{\beta}_{\mathrm{ML}} = 176.3875$.

(ii) The reliability of a component at time $t(> 0)$ is

$$\bar{F}(t) = P(X > t) = \exp(-t/\beta) \qquad (10.2.2)$$

(iii) The MLE of $\bar{F}(t)$ is

$$\widehat{\bar{F}}_{\mathrm{ML}} = \widehat{\bar{F}}_{\mathrm{ML}}(t|\beta) = \exp(-t/\widehat{\beta}_{\mathrm{ML}}) = \exp(-t/176.3875)$$
$$(10.2.3)$$

which is biased.

The bias of $\widehat{\bar{F}}_{\mathrm{ML}}$, which is a function of β, is

$$
\begin{aligned}
B(\widehat{\bar{F}}_{\mathrm{ML}}|\beta) &= E(\widehat{\bar{F}}_{\mathrm{ML}}) - \bar{F} \\
&= \int_0^\infty \{e^{-nt/x} - e^{-t/\beta}\}(\beta^n\Gamma(n))^{-1}e^{-x/\beta}x^{n-1}\,dx \\
&= \int_0^\infty \{e^{nt_*/y} - e^{-t_*}\}e^{-y}y^{n-1}/\Gamma(n)\,dy, \quad t_* = t/\beta
\end{aligned}
$$
$$(10.2.4)$$

since $n\widehat{\beta}_{\mathrm{ML}} = \sum_{i=1}^n X_i$ follows $G(n,\beta)$ distribution (with *pdf* (7.1.1)). In the present example, $n = 8$.

Figure 10.2.1 gives the plot of $B(\widehat{\bar{F}}_{\mathrm{ML}})$ as a function of $t_* = t/\beta$. Similarly, the *MSE* of $\widehat{\bar{F}}_{\mathrm{ML}}$ is given as

$$
\begin{aligned}
MSE(\widehat{\bar{F}}_{\mathrm{ML}}|\beta) &= E(\widehat{\bar{F}}_{\mathrm{ML}} - \bar{F})^2 \\
&= \int_0^\infty \{e^{nt_*/y} - e^{-t_*}\}^2 e^{-y}y^{n-1}/\Gamma(n)\,dy, \quad t_* = t/\beta
\end{aligned}
$$
$$(10.2.5)$$

The *UMVUE* of $\bar{F}(t)$ is given as

$$
\widehat{\bar{F}}_{\mathrm{U}} = \widehat{\bar{F}}_{\mathrm{U}}(t|\beta) =
\begin{cases}
\left[1 - \left\{t/(n\widehat{\beta}_{\mathrm{ML}})\right\}\right]^{n-1} & \text{if } n\widehat{\beta}_{\mathrm{ML}} > t \\
\\
0 & \text{otherwise}
\end{cases}
\qquad (10.2.6)
$$

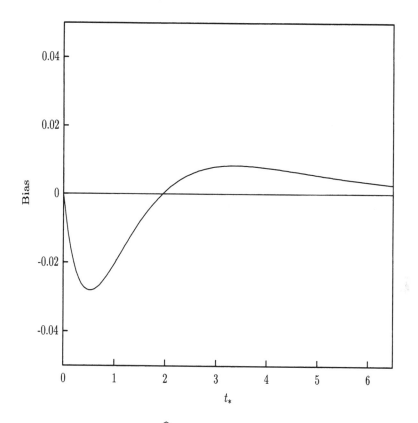

Figure 10.2.1:   Bias of $\widehat{\overline{F}}_{\text{ML}}$ as a function of $t_* = t/\beta$ with $n = 8$.

(Note that the expression in Bain (1978) (Theorem 2.2.2, page-124) is incorrect, presumably due to a typographical error.)

The *MSE* of $\widehat{\overline{F}}_U$ is given as

$$
\begin{aligned}
MSE(\widehat{\overline{F}}_{\text{U}}|\beta) &= E(\widehat{\overline{F}}_{\text{U}} - \bar{F})^2 \\
&= \int_{t_*}^{\infty} (1 - t_*/y)^{2(n-1)} e^{-y} y^{n-1}/\Gamma(n)\, dy - (e^{-t_*})^2, \\
&\hspace{4cm} t_* = t/\beta \hspace{2cm} (10.2.7)
\end{aligned}
$$

The following figure gives the plots of the *MSE* curves of $\widehat{\overline{F}}_{\text{ML}}$ and $\widehat{\overline{F}}_{\text{U}}$ as a function of $t_* = t/\beta$.

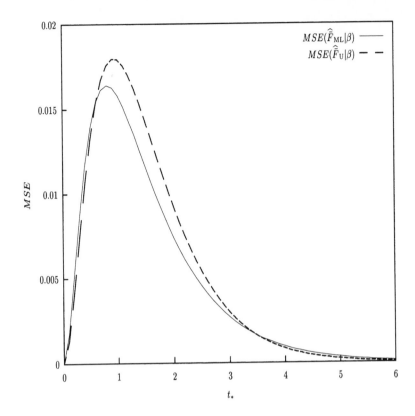

Figure 10.2.2: MSEs of $\widehat{\overline{F}}_{\mathrm{ML}}(t)$ and $\widehat{\overline{F}}_U(t)$ as a function of $t_* = t/\beta$ with $n = 8$.

In reality, since β is unknown, so is t_*; and hence the bias and MSE are unknown (except for the zero bias of UMVUE). However, these can be estimated by replacing t_* by $\widehat{t}_* = t/\widehat{\beta}_{\mathrm{ML}}$.

In Example 10.2.1, $\widehat{t}_* = t/(176.3875)$. If we want to study the component reliability at time $t = 300$ hours, then $\widehat{t}_* = 300/(176.3875) = 1.700800 \approx 1.7$. Then

$$\widehat{B}(\widehat{\overline{F}}_{\mathrm{ML}}|\beta) = \text{estimated bias of } \widehat{\overline{F}}_{\mathrm{ML}} = -0.00403, \quad \text{and}$$
$$\widehat{MSE}(\widehat{\overline{F}}_{\mathrm{ML}}|\beta) = \text{estimated } MSE \text{ of } \widehat{\overline{F}}_{\mathrm{ML}} = 0.00980$$

On the other hand, the bias of $\widehat{\overline{F}}_U$ is 0, and

$$\widehat{MSE}(\widehat{\overline{F}}_U|\beta) = \text{estimated } MSE \text{ of } \widehat{\overline{F}}_U = 0.01219$$

Case II : Two Parameter Exponential Model

Step 1 : Goodness of fit test (see Section 6.4):

(i) First we obtain the 'spacings' defined as $Y_i = (n+i-1)(X_{(i)} - X_{(i-1)})$ for $i = 2, 3, \ldots, n$. Hence

$$Y_2 = 39.97, \quad Y_3 = 276.48, \quad Y_4 = 99.65, \quad Y_5 = 110.64$$

$$Y_6 = 240.27, \quad Y_7 = 130.04 \quad \text{and} \quad Y_8 = 122.37$$

(ii) $S^* = \sum_{i=2}^{n} Y_i = 1019.42$

(iii) $Z_j^{**} = \sum_{i=2}^{j} Y_i / S^*$, $2 \leq j \leq (n-1)$, values are given respectively as (in increasing order)

    0.039209   0.310422   0.408173
    0.516709   0.752398   0.879961

(iv) $T^{**} = -2 \sum_{i=2}^{(n-1)} \ln Z_i^{**} = 12.754769$

(v) $\chi^2_{2(n-2),(\alpha/2)}$ and $\chi^2_{2(n-2),(1-\alpha/2)}$ cut off points with $\alpha = 0.05$ are respectively 23.34 and 4.40. Since T^{**} falls between these two values, at 5% level , the model $\mathbf{Exp}(\beta, \mu)$ seems plausible for the above data set.

Step 2 : Parameter estimation:

(i) The *MLE*s of the model parameters μ and β are $\hat{\mu}_{\mathrm{ML}} = X_{(1)} = \min\{X_1, \ldots, X_n\} = 48.96$ and $\hat{\beta}_{\mathrm{ML}} = \sum_{i=1}^{n}(X_i - X_{(1)})/n = 127.4275$.

(ii) The reliability of a component at time $t(> \mu)$ is

$$\bar{F}(t) = P(X > t) = P(X - \mu > t - \mu) = \exp(-(t-\mu)/\beta) \quad (10.2.8)$$

(iii) Therefore, the *MLE* of $\bar{F}(t)$ is

$$\widehat{\bar{F}}_{\mathrm{ML}} = \widehat{\bar{F}}_{\mathrm{ML}}(t|\beta, \mu) = \begin{cases} exp(-(t - \hat{\mu}_{\mathrm{ML}})/\hat{\beta}_{\mathrm{ML}}) & \text{if } t > \hat{\mu}_{\mathrm{ML}} \\ \\ 1 & \text{if } t \leq \hat{\mu}_{\mathrm{ML}} \end{cases}$$

$$= \begin{cases} exp(-(t - 48.96)/127.4275) & \text{if } t > 48.96 \\ \\ 1 & \text{if } t \leq 48.96 \end{cases} \quad (10.2.9)$$

It can be shown that the probability distribution of $\widehat{\bar{F}}_{\mathrm{ML}}$ depends on the term $(t - \mu)/\beta$, and hence the bias and MSE of $\widehat{\bar{F}}_{\mathrm{ML}}$ are functions of $(t - \mu)/\beta$ only where $(t > \mu)$.

To study the bias and MSE of $\widehat{\bar{F}}_{\mathrm{ML}}$ (see 10.2.9), it would be helpful to look at the following representation. Define $t_{**} = (t - \mu)/\beta$. Then $\widehat{\bar{F}}_{\mathrm{ML}}$ can be characterized as

$$\widehat{\bar{F}}_{\mathrm{ML}} = \begin{cases} exp\{(-nt_{**} + W_1)/W_2\} & \text{if } t_{**} > W_1 \\[2mm] 1 & \text{if } t_{**} < W_1 \end{cases} \qquad (10.2.10)$$

where W_1 and W_2 are independent random variables with $W_1 \sim \mathbf{Exp}(1)$ and $W_2 \sim G(n - 1, 1)$. Hence,

$$\begin{aligned} B(\widehat{\bar{F}}_{\mathrm{ML}}|\beta, \mu) &= E(\widehat{\bar{F}}_{\mathrm{ML}}) - \bar{F} \\[2mm] &= \int_0^\infty \int_0^\infty [\, exp\{(-nt_{**} + w_1)/w_2\} - exp(-t_{**})\,] \\[2mm] &\quad \times e^{-w_1} e^{-w_2} w_2^{n-2}/\Gamma(n-1)\, dw_1 dw_2 \\[2mm] &\quad + \int_0^\infty \int_{t_{**}}^\infty [\, 1 - exp(-t_{**})\,] \\[2mm] &\quad \times e^{-w_1} e^{-w_2} w_2^{n-2}/\Gamma(n-1)\, dw_1 dw_2 \qquad (10.2.11) \end{aligned}$$

The Figure 10.2.3 gives the plot of $B(\widehat{\bar{F}}_{\mathrm{ML}}|\beta, \mu)$ as a function of $t_{**} = (t - \mu)/\beta > 0$.

Similarly, the MSE of $\widehat{\bar{F}}_{\mathrm{ML}}$ is given as

$$\begin{aligned} MSE(\widehat{\bar{F}}_{\mathrm{ML}}|\beta, \mu) &= E(\widehat{\bar{F}}_{\mathrm{ML}} - \bar{F})^2 \\[2mm] &= \int_0^\infty \int_0^\infty [\, exp\{(-nt_{**} + w_1)/w_2\} - exp(-t_{**})\,]^2 \\[2mm] &\quad \times e^{-w_1} e^{-w_2} w_2^{n-2}/\Gamma(n-1)\, dw_1 dw_2 \\[2mm] &\quad + \int_0^\infty \int_{t_{**}}^\infty [\, 1 - exp(-t_{**})\,]^2 \\[2mm] &\quad \times e^{-w_1} e^{-w_2} w_2^{n-2}/\Gamma(n-1)\, dw_1 dw_2 \qquad (10.2.12) \end{aligned}$$

An analogue to the estimator (10.1.6) for estimating $\bar{F}(t)$ (for the two parameter exponential model) is the unbiased estimator given as

$$\begin{aligned} \widehat{\bar{F}}_U &= \widehat{\bar{F}}_U(t) \\[2mm] &= \begin{cases} 0 & \text{if } t > n\widehat{\beta}_{\mathrm{ML}} + \widehat{\mu}_{\mathrm{ML}} \\[2mm] (1 - 1/n)\Big[1 - \{(t - \widehat{\mu}_{\mathrm{ML}})/(n\widehat{\beta}_{\mathrm{ML}})\}\Big]^{n-2} & \text{if } \widehat{\mu}_{\mathrm{ML}} \le t < n\widehat{\beta}_{\mathrm{ML}} + \widehat{\mu}_{\mathrm{ML}} \\[2mm] 1 & \text{if } t < \widehat{\mu}_{\mathrm{ML}} \end{cases} \end{aligned}$$

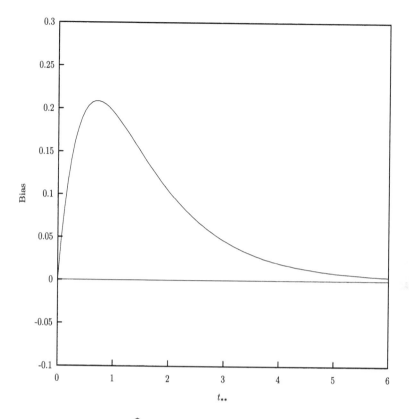

Figure 10.2.3: Bias of $\widehat{\overline{F}}_{\mathrm{ML}}(t)$ as a function of $t_{**} = (t - \mu)/\beta$ $(n = 8)$.

The MSE of $\widehat{\overline{F}}_{\mathrm{U}}$ can be obtained through the following expression. Define $g(w_1, w_2 | t_{**})$ as

$$g(w_1, w_2 | t_{**}) =$$
$$\begin{cases} 0 & \text{if } t_{**} > w_2 + w_1/n \\ (1 - 1/n)\{1 - (t_{**} - w_1/n)w_2\}^{n-2} & \text{if } w_1/n \leq t_{**} \leq w_2 + w_1/n \\ 1 & \text{if } t_{**} < w_1/n \end{cases}$$
$$(10.2.13)$$

Then the MSE of $\widehat{\overline{F}}_{\mathrm{U}}$ is

$$MSE(\widehat{\overline{F}}_{\mathrm{U}} | \beta, \mu) = \int_0^\infty \int_0^\infty [g(w_1, w_2 | t_{**}) - exp(-t_{**})]^2 \times$$
$$e^{-w_1} e^{-w_2} w_2^{n-2}/\Gamma(n-1) \ dw_1 dw_2 \quad (10.2.14)$$

The Figure 10.2.4 gives the plots of the MSE curves of $\widehat{\overline{F}}_{\text{ML}}$ and $\widehat{\overline{F}}_U$ as a functions of $t_{**} = (t - \mu)/\beta$.

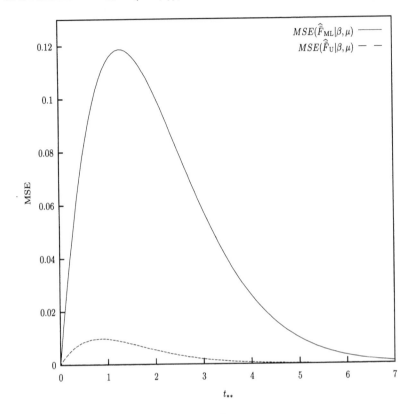

Figure 10.2.4:   MSEs of $\widehat{\overline{F}}_{\text{ML}}(t)$ and $\widehat{\overline{F}}_U(t)$ as functions of $t_{**} = (t-\mu)/\beta$ $(n = 8)$.

Remark 10.2.1 *Estimating the reliability of a component with exponential lifetime and studying the sampling distribution of these estimators is relatively easy. But this may not be so if the component lifetime follows a more general distribution as shown below. In such a case, Monte-Carlo simulation studies come handy in studying the reliability estimators.*

Gamma Distribution

Example 10.2.2.   Consider the data set of Example 7.1.1 where it is assumed that a gamma distribution might be suitable to model the lifetime (in hours) of pressure vessels. For convenience the data set has been reproduced below.

274.00	1.70	871.00	1311.00	236.00
458.00	54.90	1787.00	0.75	776.00
28.50	20.80	363.00	1661.00	828.00
290.00	175.00	970.00	1278.00	126.00

We shall estimate the reliability of a pressure vessel at time t using both the two-parameter gamma model (i.e. $G(\alpha, \beta)$ model (7.1.1)) as well as the three-parameter gamma model (i.e., $G(\alpha, \beta, \mu)$ model (7.1.9)).

Case I : Two-Parameter Gamma Model

Denoting the observations as $X_1, \ldots, X_n$ ($n = 20$), the maximum likelihood estimates of α and β are found by solving the equations $\widehat{\alpha}_{\mathrm{ML}}\widehat{\beta}_{\mathrm{ML}} = \bar{X}$ and $ln(\widehat{\beta}_{\mathrm{ML}}) + \psi(\widehat{\alpha}_{\mathrm{ML}}) = (\sum_{i=1}^{n} lnX_i)/n$, where $\bar{X} = \sum_{i=1}^{n} X_i/n$ and $\psi(\alpha) = \frac{\partial}{\partial \alpha} ln\Gamma(\alpha)$; and these are

$$\widehat{\alpha}_{\mathrm{ML}} = 0.579182; \quad \widehat{\beta}_{\mathrm{ML}} = 993.699 \qquad (10.2.15)$$

The reliability of a pressure vessel at time $t(> 0)$ is

$$
\begin{aligned}
\bar{F} &= \bar{F}(t|\alpha, \beta) = P(X > t) \\
&= \int_t^\infty exp(-x/\beta)x^{\alpha-1}/(\beta^\alpha \Gamma(\alpha)) \, dx \qquad (10.2.16)
\end{aligned}
$$

which does not have any further closed expression. The MLE of $\bar{F}$ is

$$
\begin{aligned}
\widehat{\bar{F}}_{\mathrm{ML}} &= \bar{F}(t|\widehat{\alpha}_{\mathrm{ML}}, \widehat{\beta}_{\mathrm{ML}}) \\
&= \int_t^\infty exp(-x/\widehat{\beta}_{\mathrm{ML}})x^{\widehat{\alpha}_{\mathrm{ML}}-1}/(\widehat{\beta}_{\mathrm{ML}}^{\widehat{\alpha}_{\mathrm{ML}}}\Gamma(\widehat{\alpha}_{\mathrm{ML}})) \, dx \, (10.2.17)
\end{aligned}
$$

The sampling distribution of $\widehat{\bar{F}}_{\mathrm{ML}}$ is impossible to study analytically for fixed sample sizes. For the data set in the above Example 10.2.2, with $t = 300$ hours, $\widehat{\alpha}_{\mathrm{ML}} = 0.579182$, and $\widehat{\beta}_{\mathrm{ML}} = 993.699$,

$$
\begin{aligned}
\widehat{\bar{F}}_{\mathrm{ML}} &= \frac{1}{(993.699)^{0.579182}\Gamma(0.579182)} \\
&\quad \times \int_{300}^\infty exp(-x/993.699)x^{0.579182-1} \, dx \\
&= 0.496069 \qquad (10.2.18)
\end{aligned}
$$

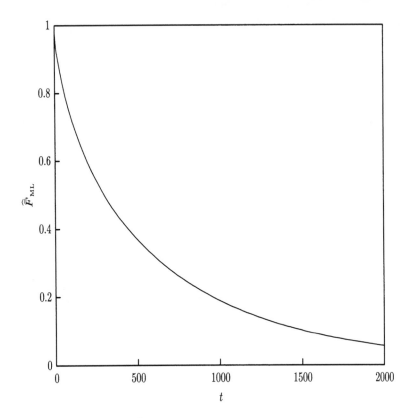

Figure 10.2.5: The $\widehat{\bar{F}}_{\mathrm{ML}}$ curve for the data set in Example 10.1.2.

The Figure 10.2.4 plots $\widehat{\bar{F}}_{\mathrm{ML}}$ (in (10.2.18)) against $t > 0$ for the above data set.

In general, one can use the Monte-Carlo simulation to study the behavior of the estimator $\widehat{\bar{F}}_{\mathrm{ML}}$ as described in Chapter 5 and given in the following.

For the sake of simplicity we generate data from a gamma distribution with $\alpha > 0$ and $\beta = 1$. Note, $\bar{F} = \int_t^\infty \exp(-x/\beta)x^{\alpha-1}/(\beta^\alpha\Gamma(\alpha))\, dx$.

Step 1 : Generate $X_1, \ldots, X_n$ *iid* from $G(\alpha, \beta)$ with a specific $\alpha > 0$ and $\beta = 1$. Fix $t > 0$.

Step 2 : Calculate $\widehat{\alpha}_{\mathrm{ML}}, \widehat{\beta}_{\mathrm{ML}}, \widehat{\bar{F}}_{\mathrm{ML}}$.

Step 3 : The error in $\bar{F}$ estimation is calculated as

$$E_{\mathrm{ML}} = (\widehat{\bar{F}}_{\mathrm{ML}} - \bar{F})$$

Step 4 : Repeat the above Step 1 through Step 3 for a large number of (say M) times. The resultant values of E_{ML} for these replications are termed as $E_{\mathrm{ML}}^{(1)}, \ldots, E_{\mathrm{ML}}^{(M)}$.

Step 5 : The bias and MSE of $\widehat{F}_{\mathrm{ML}}$ are approximated as

$$B(\widehat{F}_{\mathrm{ML}}|\alpha, \beta) \approx \frac{1}{M}\sum_{j=1}^{M}E_{\mathrm{ML}}^{(j)}$$

and

$$MSE(\widehat{F}_{\mathrm{ML}}|\alpha, \beta) \approx \frac{1}{M}\sum_{j=1}^{M}(E_{\mathrm{ML}}^{(j)})^2$$

One can carry out such a simulation for different combinations of (α, β) and t.

The Figure 10.2.6 (a) to Figure 10.2.6 (b), Figure 10.2.7 (a) to Figure 10.2.7 (b) give the bias and MSE (computed through the above simulation steps) curves of $\widehat{F}_{\mathrm{ML}}$ plotted against α (for fixed $\beta = 1$ and $t > 0$) and t (for fixed α and β).

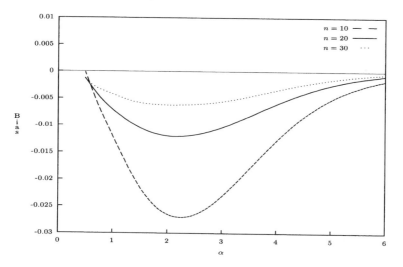

Figure 10.2.6: (a) Simulated bias of $\widehat{F}_{\mathrm{ML}}$ plotted against α ($\beta = 1, t = 1$).

The above mentioned figures, though they provide an idea about the sampling distribution of $\widehat{F}_{\mathrm{ML}}$, can do little for a specific data set. Since the parameters α, β are unknown, we can only plot the bias and MSE curves of $\widehat{F}_{\mathrm{ML}}$ for limited combinations of (α, β). A more realistic approach would

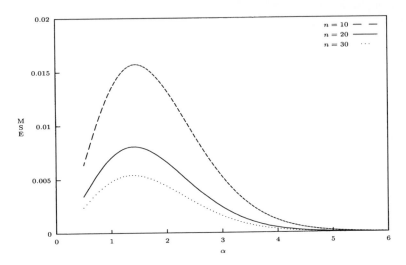

Figure 10.2.6: (b) Simulated MSE of $\widehat{\overline{F}}_{\mathrm{ML}}$ plotted against α $(\beta = 1, t = 1)$.

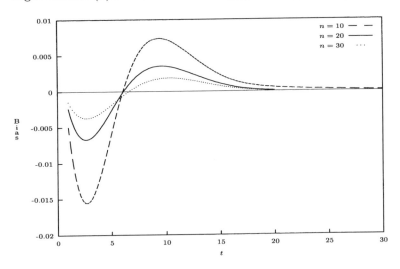

Figure 10.2.7: (a) Simulated bias of $\widehat{\overline{F}}_{\mathrm{ML}}$ plotted against $t > 0$ $(\alpha = 5, \beta = 1)$.

be to follow the bootstrap method as described in Chapter 5 to study the bias and MSE of $\widehat{\overline{F}}_{\mathrm{ML}}$. This is demonstrated below with the data set in Example 10.2.2.

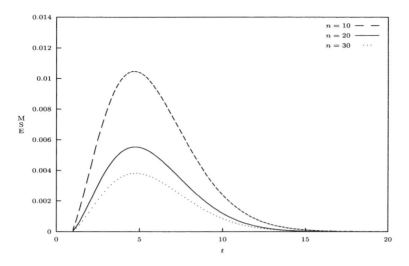

Figure 10.2.7: (b) Simulated MSE of $\widehat{\overline{F}}_{\mathrm{ML}}$ plotted against $t > 0$ ($\alpha = 5, \beta = 1$).

Semiparametric Bootstrap Method

Fix the observed sample $\{X_1, \ldots, X_n\}$, call it 'orginal sample'.

Step 1 : Draw a sample of size n with replacement from the original sample, call it a 'bootstrap sample'.

Step 2 : Calculate $\widehat{\alpha}_{\mathrm{ML}}$, $\widehat{\beta}_{\mathrm{ML}}$, and $\widehat{\overline{F}}_{\mathrm{ML}}$ based on the bootstrap sample, and call them $\widehat{\alpha}^*_{\mathrm{ML}}$, $\widehat{\beta}^*_{\mathrm{ML}}$, and $\widehat{\overline{F}}^*_{\mathrm{ML}}$ respectively.

Step 3 : The observed error in $\widehat{\overline{F}}_{\mathrm{ML}}$ for the new sample is calculated as $E^*_{\mathrm{ML}} = (\widehat{\overline{F}}^*_{\mathrm{ML}} - \widehat{\overline{F}}_{\mathrm{ML}})$.

Step 4 : Repeat the above Step 1 through Step 3 for a large number of (say, M) times. The resultant values of E^*_{ML} for these replications are termed as $E^{*(1)}_{\mathrm{ML}}, \ldots, E^{*(M)}_{\mathrm{ML}}$.

Step 5 : The estimated bias (called, semiparametric bootstrap bias) and the estimated MSE (called, semiparametric bootstrap MSE) of $\widehat{\overline{F}}_{\mathrm{ML}}$ are obtained as (abbreviated as 'SB-Bias' and 'SB-MSE' respectively)

$$SB - Bias(\widehat{\overline{F}}_{\mathrm{ML}} | \alpha, \beta) = \frac{1}{M} \sum_{j=1}^{M} E^{*(j)}_{\mathrm{ML}}$$

and

$$SB - MSE(\widehat{\overline{F}}_{\mathrm{ML}}|\alpha, \beta) = \frac{1}{M}\sum_{j=1}^{M}(E_{\mathrm{ML}}^{*(j)})^2$$

Using $M = 20,000$, the following SB-bias and SB-MSE curves have been generated for the data set in Example 10.2.2.

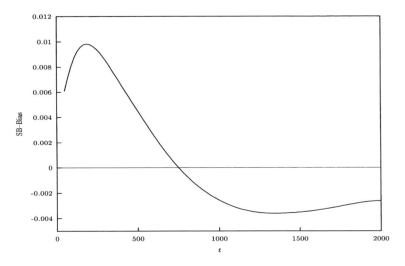

Figure 10.2.8: (a) SB-Bias of $\widehat{\overline{F}}_{\mathrm{ML}}$ (Example 10.2.2).

Remark 10.2.2 *(a) A true nonparametric bootstrap method does not require any specific probability model (parametric) for the data set. Yet we have assumed the $G(\alpha, \beta)$ model for the data in Example 10.2.2 to begin with. Then we have followed the bootstrap approach to study the behavior of the parameter estimates $\widehat{\alpha}_{ML}$ and $\widehat{\beta}_{ML}$, and hence we call it a semiparametric bootstrap (SB) method.*

(b) It has been observed that the semiparametric (or nonparametric) bootstrap method can give satisfactory results for moderate to large sample sizes (n). For small n, such bootstrap results (bias and MSE) tend to be very unstable (fluctuate too much from one set of replications to another). Hence, a parametric bootstrap is suggested as an extra tool to evaluate an estimator.

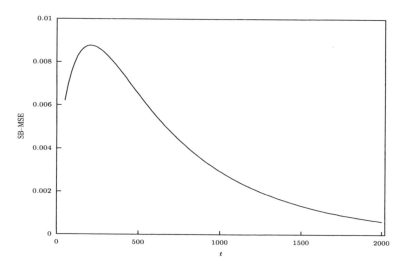

Figure 10.2.8: (b) SB-MSE of $\widehat{\overline{F}}_{\mathrm{ML}}$ (Example 10.2.2).

Parametric Bootstrap Method

For the observed sample $\{X_1, \ldots, X_n\}$ (original sample), calculate $\widehat{\alpha}_{\mathrm{ML}}$, $\widehat{\beta}_{\mathrm{ML}}$, and $\widehat{\overline{F}}_{\mathrm{ML}}$.

Step 1 : Generate a new sample (of size n) *iid* from $G(\widehat{\alpha}_{\mathrm{ML}}, \widehat{\beta}_{\mathrm{ML}})$ (i.e., the $G(\alpha, \beta)$ distribution with $(\alpha = \widehat{\alpha}_{\mathrm{ML}}, \beta = \widehat{\beta}_{\mathrm{ML}})$.

Step 2 : Calculate the *MLEs* of α, β, and $\bar{F}$ based on the new sample, and call them $\widehat{\alpha}_{\mathrm{ML}}^{**}$, $\widehat{\beta}_{\mathrm{ML}}^{**}$, and $\widehat{\overline{F}}_{\mathrm{ML}}^{**}$ respectively.

Step 3 : The observed error in $\widehat{\overline{F}}_{\mathrm{ML}}$ for the new sample is calculated as $E_{\mathrm{ML}}^{**} = (\widehat{\overline{F}}_{\mathrm{ML}}^{**} - \widehat{\overline{F}}_{\mathrm{ML}})$.

Step 4 : Repeat the above Step 1 through Step 3 for a large number (say, M) of times. The resultant values of E_{ML}^{**} for these replications are termed as $E_{\mathrm{ML}}^{**(1)}, \ldots, E_{\mathrm{ML}}^{**(M)}$.

Step 5 : The estimated bias and *MSE* (called, parametric bootstrap bias and *MSE*) of $\widehat{\overline{F}}_{\mathrm{ML}}$ are obtained as (and denoted by 'PB-Bias' and 'PB-*MSE*' respectively)

$$PB - Bias(\widehat{\overline{F}}_{\mathrm{ML}} | \alpha, \beta) = \frac{1}{M} \sum_{j=1}^{M} E_{\mathrm{ML}}^{**(j)}$$

and

$$PB - MSE(\widehat{\bar{F}}_{\text{ML}}|\alpha, \beta) = \frac{1}{M} \sum_{j=1}^{M} (E_{\text{ML}}^{**(j)})^2$$

Using $M = 20,000$, the following PB-Bias and PB-MSE curves have been generated for the data set in Example 10.2.2 and these are plotted against $t > 0$ in Figure 10.2.9 (a) and Figure 10.2.9 (b).

The other estimators $\bar{F}(t|\alpha, \beta)$ are obtained from the modified $MLEs$ (called $MMLEs$) and $MMEs$ of α and β as discussed in Section 7.3.

Using the reparametrization $\beta = 1/\lambda$ and (7.3.6), the $MMLEs$ of α and β are

$$\widehat{\alpha}_{\text{MML}} = \frac{(n-3)}{n}\widehat{\alpha}_{\text{ML}} + \frac{2}{3n}$$

and

$$\widehat{\beta}_{\text{MML}} = \widehat{\beta}_{\text{ML}}\, g_0(\widehat{\alpha}_{\text{MML}}) \qquad (10.2.19)$$

where $g_0(\alpha) = \{1 + \frac{3n\alpha}{(n-3)(n\alpha-1)}(1 + \frac{1}{9\alpha} - \frac{1}{n\alpha} + \frac{(n-1)}{27n\alpha^2})\}$.

On the other hand, the $MMEs$ of α and β are

$$\widehat{\alpha}_{\text{MM}} = \frac{n\bar{X}^2}{\sum_{i=1}^{n}(X_i - \bar{X})^2} \quad \text{and} \quad \widehat{\beta}_{\text{MM}} = \frac{\sum_{i=1}^{n}(X_i - \bar{X})^2}{n\bar{X}} \qquad (10.2.20)$$

Thus, the two new estimators of $\bar{F}(t|\alpha, \beta)$ are

$$\widehat{\bar{F}}_{\text{MML}} = \bar{F}(t|\widehat{\alpha}_{\text{MML}}, \widehat{\beta}_{\text{MML}})$$

and

$$\widehat{\bar{F}}_{\text{MM}} = \bar{F}(t|\widehat{\alpha}_{\text{MM}}, \widehat{\beta}_{\text{MM}}) \qquad (10.2.21)$$

For the data set in Example 10.2.2, with $t = 300$,

$$\widehat{\alpha}_{\text{MML}} = 0.5256, \quad \widehat{\beta}_{\text{MML}} = 1234.7$$

$$\widehat{\alpha}_{\text{MM}} = 1.0559, \quad \widehat{\beta}_{\text{MM}} = 545.061$$

and

$$\widehat{\bar{F}}_{\text{MML}} = \int_{300}^{\infty} \exp(-x/1234.7)x^{0.5256-1}/(1234.7)^{0.5256}\Gamma(0.5256))\, dx$$
$$= 0.506012$$

$$\widehat{\bar{F}}_{\text{MM}} = \int_{300}^{\infty} \exp(-x/545.061)x^{1.0559-1}/(545.061^{1.00311}\Gamma(1.0559))\, dx$$
$$= 0.603638$$

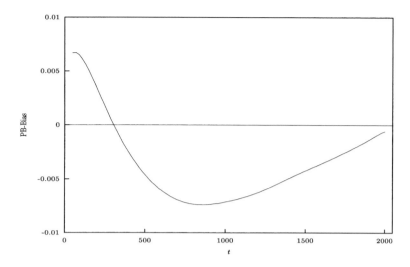

Figure 10.2.9: (a) PB-Bias of $\widehat{\overline{F}}_{\mathrm{ML}}$ (Example 10.2.2).

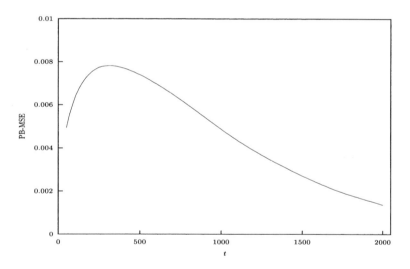

Figure 10.2.9: (b) PB-MSE of $\widehat{\overline{F}}_{\mathrm{ML}}$ (Example 10.2.2).

In the following (Figure 10.2.10 (a) and Figure 10.2.10 (b), Figure 10.2.11 (a) and Figure 10.2.11 (b)) we provide the bias and MSE curves of $\widehat{\overline{F}}_{\mathrm{MML}}$ and $\widehat{\overline{F}}_{\mathrm{MM}}$ through simulations for various n.

Remark 10.2.3 *Compare the bias and MSE of $\widehat{\overline{F}}_{ML}$ in Figure 10.2.6 (a)*

and Figure 10.2.6 (b) with those of $\widehat{\overline{F}}_{MML}$ and $\widehat{\overline{F}}_{MM}$ in Figure 10.2.10 (a)
and Figure 10.2.10 (b). Apparently, both $\widehat{\overline{F}}_{ML}$ and $\widehat{\overline{F}}_{MML}$ have similar bias
(mostly negative), whereas $\widehat{\overline{F}}_{MM}$ has opposite (mostly positive) bias. In
term of MSE, $\widehat{\overline{F}}_{MML}$ seems to have an overall best performance. Also note
that the MMEs of α and β are not functions of the sufficient statistics,
and hence $\widehat{\overline{F}}_{MM}$ is inferior to $\widehat{\overline{F}}_{ML}$ or $\widehat{\overline{F}}_{MML}$ intuitively.

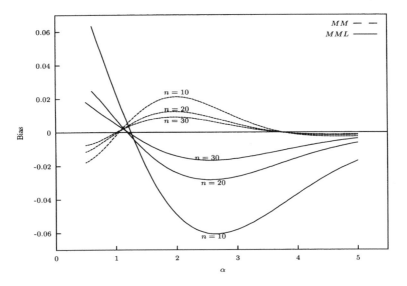

Figure 10.2.10: (a) Simulated Biases of $\widehat{\overline{F}}_{\text{MML}}$ and $\widehat{\overline{F}}_{\text{MM}}$ plotted against α ($\beta = 1, t = 1$).

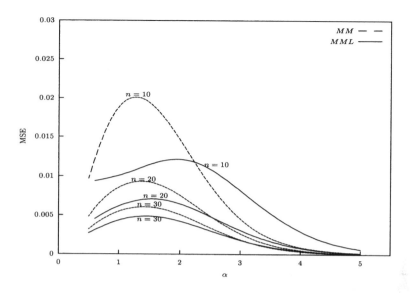

Figure 10.2.10: (b) Simulated MSEs of $\widehat{\widehat{F}}_{\text{MML}}$ and $\widehat{\widehat{F}}_{\text{MM}}$ plotted against α ($\beta = 1, t = 1$).

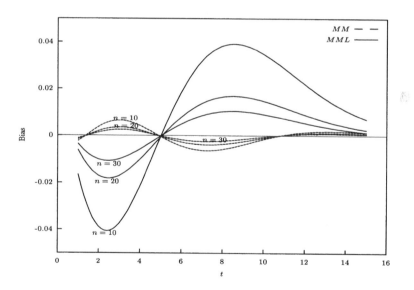

Figure 10.2.11: (a) Simulated Biases of $\widehat{\widehat{F}}_{\text{MML}}$ and $\widehat{\widehat{F}}_{\text{MM}}$ plotted against $t > 0$ ($\alpha = 5.0, \beta = 1.0$).

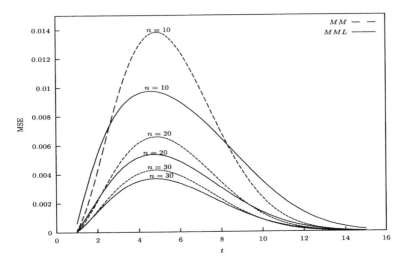

Figure 10.2.11: (b) Simulated MSEs of $\widehat{\overline{F}}_{\text{MML}}$ and $\widehat{\overline{F}}_{\text{MM}}$ plotted against $t > 0$ ($\alpha = 5.0, \beta = 1$).

The following figures (Figure 10.2.12 (a) and Figure 10.2.12 (b)) provide the semiparametric bootstrap bias and MSE of the estimators $\widehat{\overline{F}}_{\text{MML}}$ and $\widehat{\overline{F}}_{\text{MM}}$ plotted against $t > 0$. Note that the bootstrap method is dependent heavily on the specific data set under consideration. The bootstrap method is used for the data set in Example 10.2.2.

Remark 10.2.4 *One can also follow the parametric bootstrap approach to get estimated bias and MSE of the reliability estimators $\widehat{\overline{F}}_{\text{MML}}$ and $\widehat{\overline{F}}_{\text{MM}}$. But one has to be extra cautious in comparing the PB-Bias and PB-MSE curves of various estimators . Note that while SB-Bias and SB-MSE are computed based on resampling from the same 'original sample', PB-Bias and PB-MSE are computed based on sampling from different distributions ($G(\widehat{\alpha}_{MML}, \widehat{\beta}_{MML})$ and $G(\widehat{\alpha}_{MM}, \widehat{\beta}_{MM})$) and hence comparisons may be criticized.*

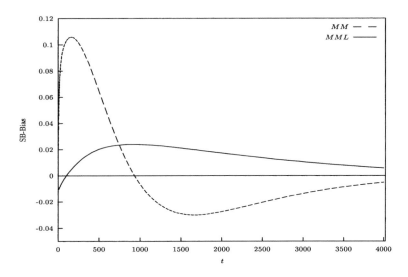

Figure 10.2.12: (a) SB-Biases of $\widehat{\overline{F}}_{\mathrm{MML}}$ and $\widehat{\overline{F}}_{\mathrm{MM}}$ (Example 10.2.2).

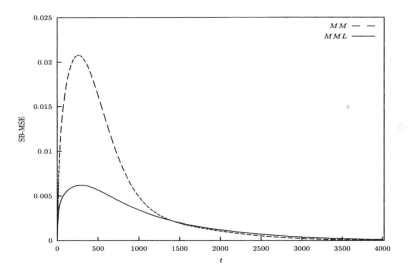

Figure 10.2.12: (b) SB-MSEs of $\widehat{\overline{F}}_{\mathrm{MML}}$ and $\widehat{\overline{F}}_{\mathrm{MM}}$ (Example 10.2.2).

Case II : Three-parameter gamma model

If we assume a three-parameter gamma model $G(\alpha, \beta, \mu)$ (as given in (7.1.9)) for the data in Example 10.2.2, the reliability of a pressure vessel at time $t(> \mu)$ is

$$\bar{F} = \bar{F}(t|\alpha, \beta, \mu) = P(X > t) \quad = \quad \int_t^\infty exp\{-(x - \mu)/\beta\}$$
$$\times (x - \mu)^{\alpha-1}/(\beta^\alpha \Gamma(\alpha)) \, dx$$
$$(10.2.22)$$

The following estimates of $\bar{F}$ are found by employing suitable estimates of the model parameters as

$$\widehat{\bar{F}}_{\text{ML}} = \begin{cases} 1 & \text{if } t \leq \widehat{\mu}_{\text{ML}} \\ \bar{F}(t|\widehat{\alpha}_{\text{ML}}, \widehat{\beta}_{\text{ML}}, \widehat{\mu}_{\text{ML}}) & \text{if } t > \widehat{\mu}_{\text{ML}} \end{cases}$$

(using the MLEs in (10.2.23));

$$\widehat{\bar{F}}_{\text{MML}} = \begin{cases} 1 & \text{if } t \leq \widehat{\mu}_{\text{MML}} \\ \bar{F}(t|\widehat{\alpha}_{\text{MML}}, \widehat{\beta}_{\text{MML}}, \widehat{\mu}_{\text{MML}}) & \text{if } t > \widehat{\mu}_{\text{MML}} \end{cases}$$

(using the MMLEs in (10.2.23));

$$\widehat{\bar{F}}_{\text{MM}} = \begin{cases} 1 & \text{if } t \leq \widehat{\mu}_{\text{MM}} \\ \bar{F}(t|\widehat{\alpha}_{\text{MM}}, \widehat{\beta}_{\text{MM}}, \widehat{\mu}_{\text{MM}}) & \text{if } t > \widehat{\mu}_{\text{MM}} \end{cases}$$

(using the MMEs in (10.2.23)).

For the data set in Example 10.2.2, using $t = 300$ hours,

$$\widehat{\bar{F}}_{\text{ML}} = 0.667; \ \widehat{\bar{F}}_{\text{MML}} = 0.644; \ \widehat{\bar{F}}_{\text{MM}} = 0.652 \qquad (10.2.23)$$

Also, the above three estimators have been plotted against $t > 0$ for the above data set in Figure 10.2.13.

Similar to the case of a two-parameter gamma model, simulated bias and MSE of $\widehat{\bar{F}}_{\text{ML}}$, $\widehat{\bar{F}}_{\text{MML}}$ and $\widehat{\bar{F}}_{\text{MM}}$ are plotted against $t > 0$ as given in Figure 10.2.14 (a) and Figure 10.2.14 (b).

The data set in Example 10.2.2 is now used to get the semiparametric bootstrap bias and MSE (i.e., SB-bias and SB-MSE) of the estimates $\widehat{\bar{F}}_{\text{ML}}$, $\widehat{\bar{F}}_{\text{MML}}$, and $\widehat{\bar{F}}_{\text{MM}}$ for various values of $t > 0$ (and then plotted against t) as given in Figure 10.2.16 (a) and Figure 10.2.16 (b). Again, $\widehat{\bar{F}}_{\text{MML}}$ seems to perform best among all the three estimators.

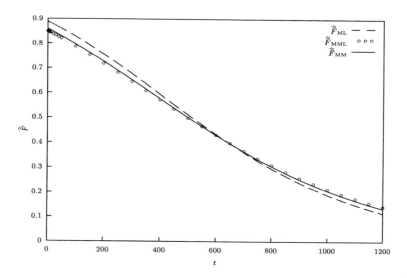

Figure 10.2.13: The $\widehat{\widehat{F}}_{\text{ML}}$, $\widehat{\widehat{F}}_{\text{MML}}$, and $\widehat{\widehat{F}}_{\text{MM}}$ curves for the data set in Example 10.2.2 with the $G(\alpha, \beta, \mu)$ model.

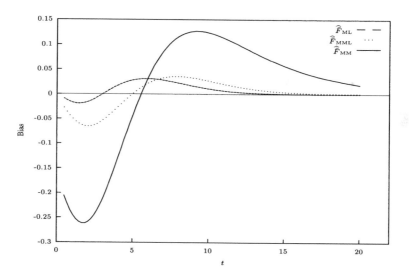

Figure 10.2.14: (a) Simulated Biases of $\widehat{\widehat{F}}_{\text{ML}}$, $\widehat{\widehat{F}}_{\text{MML}}$, and $\widehat{\widehat{F}}_{\text{MM}}$ plotted against $t > 0$ ($\alpha = 5.0, \beta = 1.0, \mu = 0.0$) for $n = 10$ with a $G(\alpha, \beta, \mu)$ model.

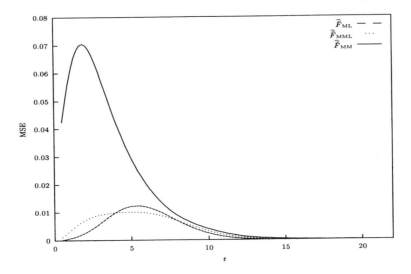

Figure 10.2.14: (b) Simulated MSEs of $\widehat{\overline{F}}_{\mathrm{ML}}$, $\widehat{\overline{F}}_{\mathrm{MML}}$, and $\widehat{\overline{F}}_{\mathrm{MM}}$ plotted against $t > 0$ ($\alpha = 5.0, \beta = 1.0, \mu = 0.0$) for $n = 10$ with a $G(\alpha, \beta, \mu)$ model.

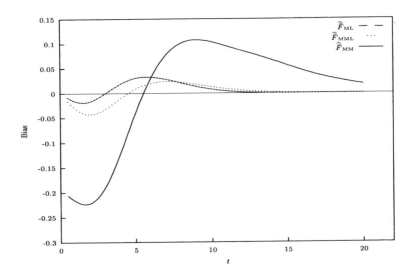

Figure 10.2.15: (a) Simulated Biases of $\widehat{\overline{F}}_{\mathrm{ML}}$, $\widehat{\overline{F}}_{\mathrm{MML}}$, and $\widehat{\overline{F}}_{\mathrm{MM}}$ plotted against $t > 0$ ($\alpha = 5.0, \beta = 1.0, \mu = 0.0$) for $n = 20$ with a $G(\alpha, \beta, \mu)$ model.

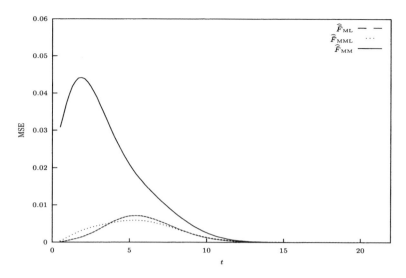

Figure 10.2.15: (b) Simulated MSEs of $\widehat{\bar{F}}_{\mathrm{ML}}$, $\widehat{\bar{F}}_{\mathrm{MML}}$, and $\widehat{\bar{F}}_{\mathrm{MM}}$ plotted against $t > 0$ ($\alpha = 5.0, \beta = 1.0, \mu = 0.0$) for $n = 20$ with a $G(\alpha, \beta, \mu)$ model.

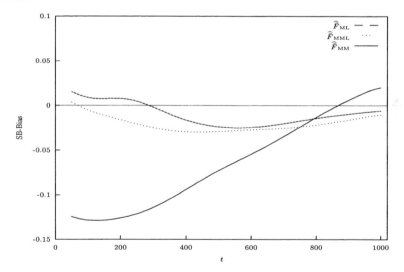

Figure 10.2.16: (a) SB-Biases of $\widehat{\bar{F}}_{\mathrm{ML}}$, $\widehat{\bar{F}}_{\mathrm{MML}}$, and $\widehat{\bar{F}}_{\mathrm{MM}}$ using a $G(\alpha, \beta, \mu)$ model for the data in Example 10.2.2.

Weibull Distribution

Apart from exponential and gamma distributions, Weibull distribution is widely used to model lifetime data sets. In the following we study the

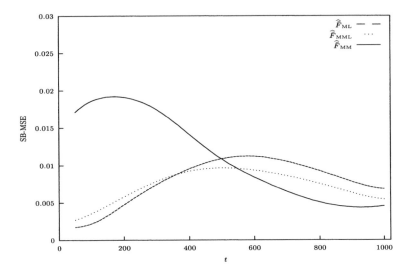

Figure 10.2.16: (b) SB-MSEs of $\widehat{F}_{\mathrm{ML}}$, $\widehat{F}_{\mathrm{MML}}$, and $\widehat{F}_{\mathrm{MM}}$ using a $G(\alpha, \beta, \mu)$ model for the data in Example 10.2.2.

estimation of a system reliability using a Weibull distribution.

Example 10.2.3. Consider the data set of Example 8.1.1 where it is assumed that a Weibull distribution could be used to model the endurance (in m.o.r.) of ball bearings. The sample observations from 23 ball bearings are:

17.88	41.52	48.48	54.12	68.64	84.12	105.12	128.04
28.92	42.12	51.84	55.56	68.64	93.12	105.84	173.40
33.00	45.60	51.96	67.80	68.88	98.64	127.92

We shall estimate the reliability of a ball bearing at t m.o.r. using both a two-parameter Weibull model (i.e., $W(\alpha, \beta)$ model (8.1.1)) as well as a three-parameter Weibull model (i.e., $W(\alpha, \beta, \mu)$ model (8.1.8)). We will see that in terms of simpler closed expression of the *cdf* Weibull model has certain advantages over a gamma model.

Case I : Two-parameter Weibull model

With observations $X_1, X_2, \ldots, X_n$ ($n = 23$ in Example 10.2.3), the maximum likelihood estimates of α and β are found by solving the equations

$$\widehat{\beta}_{\text{ML}} = \left\{\sum_{i=1}^{n} X_i^{\widehat{\alpha}_{\text{ML}}}/n\right\}^{1/\widehat{\alpha}_{\text{ML}}}$$

$$\widehat{\alpha}_{\text{ML}} = \left[\left\{\sum_{i=1}^{n} X_i^{\widehat{\alpha}_{\text{ML}}} \ln X_i\right\} / \left\{\sum_{i=1}^{n} X_i^{\widehat{\alpha}_{\text{ML}}}\right\} - \left\{\sum_{i=1}^{n} \ln X_i/n\right\}\right]^{-1}$$

For the above data set, we have

$$\widehat{\alpha}_{\text{ML}} = 2.10 \quad \text{and} \quad \widehat{\beta}_{\text{ML}} = 81.86 \tag{10.2.24}$$

The reliability at time (or m.o.r. for the above data set) $t > 0$ is

$$\bar{F} = \bar{F}(t|\alpha, \beta) = P(X > t) = exp\left\{-(t/\beta)^{\alpha}\right\} \tag{10.2.25}$$

Thus, the MLE of $\bar{F}$ is

$$\widehat{\bar{F}}_{\text{ML}} = \bar{F}(t|\widehat{\alpha}_{\text{ML}}, \widehat{\beta}_{\text{ML}}) = exp\{-(t/\widehat{\beta}_{\text{ML}})^{\widehat{\beta}_{\text{ML}}}\} \tag{10.2.26}$$

which for the data set in Example 10.2.3, with $t = 100$ m.o.r., yields

$$\widehat{\bar{F}}_{\text{ML}} = 0.21817 \tag{10.2.27}$$

The graph of $\widehat{\bar{F}}_{\text{ML}}$ is plotted against $t > 0$ in the following figure.

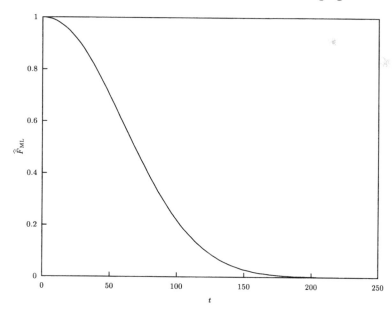

Figure 10.2.17: The $\widehat{\bar{F}}_{\text{ML}}$ curve for the data set in Example 10.2.3.

On the other hand, the method of moments estimates of α and β are obtained from

$$\frac{\left(\Gamma\left(1+\frac{1}{\alpha_{MM}}\right)\right)^2}{\Gamma\left(1+\frac{2}{\alpha_{MM}}\right)} = \frac{n\bar{X}^2}{\sum_{i=1}^{n} X_i^2} \quad \text{and} \quad \widehat{\beta}_{MM} = \frac{\bar{X}}{\Gamma\left(1+\frac{1}{\alpha_{MM}}\right)}$$

which for the above data set (Example 10.2.3) gives

$$\widehat{\alpha}_{MM} = 2.066, \quad \widehat{\beta}_{MM} = 81.5342$$

The resultant estimator of $\bar{F}$ is

$$\widehat{\bar{F}}_{MM} = \bar{F}(t|\widehat{\alpha}_{MM}, \widehat{\beta}_{MM}) = exp\left\{-(t/81.5342)^{2.066}\right\} \qquad (10.2.28)$$

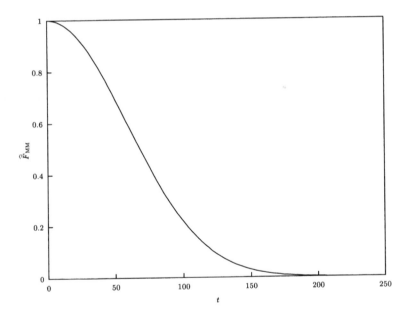

Figure 10.2.18: The $\widehat{\bar{F}}_{MM}$ curve for the data set in Example 10.2.3.

To study how the estimates $\widehat{\bar{F}}_{ML}$ and $\widehat{\bar{F}}_{MM}$ perform under the above mentioned $W(\alpha, \beta)$ model, we carry out a simulation study as described below.

Step-1 : Generate $X_1, X_2, \ldots, X_n$ *iid* from $W(\alpha, \beta)$ with a specific $\alpha > 0$ and $\beta = 1$. Fix $t > 0$.

Step-2 : Calculate $\widehat{\alpha}_{\mathrm{ML}}$, $\widehat{\beta}_{\mathrm{ML}}$, $\widehat{\bar{F}}_{\mathrm{ML}}$ and $\widehat{\alpha}_{\mathrm{MM}}$, $\widehat{\beta}_{\mathrm{MM}}$, $\widehat{\bar{F}}_{\mathrm{MM}}$.

Step-3 : The error in estimation by each method is calculated as

$$E_{\mathrm{ML}} = (\widehat{\bar{F}}_{\mathrm{ML}} - \bar{F}) \quad \text{and} \quad E_{\mathrm{MM}} = (\widehat{\bar{F}}_{\mathrm{MM}} - \bar{F})$$

Step-4 : Repeat the above Step 1 through Step 3 for a large number of (say M) times. The resultant values of E_{ML} and E_{MM} for these replications are termed as $E_{\mathrm{ML}}^{(1)}, \ldots, E_{\mathrm{ML}}^{(M)}$ and $E_{\mathrm{MM}}^{(1)}, \ldots, E_{\mathrm{MM}}^{(M)}$, respectively.

Step-5 : The bias and MSE of $\widehat{\bar{F}}_{\mathrm{ML}}$ and $\widehat{\bar{F}}_{\mathrm{MM}}$ are approximated as

$$B(\widehat{\bar{F}}_{\mathrm{ML}}|\alpha, \beta) \approx \frac{1}{M} \sum_{j=1}^{M} E_{\mathrm{ML}}^{(j)}; \quad B(\widehat{\bar{F}}_{\mathrm{MM}}|\alpha, \beta) \approx \frac{1}{M} \sum_{j=1}^{M} E_{\mathrm{MM}}^{(j)}$$

and

$$MSE(\widehat{\bar{F}}_{\mathrm{ML}}|\alpha, \beta) \approx \frac{1}{M} \sum_{j=1}^{M} (E_{\mathrm{ML}}^{(j)})^2; \quad MSE(\widehat{\bar{F}}_{\mathrm{MM}}|\alpha, \beta) \approx \frac{1}{M} \sum_{j=1}^{M} (E_{\mathrm{MM}}^{(j)})^2$$

$$(10.2.29)$$

First we plot the bias curves against α for $\beta = 1.0$ and $t = 1$, and then against t for $\beta = 1.0$ and $\alpha = 5.0$. These are shown in Figure 10.2.19 (a) and Figure 10.2.19 (b).

Similarly the MSE curves are plotted in Figure 10.2.20 (a) and Figure 10.2.20 (b).

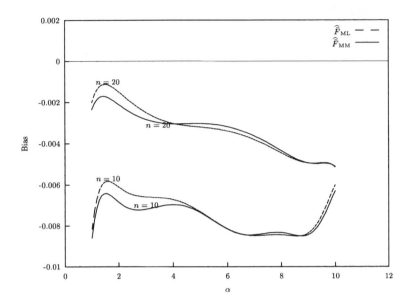

Figure 10.2.19: (a) Simulated Biases of $\widehat{\overline{F}}_{\mathrm{ML}}$ and $\widehat{\overline{F}}_{\mathrm{MM}}$ plotted against α with $n = 10$ and 20.

Following the steps of semiparametric bootstrap discussed earlier we now study $\widehat{\overline{F}}_{\mathrm{ML}}$ and $\widehat{\overline{F}}_{\mathrm{MM}}$ for the data set in Example 10.2.3.

Semiparametric Bootstrap Method

Fix the observe sample $\{X_1, \ldots, X_n\}$ with $n = 23$. This is our 'original sample'.

Step-1 : Draw a sample of size n with replacement from the original sample, and call it a 'bootstrap sample'.

Step-2 : Calculate $\widehat{\alpha}_{\mathrm{ML}}$, $\widehat{\beta}_{\mathrm{ML}}$, and $\widehat{\overline{F}}_{\mathrm{ML}}$ based on the bootstrap sample, and call them $\widehat{\alpha}^*_{\mathrm{ML}}$, $\widehat{\beta}^*_{\mathrm{ML}}$, and $\widehat{\overline{F}}^*_{\mathrm{ML}}$, respectively. Similarly get $\widehat{\alpha}^*_{\mathrm{MM}}$, $\widehat{\beta}^*_{\mathrm{MM}}$, and $\widehat{\overline{F}}^*_{\mathrm{MM}}$.

Step-3 : The observed error in $\widehat{\overline{F}}_{\mathrm{ML}}$ for the bootstrap sample is $E^*_{\mathrm{ML}} = (\widehat{\overline{F}}^*_{\mathrm{ML}} - \widehat{\overline{F}}_{\mathrm{ML}})$. Similarly $E^*_{\mathrm{MM}} = (\widehat{\overline{F}}^*_{\mathrm{MM}} - \widehat{\overline{F}}_{\mathrm{MM}})$.

Step-4 : Repeat the above Step 1 through Step 3 for a large number of (say M) times. The resultant values of E^*_{ML} for these replicated values are $E^{*(1)}_{\mathrm{ML}}, \ldots, E^{*(M)}_{\mathrm{ML}}$. Similarly get $E^{*(1)}_{\mathrm{MM}}, \ldots, E^{*(M)}_{\mathrm{MM}}$.

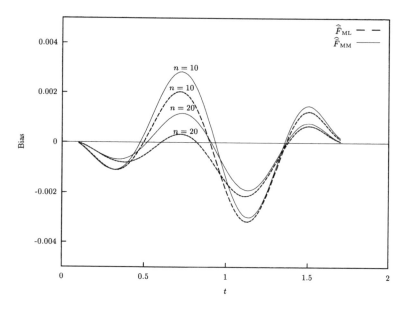

Figure 10.2.19: (b) Simulated Biases of $\widehat{\widehat{F}}_{\mathrm{ML}}$ and $\widehat{\widehat{F}}_{\mathrm{MM}}$ plotted against t with $n = 10$ and 20.

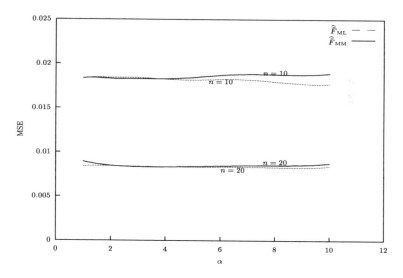

Figure 10.2.20: (a) Simulated MSEs of $\widehat{\widehat{F}}_{\mathrm{ML}}$ and $\widehat{\widehat{F}}_{\mathrm{MM}}$ plotted against α ($\beta = 1$, $t = 1.0$) with $n = 10$ and 20.

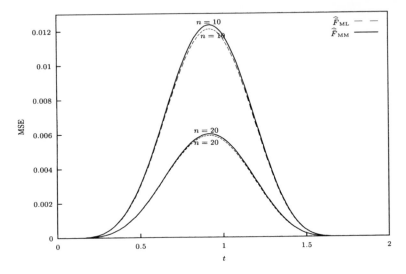

Figure 10.2.20: (b) Simulated MSEs of $\widehat{\bar{F}}_{\mathrm{ML}}$ and $\widehat{\bar{F}}_{\mathrm{MM}}$ plotted against $t > 0$ ($\alpha = 5.0$, $\beta = 1.0$) with $n = 10$ and 20.

Step-5 : The semiparametric bootstrap bias and MSE are

$$SB - Bias(\widehat{\bar{F}}_{\mathrm{ML}}|\alpha,\beta) = \frac{1}{M}\sum_{j=1}^{M}E_{\mathrm{ML}}^{*(j)}$$

and

$$SB - MSE(\widehat{\bar{F}}_{\mathrm{ML}}|\alpha,\beta) = \frac{1}{M}\sum_{j=1}^{M}(E_{\mathrm{ML}}^{*(j)})^2$$

Similarly get SB-Bias $(\widehat{\bar{F}}_{\mathrm{MM}}|\alpha,\beta))$ and SB-MSE $(\widehat{\bar{F}}_{\mathrm{MM}}|\alpha,\beta))$.

The semiparametric bias and MSE of $\widehat{\bar{F}}_{\mathrm{ML}}$ and $\widehat{\bar{F}}_{\mathrm{MM}}$ are plotted against t as shown in Figure 10.2.21 (a) and Figure 10.2.21 (b).

Remark 10.2.5 *From the above simulation studies it is clear that over-all performance of $\widehat{\bar{F}}_{ML}$ is better than that of $\widehat{\bar{F}}_{MM}$. This is not surprising for small to moderate sample sizes since the MLE $\widehat{\bar{F}}_{ML}$ of $\bar{F} = \bar{F}(t|\alpha,\beta)$ depends on $\widehat{\alpha}_{ML}$ and $\widehat{\beta}_{ML}$ which in turn depend on the data through the minimal sufficient statistics thereby ensuring that no information is lost. On the other hand, the MME $\widehat{\bar{F}}_{MM}$ of $\bar{F} = \bar{F}(t|\alpha,\beta)$ depends only on the sample moments (through $\widehat{\alpha}_{MM}$ and $\widehat{\beta}_{MM}$) which may not retain all the information necessary for (α,β) and hence for $\bar{F}$. However, for large*

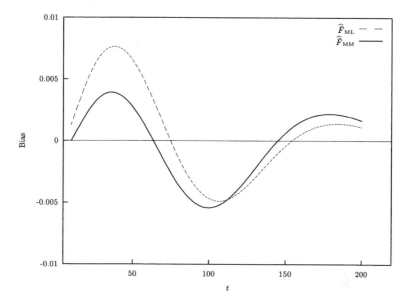

Figure 10.2.21: (a) SB-Biases of $\widehat{\overline{F}}_{\mathrm{ML}}$ and $\widehat{\overline{F}}_{\mathrm{MM}}$ (Example 10.2.3).

sample sizes this gap between the information in the data and the infor-
mation used through sample moments diminishes, and as a result $\widehat{\overline{F}}_{MM}$
can perform as good as $\widehat{\overline{F}}_{ML}$.

Parametric bootstrap method

In the following we provide the parametric bootstrap bias and MSE
of $\widehat{\overline{F}}_{\mathrm{ML}}$ for the data set in Example 10.2.3.

Step-1 : Obtain $\widehat{\alpha}_{\mathrm{ML}} = 2.10$ and $\widehat{\beta}_{\mathrm{ML}} = 81.86$ from the original data set
of size $n = 23$. Also, $\widehat{\overline{F}}_{\mathrm{ML}} = \bar{F}(t|\widehat{\alpha}_{\mathrm{ML}}, \widehat{\beta}_{\mathrm{ML}})$.

Step-2 : Generate new data $\{X_1^{**}, \ldots, X_n^{**}\}$ $(n = 23)$ *iid* from
$W(2.10, 81.86)$. Based on this new data recalculate $\widehat{\alpha}_{\mathrm{ML}}$ and $\widehat{\beta}_{\mathrm{ML}}$,
and call them $\widehat{\alpha}_{\mathrm{ML}}^{**}$ and $\widehat{\beta}_{\mathrm{ML}}^{**}$, respectively. Then $\widehat{\overline{F}}_{\mathrm{ML}}^{**} = \bar{F}(t|\widehat{\alpha}_{\mathrm{ML}}^{**}, \widehat{\beta}_{\mathrm{ML}^{**}})$.

Step-3 : The observed error in estimating $\bar{F}$ by $\widehat{\overline{F}}_{\mathrm{ML}}$ for the new sample
is $E_{\mathrm{ML}}^{**} = (\widehat{\overline{F}}_{\mathrm{ML}}^{**} - \widehat{\overline{F}}_{\mathrm{ML}})$.

Step-4 : Repeat the above Step 2 through Step 3 a large number of (say
M) times. The resultant values of E_{ML}^{**} for these replications are
$E_{\mathrm{ML}}^{**(1)}, \ldots, E_{\mathrm{ML}}^{**(M)}$.

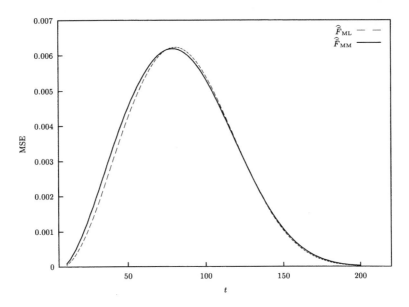

Figure 10.2.21: (b) SB-MSEs of $\widehat{\overline{F}}_{\mathrm{ML}}$ and $\widehat{\overline{F}}_{\mathrm{MM}}$ (Example 10.2.3).

Step-5 : The parametric bootstrap bias ('PB-bias') and MSE ('PB-MSE') of $\widehat{\overline{F}}_{\mathrm{ML}}$ are

$$PB - Bias(\widehat{\overline{F}}_{\mathrm{ML}}|\alpha, \beta) = \frac{1}{M} \sum_{j=1}^{M} E_{\mathrm{ML}}^{**(j)}$$

and

$$PB - MSE(\widehat{\overline{F}}_{\mathrm{ML}}|\alpha, \beta) = \frac{1}{M} \sum_{j=1}^{M} (E_{\mathrm{ML}}^{**(j)})^2$$

Using $M = 20,000$, PB-Bias and PB-MSE are plotted in Figure 10.2.22 (a) and Figure 10.2.22 (b) against t.

Case-II : **Three-Parameter Weibull Model**

We now assume a three-parameter Weibull model $W(\alpha, \beta, \mu)$ (as given in (8.1.8)) for the data in Example 10.2.3. The reliability of ball-bearing at t m.o.r. $(t > \mu)$ is (from (8.1.9))

$$\bar{F} = \bar{F}(t|\alpha, \beta, \mu) = P(X > t) = exp\{-((x - \mu)/\beta)^{\alpha}\} \qquad (10.2.30)$$

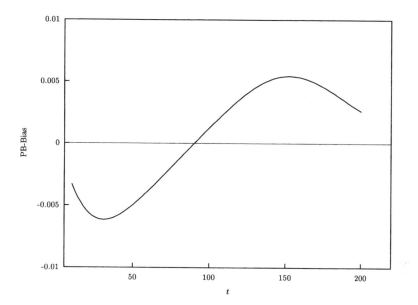

Figure 10.2.22: (a) PB-Bias of $\widehat{\overline{F}}_{\mathrm{ML}}$ (Example 10.2.3).

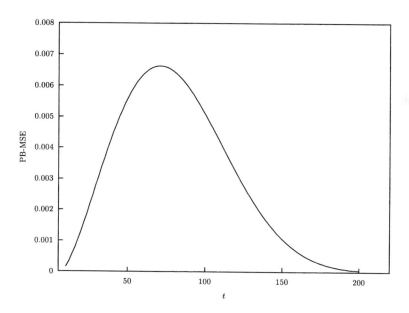

Figure 10.2.22: (b) PB-MSE of $\widehat{\overline{F}}_{\mathrm{ML}}$ (Example 10.2.3).

The following estimates of $\bar{F}$ are obtained by using suitable estimates of α, β, and μ, and these are plotted in Figure 10.2.23

$$\widehat{\bar{F}}_{\mathrm{ML}} = \begin{cases} 1 & \text{if } t \leq \widehat{\mu}_{\mathrm{ML}} \\ \bar{F}(t \mid \widehat{\alpha}_{\mathrm{ML}}, \widehat{\beta}_{\mathrm{ML}}, \widehat{\mu}_{\mathrm{ML}}) & \text{if } t > \widehat{\mu}_{\mathrm{ML}} \end{cases}$$

$$\widehat{\bar{F}}_{\mathrm{MM}} = \begin{cases} 1 & \text{if } t \leq \widehat{\mu}_{\mathrm{MM}} \\ \bar{F}(t \mid \widehat{\alpha}_{\mathrm{MM}}, \widehat{\beta}_{\mathrm{MM}}, \widehat{\mu}_{\mathrm{MM}}) & \text{if } t > \widehat{\mu}_{\mathrm{MM}} \end{cases}$$

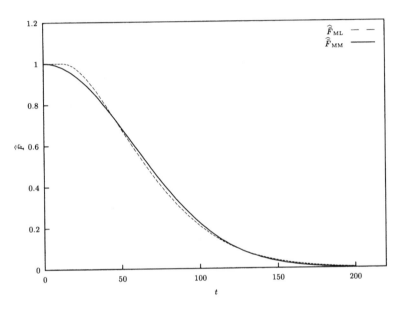

Figure 10.2.23: $\widehat{\bar{F}}_{\mathrm{ML}}$ and $\widehat{\bar{F}}_{\mathrm{MM}}$ plotted against t (Example 10.2.3).

Similar to the case of two-parameter Weibull model, simulated bias and MSE of $\widehat{\bar{F}}_{\mathrm{ML}}$ and $\widehat{\bar{F}}_{\mathrm{MM}}$ are plotted against t for fixed α, β, μ (in Figure 10.2.24 (a) and Figure 10.2.25 (a)), and against α for fixed t, β, μ (in Figures 10.2.24 (b) and Figure 10.2.25 (b)).

Remark 10.2.6 *From our extensive simulation study it appears that $\widehat{\bar{F}}_{\mathrm{ML}}$ has overall better performance than $\widehat{\bar{F}}_{\mathrm{MM}}$ for estimating $\bar{F}$ using both a two-parameter as well as a three-parameter Weibull model.*

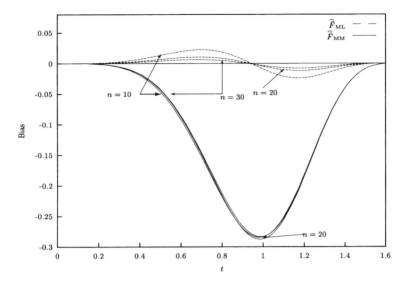

Figure 10.2.24:   (a) Simulated Biases of $\widehat{\widehat{F}}_{\text{ML}}$ and $\widehat{\widehat{F}}_{\text{MM}}$ plotted against $t > \mu$ ($\alpha = 5.0$, $\beta = 1.0$, $\mu = 0.0$).

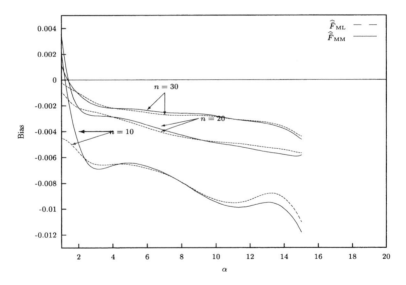

Figure 10.2.24:   (b) Simulated Biases of $\widehat{\widehat{F}}_{\text{ML}}$ and $\widehat{\widehat{F}}_{\text{MM}}$ plotted against α ($t = 1.0$, $\beta = 1.0$, $\mu = 0.0$).

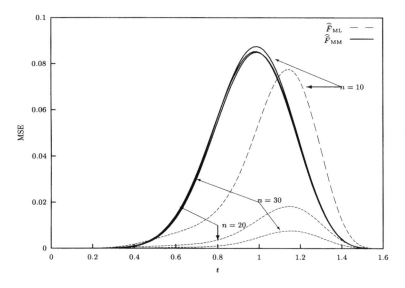

Figure 10.2.25:   (a) Simulated MSEs of $\widehat{\widehat{F}}_{\mathrm{ML}}$ and $\widehat{\widehat{F}}_{\mathrm{MM}}$ plotted against $t > \mu$ ($\alpha = 5.0$, $\beta = 1.0$, $\mu = 0.0$).

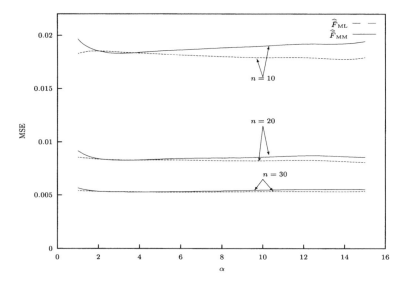

Figure 10.2.25:   (b) Simulated MSEs of $\widehat{\widehat{F}}_{\mathrm{ML}}$ and $\widehat{\widehat{F}}_{\mathrm{MM}}$ plotted against $t > \mu$ ($\alpha = 5.0$, $\beta = 1.0$, $\mu = 0.0$).

Remark 10.2.7 *Again, for the data set in Example 10.2.3, one can carry out bootstrap bias and MSE estimation of $\widehat{\bar{F}}_{ML}$ and $\widehat{\bar{F}}_{MM}$ for a three-parameter Weibull model, but we skip it for brevity.*

10.3   Reliability of a Series System with Componentwise Data

Consider a system which consists of k independent components, say $C-1, C-2, \ldots, C-k$, connected in series. The whole system works as long as all components work, and the whole system fails as soon as any one of the components fails.

Figure 10.3.1: A Series System consisting of k components.

Let the lifetime of the component $C-i$ ($1 \leq i \leq k$) follow a probability distribution, independent of the other components, which can be modelled by a *pdf* $f_i(\cdot)$ with corresponding *cdf* $F_i(\cdot)$; i.e., if X_i (≥ 0) represents the lifetime of $C-i$, then

$$P(X_i \leq t) \;=\; \int_0^t f_i(x)\,dx = F_i(t)$$
$$\text{i.e., } P(X_i > t) \;=\; 1 - F_i(t) = \bar{F}_i(t) \tag{10.3.1}$$

The series system fails as soon as any one of the components fails, and hence the reliability of the series system at time t, denoted by $\bar{F}_s(t)$, is

$$\bar{F}_s(t) \;=\; P(\text{Series System life } > t)$$
$$=\; P(X_1 > t, \ldots, X_k > t) \tag{10.3.2}$$

On the assumption that the failure time distributions of the components are independent,

$$\bar{F}_s(t) = P(X_1 > t) \ldots P(X_k > t) = \prod_{i=1}^{k} \bar{F}_i(t) \tag{10.3.3}$$

where $\bar{F}_i(t)$ is the reliability of $C - i$ at time t. Note that:

(i) Since each $\bar{F}_i(t) \leq 1$, $\bar{F}_s(t) \leq \min_{1 \leq i \leq k} \bar{F}_i(t)$; i.e, the reliability of
     a series system is much smaller than the reliability of the individual
     components

(ii) If the components have identical probability distribution then $\bar{F}_s(t) = (\bar{F}(t))^k$, where $\bar{F}(\cdot)$ is the common reliability function of the indi-
     vidual components

From the expression (10.3.3) it is clear that if each $\bar{F}_i(t)$ is not simple
(or mathematically trackable), then $\bar{F}_s(t)$ gets even more complicated.
This is probably one reason why most of the published research has fo-
cused on on $\bar{F}_s(t)$ with each X_i following an exponential distribution. It
is also customary to assume that the distributions of X_i's belong to the
same family of probability distributions. In the following we will study
$\bar{F}_s(t)$ for a few families of distributions for individual X_i's.

As mentioned earlier (in Section 10.1), we assume that we have data
on the lifetime of each component (instead of the whole system); i.e., we
have

$$X_{i1}, \ldots, X_{in_i} \quad iid \text{ with } pdf \quad f_i(x),\ 1 \leq i \leq k \qquad (10.3.4)$$

where $n = n_1 + \ldots + n_k = $ total sample size.

Components Following an Exponential Distribution

If we assume that each X_i follows an **Exp**(β_i) distribution, then $\bar{F}_i(t) = exp(-t/\beta_i)$; i.e.,

$$\bar{F}_s(t) = exp\left\{ -t\left(\frac{1}{\beta_1} + \frac{1}{\beta_2} + \ldots + \frac{1}{\beta_k}\right) \right\} \qquad (10.3.5)$$

Remark 10.3.1 *If the components are all identically distributed with a
common* **Exp**(β) *distribution, then* $\bar{F}_s(t) = exp\{-kt/\beta\}$. *The problem
then becomes drawing inferences on the reliability function* $exp(-t_*/\beta)$,
$t_* = kt$, *based on n iid observations with pdf* $(\frac{1}{\beta})exp(-x/\beta)$; *and one can
follow the method discussed in Section 10.2 with* **Exp**(β) *model.*

For the independent (and possibly nonidentical) exponential components, one can summarize the data by looking at the minimal sufficient

statistics $\mathbf{T} = (T_1, \ldots, T_k)$, where

$$T_i = \sum_{j=1}^{n_i} X_{ij} = \text{total observed lifetime of the prototypes of C-}i,$$

$$i = 1, 2, \ldots, k \qquad (10.3.6)$$

Also, it can be shown that T_i follows $G(n_i, \beta_i)$ distribution, $1 \leq i \leq k$; and they are independent. Since the MLE of β_i is $\widehat{\beta}_{i(\text{ML})} = T_i/n_i$, the MLE of $\bar{F}_s(t)$ is

$$\widehat{\bar{F}}_{s(\text{ML})}(t) = exp\{-t \sum_{i=1}^{k} (n_i/T_i)\} \qquad (10.3.7)$$

However, the UMVUE of $\bar{F}_s(t)$ is

$$\widehat{\bar{F}}_{s(\text{U})} = \prod_{i=1}^{k} \left\{ \left(1 - t/T_i\right)^{n_i-1} I\left(T_i > t\right) \right\} \qquad (10.3.8)$$

where $I(\cdot)$ denotes an indicator function. In a recent study Chandhuri, Pal, and Sarkar (1998) derived the asymptotic bias and MSE expressions of $\widehat{\bar{F}}_{s(\text{ML})}$ and $\widehat{\bar{F}}_{s(\text{U})}$, and it was shown that $\widehat{\bar{F}}_{s(\text{ML})}$ has smaller first order (i.e., in terms of expansion up to order $O(n_i^{-1})$) MSE than that of $\widehat{\bar{F}}_{s(\text{U})}$. The first order bias and MSE of $\widehat{\bar{F}}_{s(\text{ML})}$ (hence called 'FO-Bias' and 'FO-MSE') are

$$FO\text{-}Bias(\widehat{\bar{F}}_{s(\text{ML})} \mid \beta_1, \ldots, \beta_k) \approx \left(\bar{F}_s(t) \right) \sum_{i=1}^{k} t(t - 2\beta_i)/(2n_i\beta_i^2)$$

$$FO\text{-}MSE(\widehat{\bar{F}}_{s(\text{ML})} \mid \beta_1, \ldots, \beta_k) \approx \left(\bar{F}_s(t) \right)^2 t^2 \sum_{i=1}^{k} 1/(n_i\beta_i^2) \qquad (10.3.9)$$

On the other hand, the exact bias and MSE are

$$Bias(\widehat{\bar{F}}_{s(\text{ML})} \mid \beta_1, \ldots, \beta_k) =$$

$$\prod_{i=1}^{k} \left[\int_0^\infty \left\{ \Gamma(n_i)\beta_i^{n_i} \right\}^{-1} exp\left\{ - (tn_i/t_i) - (t_i/\beta_i) \right\} t_i^{n_i-1} \, dt_i \right] - \bar{F}_s(t)$$

and

$$MSE(\widehat{\bar{F}}_{s(\text{ML})} \mid \beta_1, \ldots, \beta_k) =$$

$$\prod_{i=1}^{k} \left[\int_0^\infty \left\{ \Gamma(n_i)\beta_i^{n_i} \right\}^{-1} exp\left\{ - (2tn_i/t_i) - (t_i/\beta_i) \right\} t_i^{n_i-1} \, dt_i \right]$$

$$-2\bar{F}_s(t) \cdot Bias(\widehat{\bar{F}}_{s(\text{ML})} \mid \beta_1, \ldots, \beta_k) - (\bar{F}_s(t))^2 \qquad (10.3.10)$$

(n_1, n_2)	(β_1, β_2)	FO-Bias	Bias	FO-MSE	MSE
	$(1.0, 0.1)$	0.0001	0.0002	0.0000	0.0000
	$(1.0, 0.5)$	-0.0050	-0.0035	0.0025	0.0019
$(5, 5)$	$(1.0, 1.0)$	-0.0270	-0.0220	0.0073	0.0063
	$(1.0, 2.0)$	-0.0390	-0.0340	0.0124	0.0113
	$(1.0, 0.1)$	0.0000	0.0000	0.0001	0.0000
	$(1.0, 0.5)$	-0.0050	-0.0039	0.0015	0.0013
$(5, 10)$	$(1.0, 1.0)$	-0.0203	-0.0167	0.0055	0.0048
	$(1.0, 1.5)$	-0.0273	-0.0229	0.0087	0.0077
	$(1.0, 2.0)$	-0.0307	-0.0259	0.0112	0.0099

Table 10.3.1:   A Comparison of First Order Bias and MSE of $\widehat{\bar{F}}_{s(\text{ML})}$ with the Exact Ones ($k = 2$).

For a complex system it is often desired to have a $(1 - \alpha)$ level lower confidence bound for the system reliability, i.e., one seeks a bound say, $\widehat{\bar{F}}_{s*}$ such that

$$(1 - \alpha) \leq P(\widehat{\bar{F}}_{s*} \leq \bar{F}_s(t) \leq 1) \qquad (10.3.11)$$

One can find a two-sided interval estimate of $\bar{F}_s(t)$, but since $\bar{F}_s(t)$ is always bounded above by 1, a lower confidence bound as given in (10.3.11) is enough for the interval estimate. Since $\bar{F}_s(t)$ is a monotone function of $\sum_{i=1}^{k}(1/\beta_i)$, it is sufficient to find a $(1 - \alpha)$ level upper confidence bound for it. Arazyan and Pal (2005) derived a conditional test procedure for testing a hypothesis on $\sum_{i=1}^{k}(1/\beta_i)$, and then inverted the acceptance region to get a $(1 - \alpha)$ level upper confidence bound for $\sum_{i=1}^{k}(1/\beta_i)$.

The following notations will be useful in deriving an interval estimate for $\sum_{i=1}^{k}(1/\beta_i)$, and subsequently that for $\bar{F}_s(t)$.

Notations: $\lambda = \sum_{i=1}^{k}(1/\beta_i), \quad n = \sum_{i=1}^{k} n_i; \quad k_i = n/n_i$

$$\bar{T} = \sum_{i=1}^{k} n_i T_i / n; \text{ and } U_i = n(\bar{T} - T_i)/n_i^2, \ i = 1, 2, \ldots, k$$

Result 10.3.1 *The conditional pdf of $\bar{T}$ given $(U_1, \ldots, U_{k-1})$ is $h_\lambda(\bar{t} \mid u_1, \ldots, u_{k-1})$ where*

$$h_\lambda(\bar{t} \mid u_1, \ldots, u_{k-1}) \quad \propto \quad \prod_{i=1}^{k-1} \left(\bar{t} - \frac{n_i^2}{n} u_i\right)^{(n_i-1)} \left(\bar{t} + \frac{1}{nn_k} \sum_{i=1}^{k-1} n_i^3 u_i\right)^{(n_k-1)}$$
$$\times exp(-\lambda \bar{t})$$

where $\bar{t} > \max \left\{ -\frac{1}{nn_k} \sum_{i=1}^{k-1} n_i^2 u_i, \frac{n_1^2}{n} u_1, \ldots, \frac{n_{k-1}^2}{n} u_{k-1} \right\} = t_0$ (say).

Result 10.3.2 *To test $H_0 : \lambda = \lambda_0$, a conditional test procedure is: reject H_0 if (given $U_1, \ldots, U_{k-1}$)*

$$\bar{T} > c \quad for \quad H_1 : \lambda < \lambda_0$$
$$\bar{T} < c \quad for \quad H_1 : \lambda > \lambda_0$$

The cut-off point c is such that

$$\alpha = level\ of\ the\ test = \begin{cases} \int_c^\infty h_{\lambda_0}(\bar{t} \mid u_1, \ldots, u_{k-1})\, d\bar{t} & for\ H_1 : \lambda < \lambda_0 \\[2mm] \int_{t_0}^c h_{\lambda_0}(\bar{t} \mid u_1, \ldots, u_{k-1})\, d\bar{t} & for\ H_1 : \lambda > \lambda_0 \end{cases}$$

Also, the cut-off point $c = c(\lambda_0 \mid \alpha, u_1, \ldots, u_{k-1})$ is a decreasing function of λ_0.

The above procedure is now implemented, as a demonstration, for a two component series system with componentwise data presented by Liberman and Ross (1971).

Example 10.3.1. Independent random samples of size 4 are obtained on each of two components as given below

Component-1 (or C-1): 2.0 7.1 6.0 3.0
Component-2 (or C-2): 5.0 4.0 2.0 3.1 .

Assuming exponential lifetime distribution ($\mathbf{Exp}(\beta_i)$) for each component, draw inferences on the reliability of a series system with the above two components.

For the above data set, we have $T_1 = 18.1$ and $T_2 = 14.1$. Since $\widehat{\lambda}_{\mathrm{ML}} = 0.504882$ the MLE of $\bar{F}_s(t)$ is

$$\widehat{\bar{F}}_{s(\mathrm{ML})} = exp\{-t(0.504882)\} \tag{10.3.12}$$

We now get the cut-off point $c = c(\lambda_0|\alpha, u_1, \ldots, u_{k-1})$ through numerical integration for selected values of λ_0. As a rule of thumb one can select values of λ_0 over the interval $(LB, LB + 3SE)$, where $LB =$ lower bound $= \max_i(1/\hat{\beta}_{i(\text{ML})}) = \max_i(n_i/T_i)$, and SE is the standard error of $\hat{\lambda}_{\text{ML}} = \sum_{i=1}^{k}(1/\hat{\beta}_{i(\text{ML})}) = \sum_{i=1}^{k}(n_i/T_i)$. (All we need is a rough estimate of SE so that we have an interval for λ_0.) For large n_i values,

$$SE = \sqrt{Var(\hat{\lambda}_{\text{ML}})} \approx \left\{ \sum_{i=1}^{k}(n_i\beta_i^2)^{-1} \right\}^{1/2} \qquad (10.3.13)$$

Using (10.3.13), we now obtain the values of $c = c(\lambda_0|\alpha, U_1)$ where $U_1 = -1.0$ and $\lambda_0 = 0.2837(0.01)1.185$.

We need a $(1 - \alpha)$ level upper confidence bound for λ. So, we test $H_0 : \lambda = \lambda_0$ vs. $H_A : \lambda < \lambda_0$, and the cut-off point $c = c(\lambda_0|\alpha, U_1)$ is such that

$$\alpha = \int_c^\infty h_{\lambda_0}(\bar{t}|u_1 = -1.0) \, d\bar{t} \qquad (10.3.14)$$

where $h_{\lambda_0}(\bar{t}|u_1 = -1.0)$ is available from Result 10.3.1. Also, $\bar{T} = 16.1$.

For $\alpha = 0.05$, the values of c satisfying (10.3.14) can be approximated fairly well by the nonlinear function

$$c(\lambda_0|\alpha = 0.05, u_1 = -1.0) \approx c_*(\lambda_0) = b_1 exp(-b_2\lambda_0) + b_3 \qquad (10.3.15)$$

with $b_1 = 97.33$, $b_2 = 3.83$ and $b_3 = 9.0$. By inverting $c_*(\lambda_0)$ we get a $(1 - \alpha) = 0.95$ level upper confidence bound for λ as $(0, c_*^{-1}(\bar{T}))$. [Since the conditional acceptance region for H_0 is $\bar{T} \leq c \approx c_*(\lambda_0)$, and c_* is decreasing in λ_0, we have $\lambda_0 \leq c_*^{-1}(\bar{T})$.] So, our 95% upper confidence bound for λ is 0.6827, which gives the lower 95% confidence bound for $\bar{F}_s(t)$ as $\hat{\bar{F}}_{s*} = exp(-0.6827t)$.

The following Figure 10.3.2 plots $\hat{\bar{F}}_{s(\text{ML})} = exp(-0.504882t)$ and its 95% lower confidence bound $\hat{\bar{F}}_{s*} = exp(-0.6827t)$ against $t > 0$.

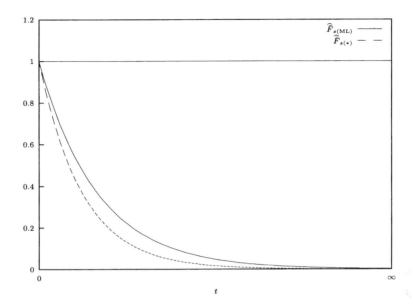

Figure 10.3.2: Graphs of $\widehat{\overline{F}}_{s(\text{ML})}$ and its 95% lower confidence bound $\widehat{\overline{F}}_{s*}$ plotted against t (Example 10.3.1).

The general methodology is given through the following steps:

Step-1 : Get the minimal sufficient statistic $\mathbf{T} = (T_1, \ldots, T_k)$ where
$T_i = \sum_{j=1}^{k} X_{ij}$, $i = 1, 2, \ldots, k$.

Step-2 : (a) $\widehat{\lambda}_{\text{ML}} = $ (MLE of $\lambda = \sum_{i=1}^{k}(1/\beta_i)) = \sum_{i=1}^{k}(n_i/T_i)$.
(b) Obtain LB(lower bound) and UB(upper bound) for λ as $LB = \max_i(n_i/T_i)$ and $UB = LB + \{\sum_{i=1}^{k}(n_i\beta_i^2)^{-1}\}^{1/2}$, with β_i replaced by $\widehat{\beta}_{i(\text{ML})}$.

Step-3 : For $\lambda_0 = (LB)(0.1)(UB)$, calculate $c = c(\lambda_0 | \alpha, u_1, \ldots, u_{k-1})$ such that

$$\alpha = \frac{\int_c^{\infty} h^*(\bar{t} | u_1, \ldots, u_{k-1}) \, d\bar{t}}{\int_{t_0}^{\infty} h^*(\bar{t} | u_1, \ldots, u_{k-1}) \, d\bar{t}}; \quad \text{where}$$

$$h^*(\bar{t} | u_1, \ldots, u_{k-1}) =$$
$$\prod_{i=1}^{k-1}(\bar{t} - \frac{n_i^2}{n} u_i)^{n_i-1}(\bar{t} + \frac{1}{nn_k}\sum_{i=1}^{k-1} n_i^3 u_i)^{(n_k-1)} exp(-\lambda \bar{t})$$

The expression for t_0 is given in Result 10.3.1.

Step-4 : Plot the c values against λ_0. Note that c, as a function of λ_0, is decreasing. Approximate $c(\lambda_0|\alpha, u_1, \ldots, u_{k-1})$ by $c_*(\lambda_0) = b_1 exp(-b_2\lambda_0) + b_3$, and obtain the values of b_1, b_2, and b_3.

Step-5 : The $(1 - \alpha)$ level upper confidence bound for λ is obtained as $\lambda \leq c_*^{-1}(\bar{T}) = -\left[ln\{(\bar{T} - b_3)/b_1\}\right]/b_2$. The $(1 - \alpha)$ level lower confidence bound for $\bar{F}_s(t)$ is $\bar{F}_s(t) \geq exp\{-tc_*^{-1}(\bar{T})\} = \widehat{\bar{F}}_{s*}$.

Components Following Gamma Distribution

If we assume that each X_i, i.e., the lifetime of the ith component (C-i) follows a $G(\alpha_i, \beta_i)$ model (see (7.1.1)) then $\bar{F}_i(t) = \int_t^\infty \{G(\alpha_i, \beta_i) \ pdf \ at \ x\}dx$, and the reliability of the whole system at time t is

$$\bar{F}_s(t) = \prod_{i=1}^{k} \left[\int_t^\infty \left\{ G(\alpha_i, \beta_i) \ pdf \ at \ x \right\} dx \right] \qquad (10.3.16)$$

Further, the MLE of $\bar{F}_s(t)$ is

$$\widehat{\bar{F}}_{s(\text{ML})}(t) = \prod_{i=1}^{k} \left[\int_t^\infty \left\{ G(\widehat{\alpha}_{i(\text{ML})}, \widehat{\beta}_{i(\text{ML})}) \ pdf \ at \ x \right\} dx \right] \qquad (10.3.17)$$

(One can use $\widehat{\alpha}_{i(\text{MML})}$ and $\widehat{\beta}_{i(\text{MML})}$ instead of $\widehat{\alpha}_{i(\text{ML})}$ and $\widehat{\beta}_{i(\text{ML})}$. The resultant estimate of $\bar{F}_s(t)$ will be denoted by $\widehat{\bar{F}}_{s(\text{MML})}$.)

The sampling distribution of $\widehat{\bar{F}}_{s(\text{ML})}$ (or that of $\widehat{\bar{F}}_{s(\text{MML})}$) is not possible to derive due to the complexity involved. Even if one tries to study the bias and MSE of $\widehat{\bar{F}}_{s(\text{ML})}$ through simulation, a complete picture is hard to comprehend if $k \geq 3$ due to the large number of parameters involved. When n_i's are sufficiently large, one can follow the bootstrap method to study the sampling distribution of $\widehat{\bar{F}}_{s(\text{ML})}$. Also, for large n_i's, $\widehat{\bar{F}}_{s(\text{ML})}$ is approximately unbiased and follows an asymptotic normal distribution which can be used for interval estimating as given below.

Define η_{i1}, η_{i2}, and η_{i3} (which are functions of α_i, β_i, and t) for $i = 1, 2, \ldots, k$ as

$$\eta_{i1} = \int_0^t \left\{ G(\alpha_i, \beta_i) \ pdf \ at \ x \right\} dx = 1 - \bar{F}_i(t) \qquad (10.3.18)$$

$$\eta_{i2} = \int_0^t \left(ln x \right) \left\{ G(\alpha_i, \beta_i) \ pdf \ at \ x \right\} dx \qquad (10.3.19)$$

$$\eta_{i3} = \int_0^t \left(lnx\right)^2 \left\{G(\alpha_i, \beta_i) \ pdf \ at \ x\right\} dx \qquad (10.3.20)$$

Define σ_i^{*2} (which is a function of α_i, β_i, and t) as

$$
\begin{aligned}
\sigma_i^{*2} \;=\; & \left\{n_i(\alpha_i\psi'(\alpha_i) - 1)\right\}^{-1}\left[\alpha_i\left\{\eta_{i2} - (ln\beta_i + \psi(\alpha_i))\eta_{i1}\right\}^2\right. \\
& +\psi'(\alpha_i)\left\{(\alpha_i - \beta_i^{-2})\eta_{i1}\right\}^2 \\
& \left. +2\left\{\eta_{i2} - (ln\beta_i + \psi(\alpha_i))\eta_{i1}\right\}\left\{(\alpha_i - \beta_i^{-2})\eta_{i1}\right\}\right] \quad (10.3.21)
\end{aligned}
$$

where $\psi(\alpha_i) = \Gamma'(\alpha_i)/\Gamma(\alpha_i)$ and $\psi'(\alpha_i) = (\partial/\partial\alpha_i)\psi(\alpha_i)$ are di-gamma and tri-gamma functions respectively. Using the standard asymptotic properties of $\widehat{\alpha}_{i(\mathrm{ML})}$ and $\widehat{\beta}_{i(\mathrm{ML})}$, $i = 1, 2, \ldots, k$, it can be shown that as each $n_i \to \infty$,

$$\left[ln\widehat{\bar{F}}_{s(\mathrm{ML})}\right] \xrightarrow{d} N\left(ln\bar{F}_s(t), \sum_{i=1}^k \left(\sigma_i^{*2}/\eta_{i1}^2\right)\right) \qquad (10.3.22)$$

Since σ_i^{*2} and η_{i1}^2 are unknown (they have unknown parameters α_i's and β_i's in them), obtain their estimates $\widehat{\sigma}_i^{*2}$ and $\widehat{\eta}_{i1}^2$ respectively by replacing α_i and β_i by $\widehat{\alpha}_{i(\mathrm{ML})}$ and $\widehat{\beta}_{i(\mathrm{ML})}$. For convenience, define

$$\widehat{\delta}^2 = \sum_{i=1}^k \left(\widehat{\sigma}_i^{*2}/\widehat{\eta}_{i1}^2\right) \qquad (10.3.23)$$

Then, an approximate $(1 - \alpha)$ level confidence bound for $\bar{F}_s(t)$ is found as (using (10.3.23))

$$\widehat{\bar{F}}_{s*} = \widehat{\bar{F}}_{s(\mathrm{ML})}exp\{-z_\alpha\widehat{\delta}\} \qquad (10.3.24)$$

where z_α is the right tail $\alpha-$probability cut-off point of the standard normal distribution.

Though the confidence bound in (10.3.24) is meant for large n_i values, it can still be used as a crude bound for small sample sizes. In the following we compute the value of $\widehat{\bar{F}}_{s*}$ for a small data set used by Draper and Guttman (1972).

Example 10.3.2. Independent samples of size 3 and 4 are obtained on lifetime (in 1000 hours) of each of two components of a series system with the following results:

Component-1 :   28.56,   22.42,   180.24
Component-2 :   35.76,   226.58,   160.08;   556.06

Assuming a $G(\alpha_i, \beta_i)$ model for component-i lifetime ($i = 1, 2$) infer about the system reliability at time $t = 100$.

Step-1 : Using the observations $\{28.56, 22.42, 180.24\}$ obtain $\widehat{\alpha}_{1(\text{ML})} = 1.228$, $\widehat{\beta}_{1(\text{ML})} = 62.742$ (solving the system of equations given in Chapter 7). Similarly, using the observations $\{35.76, 226.58, 160.08, 556.06\}$ obtain $\widehat{\alpha}_{2(\text{ML})} = 1.39$, $\widehat{\beta}_{2(\text{ML})} = 175.823$.

Step-2 : Plot $\widehat{\overline{F}}_{s(\text{ML})}$ (given in (10.3.17)) against $t > 0$. At $t = 100$, $\widehat{\overline{F}}_{s(\text{ML})} = 0.1986$.

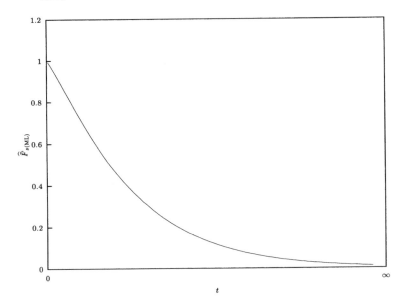

Figure 10.3.3: Plot of $\widehat{\overline{F}}_{s(\text{ML})}$ against $t > 0$ (Example 10.3.2).

Step-3 : Using $t = 100$, we have $\widehat{\eta}_{11} = 0.7284$, $\widehat{\eta}_{21} = 0.2688$; $\widehat{\eta}_{12} = 2.5013$, $\widehat{\eta}_{22} = 1.0165$; $\widehat{\eta}_{13} = 9.2758$, $\widehat{\eta}_{23} = 4.0058$. Also, $\widehat{\delta}^2 = 1.0816$.

Step-4 : The lower confidence bound for $\overline{F}_s(t)$ at $t = 100$ is $\widehat{\overline{F}}_{s*} = \widehat{\overline{F}}_{s(\text{ML})} \; exp\{-z_\alpha \widehat{\delta}\} = 0.1986 \; exp\{-1.645(1.04)\} = 0.03589$ for $\alpha = 0.05$.

Step-5 : Plot $\widehat{\overline{F}}_{s*}$ (in Step-4) against α, $0 < \alpha < 1$. (Here, $(1-\alpha)$ is the confidence level.)

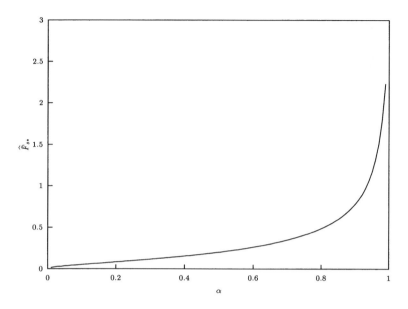

Figure 10.3.4: Plot of $\widehat{\overline{F}}_{s*}$ against α (Example 10.3.2).

10.4   Reliability of a Parallel System with Componentwise Data

In this Section we consider a system consisting of $k-$independent components say C-1, C-2, ... , C-k, connected parallelly. The whole system works as long as at least one component works, and the whole system fails as soon as all the components fail.

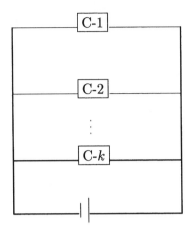

Figure 10.4.1: A parallel system consisting of k components.

If the life time X_i of C-i follows a *pdf* $f_i(\cdot)$ (with *cdf* $F_i(\cdot)$), then assuming that the components are all independent, the reliability of the parallel system at time t, denoted by $\bar{F}_p(t)$, is

$$
\begin{aligned}
\bar{F}_p(t) &= P(\text{Parallel System life } > t) \\
&= 1 - P(\text{Parallel System life } \leq t) \\
&= 1 - P(X_1 \leq t, \ldots, X_k \leq t) \\
&= 1 - \prod_{i=1}^{k} F_i(t) = 1 - \prod_{i=1}^{k}(1 - \bar{F}_i(t)) \qquad (10.4.1)
\end{aligned}
$$

Note that

(a) $\bar{F}_p(t) \geq \max_{1 \leq i \leq k} \bar{F}_i(t)$; i.e., the reliability of a parallel system is larger than the reliability of individual components

(b) If the components have identical probability distribution (with common *cdf* $F(\cdot)$ and reliability function $\bar{F}(\cdot)$), then $\bar{F}_p(t) = 1 - (1 - \bar{F}(t))^k = 1 - (F(t))^k$

Again, similar to the series system case, the expression (10.4.1) gets very complicated for most of the probability distributions. So most of the work that is available in the literature is only for the exponential distribution. We adopt the componentwise data set as presented in (10.3.4).

Components Following an Exponential Distribution

If we follow that each X_i follows an **Exp**(β_i) distribution, then $\bar{F}_i(t) = exp(-t/\beta_i)$; i.e.,

$$\bar{F}_p(t) = 1 - \prod_{i=1}^{k}\left\{1 - exp(-t/\beta_i)\right\} \qquad (10.4.2)$$

For (i) $k = 2$,

$$\bar{F}_p(t) = (e^{-t/\beta_1} + e^{-t/\beta_2}) - e^{-t(1/\beta_1 + 1/\beta_2)}$$

(ii) $k = 3$,

$$\bar{F}_p(t) = \left(e^{-t/\beta_1} + e^{-t/\beta_2} + e^{-t/\beta_3}\right) - \left(e^{-t(1/\beta_1 + 1/\beta_2)} + e^{-t(1/\beta_2 + 1/\beta_3)} + e^{-t(1/\beta_1 + 1/\beta_3)}\right) + e^{-t(1/\beta_1 + 1/\beta_2 + 1/\beta_3)}$$

In general, the expansion of (10.4.2) looks like

$$\bar{F}_p(t) = \sum_{l=1}^{k}(-1)^{l-1} \sum_{\substack{(i_1, \ldots, i_l) \in J \\ J = \{1, 2, \ldots, k\}}} exp\left\{-t(1/\beta_{i_1} + 1/\beta_{i_2} + \ldots + 1/\beta_{i_l})\right\}$$

$$(10.4.3)$$

Using the componentwise data and the notations used in (10.3.6), the MLE of $\bar{F}_p(t)$ is

$$\widehat{\bar{F}}_{p(ML)}(t) = \sum_{l=1}^{k}(-1)^{l-1} \sum_{\substack{(i_1, \ldots, i_l) \in J \\ J = \{1, 2, \ldots, k\}}} exp\left\{-t\sum_{i=1}^{l}(n_{i_j}/T_{i_j})\right\} \qquad (10.4.4)$$

where n_{i_j} and T_{i_j} are respectively the n_i and T_i values corresponding to subscript value i_j.

Similar to the case of the series system, the UMVUE of $\bar{F}_p(t)$ can be found as

$$\widehat{\bar{F}}_{p(U)} = \sum_{l=1}^{k}(-1)^{l-1} \sum_{\substack{(i_1, \ldots, i_l) \in J \\ J = \{1, 2, \ldots, k\}}} \prod_{j=1}^{l}\left\{(1 - \frac{t}{T_{i_j}})^{n_{i_j}-1}I(T_{i_j} > t)\right\}$$

$$(10.4.5)$$

For the sake of simplicity we focus on the case of $k = 2$ (i.e., a two-component parallel system with exponential lifetime for each component).

With $k = 2$,

$$\widehat{\bar{F}}_{p(\mathrm{ML})} = \left\{ e^{-tn_1/T_1} + e^{-tn_2/T_2} \right\} - e^{t((n_1/T_1)+(n_1/T_1))} \qquad (10.4.6)$$

To derive the bias and MSE of $\widehat{\bar{F}}_{p(\mathrm{ML})}$ in (10.4.6), it is convenient to adopt the following notations. Let

$$
\begin{aligned}
t_{i*} &= t/\beta_i, \quad i = 1, 2 \\
E_{1i} &= \int_0^\infty exp\left\{(-n_i t_{i*}/u_i) - u_i\right\} du_i/\Gamma(n_i), \quad i = 1, 2
\end{aligned}
$$
$$(10.4.7)$$

Then the bias of $\widehat{\bar{F}}_{p(\mathrm{ML})}$ is

$$B(\widehat{\bar{F}}_{p(\mathrm{ML})}|\beta_1, \beta_2) = \sum_{i=1}^{2} \left(E_{1i} - exp(-t_{i*}) \right) - \left\{ \prod_{i=1}^{2} E_{1i} - \prod_{i=1}^{2} exp(-t_{i*}) \right\}$$
$$(10.4.8)$$

In order to write the MSE expansion of $\widehat{\bar{F}}_{p(\mathrm{ML})}$ define

$$E_{2i} = \int_0^\infty exp\left((-2n_i t_{i*}/u_i) - u_i\right) du_i/\Gamma(n_i), \quad i = 1, 2 \qquad (10.4.9)$$

Then,

$$
\begin{aligned}
MSE(\widehat{\bar{F}}_{p(\mathrm{ML})}|\beta_1, \beta_2) &= \sum_{i=1}^{2} E_{2i} + 2 \prod_{i=1}^{2} E_{1i} + \prod_{i=1}^{2} E_{2i} - 2E_{21}E_{12} \\
&\quad - 2E_{11}E_{22} - 2\bar{F}_p(t)B(\widehat{\bar{F}}_{p(\mathrm{ML})}|\beta_1, \beta_2) - \left(\bar{F}_p(t)\right)^2
\end{aligned}
$$

Note that, using the notations $t_{i*} = t/\beta_i, \quad i = 1, 2$,

$$\bar{F}_p(t) = \sum_{i=1}^{2} exp(-t_{i*}) - \prod_{i=1}^{2} exp(-t_{i*}) \qquad (10.4.10)$$

Thus, both the bias and the MSE of $\widehat{\bar{F}}_{p(\mathrm{ML})}$ are dependent on β_1 and β_2 only through t_{1*} and t_{2*}.

In the following we plot three dimensional diagrams of $B(\widehat{\bar{F}}_{p(\mathrm{ML})}|\beta_1, \beta_2)$ and $MSE(\widehat{\bar{F}}_{p(\mathrm{ML})}|\beta_1, \beta_2)$ against t_{1*} and t_{2*}.

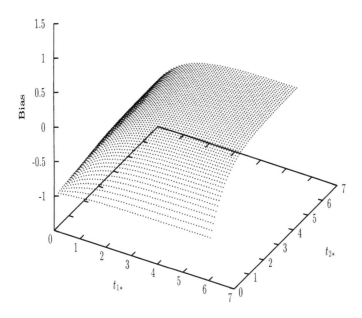

Figure 10.4.2:   (a) Bias of $\widehat{\overline{F}}_{p(\mathrm{ML})}$ for $k = 2$ and $(n_1, n_2) = (5, 10)$.

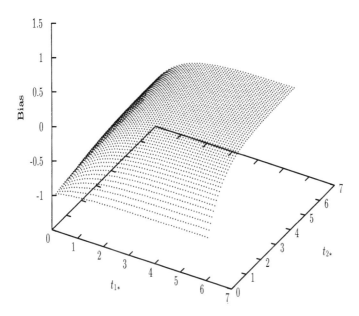

Figure 10.4.2:   (b) Bias of $\widehat{\overline{F}}_{p(\mathrm{ML})}$ for $k = 2$ and $(n_1, n_2) = (10, 10)$.

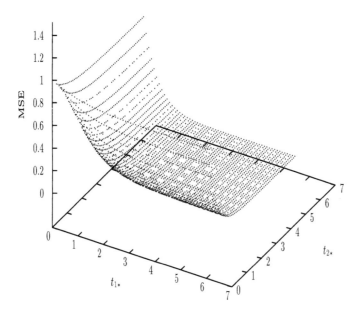

Figure 10.4.3: (a) MSE of $\widehat{\overline{F}}_{p(\mathrm{ML})}$ for $k = 2$ and $(n_1, n_2) = (5, 10)$.

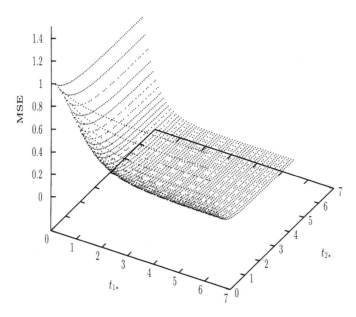

Figure 10.4.3: (b) MSE of $\widehat{\overline{F}}_{p(\mathrm{ML})}$ for $k = 2$ and $(n_1, n_2) = (10, 10)$.

Components Following Gamma Distribution

If we assume that X_i, i.e., the lifetime of the ith component (C-i) follows a $G(\alpha_i, \beta_i)$ model (in (7.1.1)) then $\bar{F}_i(t) = \int_t^\infty \{G(\alpha_i, \beta_i) \; pdf \; \text{at} \; x\} \; dx$, and the reliability of the whole parallel at time t is

$$\bar{F}_p(t) = 1 - \prod_{i=1}^{k} \left[\int_0^t \{G(\alpha_i, \beta_i) \; pdf \; \text{at} \; x\} dx \right] \qquad (10.4.11)$$

The MLE of $\bar{F}_p(t)$ is

$$\widehat{\bar{F}}_{p(ML)} = 1 - \prod_{i=1}^{k} \left[\int_0^t \{G(\widehat{\alpha}_{i(ML)}, \widehat{\beta}_{i(ML)}) \; pdf \; \text{at} \; x\} dx \right] \qquad (10.4.12)$$

One can also use $\widehat{\alpha}_{i(MML)}$ and $\widehat{\beta}_{i(MML)}$ instead of $\widehat{\alpha}_{i(ML)}$ and $\widehat{\beta}_{i(ML)}$. The resultant estimate of $\bar{F}_p(t)$ will be denoted by $\widehat{\bar{F}}_{p(MML)}(t)$.

Again, similar to the series system, the sampling distribution of $\widehat{\bar{F}}_{p(ML)}(t)$ (or $\widehat{\bar{F}}_{p(MML)}(t)$) is nearly impossible. However, for large n_i's $\widehat{\bar{F}}_{p(ML)}(t)$ (and $\widehat{\bar{F}}_{p(MML)}(t)$) is (are) approximately unbiased and has (have) asymptotic normal distribution. We will use the notations adopted in (10.3.18) to (10.3.21). It can be shown that as each $n_i \longrightarrow \infty$,

$$ln(1 - \widehat{\bar{F}}_{p(ML)}) \xrightarrow{d} N\left(ln \prod_{i=1}^{k} \eta_i, \sum_{i=1}^{k} (\sigma_i^{*2}/\eta_{i1}^2) \right) \qquad (10.4.13)$$

Note $\widehat{\delta}^2$ as defined in (10.3.23). Then an approximate $(1 - \alpha)$ level lower confidence bound for $\bar{F}_p(t)$ is found as

$$\widehat{\bar{F}}_{p*} = 1 - (1 - \widehat{\bar{F}}_{p(ML)}) exp\{z_\alpha \widehat{\delta}\} \qquad (10.4.14)$$

where z_α is the right tail α-probability cut-off point of the standard normal distribution.

Example 10.4.1. We use the data set of Example 10.3.2 for two components assuming that they are parallelly connected and each follows a gamma model.

We already have (from Step-2 of Example 10.3.2) $\widehat{\alpha}_{1(ML)} = 1.228$, $\widehat{\beta}_{1(ML)} = 62.742$; $\widehat{\alpha}_{2(ML)} = 1.390$, $\widehat{\beta}_{2(ML)} = 175.823$.

Also, using $t = 25$, we have $\widehat{\eta}_{11} = 0.2324$, $\widehat{\eta}_{21} = 0.0496$; $\widehat{\eta}_{12} = 0.5401$, $\widehat{\eta}_{22} = 0.1227$; $\widehat{\eta}_{13} = 1.4253$, $\widehat{\eta}_{23} = 0.3304$; and $\widehat{\delta}^2 = 3.100$. So $\widehat{\bar{F}}_{p(ML)} =$

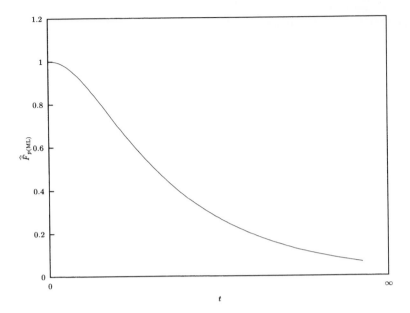

Figure 10.4.4: Plot of $\widehat{\bar{F}}_{p(\mathrm{ML})}$ as a function of t.

0.9884. For the above specific data set, the plot of the expression $\widehat{\bar{F}}_{p(\mathrm{ML})}$ as a function of t is given in Figure 10.4.4.

The lower confidence bound for $\bar{F}_p(t)$ at $t = 25$ is

$$\widehat{\bar{F}}_{p*} = 1 - (1 - \widehat{\bar{F}}_{p(\mathrm{ML})})exp(-z_\alpha\widehat{\delta}) \;\; = \;\; 1 - 0.0116\ exp(z_\alpha 1.76)$$
$$= \;\; 0.7914 \ \ (\text{using } \alpha = 0.05)$$

[We would like to point out that the above lower bound of the parallel system reliability is not very effective for small sample sizes, and can be meaningful only for moderate to large sample sizes.]

Remark 10.4.1 *Recently Chandhury, Hu, and Afshar (2001) have pro-vided a nice algorithm to calculate the confidence bounds for system re-liability which is applicable for general systems (beyond the series and parallel systems discussed here).*

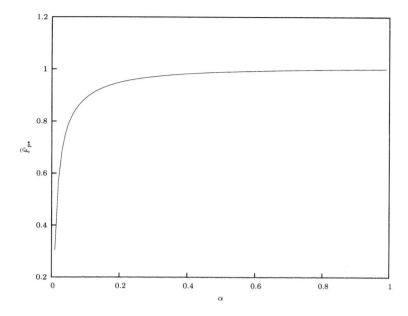

Figure 10.4.5: Plot of the lower confidence bound $\widehat{\overline{F}}_{p*}$ against α.

Bibliography

[1] Ahsanullah (1977). A Characteristic Property of the Exponential Distribution. *Annals of Statistics*, Vol. 5, 580–582.

[2] Anderson, C. W. and Ray, W. D. (1975). Improved Maximum Likelihood Estimators for the Gamma Distribution. *Communications in Statistics*, Series A (Theory and Methods), Vol. 4, No. 5, 437–448.

[3] Arazyan, A. Y. and Pal, N. (2005). On the Reliability of a Multi-component Series System. *Journal of Statistical Computation and Simulation*, Vol. 75 (to appear).

[4] Aziz, P. M. (1955). Application of the Statistical Theory of Extreme Values to the Analysis of Maximum Pit Depth Data for Aluminum. *Corrosion.* 12, 495–506.

[5] Bain, L. J. (1978). Statistical Analysis of Reliability and Life-Testing Models: Theory and Methods. Marcel Dekker Inc., New York.

[6] Ballerini, R. (1987). Another Characterization of the Type-I Extreme Value Distribution. *Statistics & Probability Letters*, 5, 83–85.

[7] Barros, V. R. and Estevan, E. A. (1983). On the Evaluation of Wind Power from Short Wind Records, *Journal of Climate and Applied Meteorology*, 22, 1116–1123.

[8] Berman, M. (1981). The Maximum Likelihood Estimators of the Parameters of the Gamma Distribution Are Always Positively Biased. *Communications in Statistics*, Series A (Theory and Methods), Vol. 10, No. 7, 693–697.

307

[9] Blom, G. (1958). *Statistical Estimates and Transformed Beta Variables*, Wiley, New York.

[10] Bougeault, P. (1982). Cloud-ensemble Relations Based on the Ability Distribution for the Higher Order Models of the Planet Layer. *Journal of the Atmospheric Sciences*, Vol. 39, 2691–2700.

[11] Carlin, J. and Haslett, J. (1982). The Probability Distribution of Wind Power from a Dispersed Array of Wind Turbine Generators, *Journal of Applied Meteorology*, 21, 303–313.

[12] Chandhuri, G., Pal, N., and Sarkar, J. (1998). Estimation of the Reliability Function of a Series System: Some Decision Theoretic Results. *Statistics*, 32, 59–74.

[13] Chandhuri, G., Hu, K., and Afshar, N. (2001). A New Approach to System Reliability, *IEEE Transactions on Reliability*, 50, No. 1, 75–84.

[14] Coe, R. and Stern, R. D. (1982). Fitting Models to Daily Rainfall Data. *Journal of the Applied Meteorology*, 21, 1024–1031.

[15] Coles, S. G. (1989). On Goodness-of-Fit Tests for the Two-Parameter Weibull Distribution Derived from the Stabilized Probability Plot, *Biometrika*, 76(3), 593–598.

[16] Cox, D. R. and Oakes, D. (1984). *Analysis of Survival Data*, London: Chapman and Hall.

[17] Dahiya, R. C. and Gurland, J. (1972). Goodness-of-Fit Tests for the Gamma and Exponential Distributions. *Technometrics*, Vol. 14, No. 3, 791–801.

[18] DasGupta, R. (1993). On the Distribution of Eccentricity, *Sankhyā: The Indian Journal of Statistics*, Series A, 55, Part-2, 226–232.

[19] DasGupta, R. and Bhanmik, D. K. (1995). Upper and Lower Tolerance Limits of Atmospheric Ozone Level and Extreme Value Distribution. *Sankhyā: The Indian Journal of Statistics*, Series B, Volume 57, Part-2, 182–199.

[20] DasGupta, R., Ghosh, J. K., and Rao, N. T. V. (1981). A Cutting Model and Distribution of Ovality and Related Topics. *Proceedings of Indian Statistical Institute Golden Jubilee Conferences*, 182–204.

[21] Desu (1971). A Characterization of the Exponential Distribution by Order Statistics. *Annals of Mathematical Statistics*, Vol. 42, No. 2, 837–838.

[22] Downton, F. (1966). Linear Estimates of Parameters in the Extreme Value Distribution. *Technometrics*, Vol. 8, No. 1, 3–17.

[23] Draper, N. and Guuttman, I. (1972). The Reliability of Independent Exponential Series Systems — A Bayesian Approach. *Technical Report #317*, Department of Statistics, University of Wisconsin-Madison.

[24] Dubey, S. D. (1965). Asymptotic Properties of Several Estimators of Weibull Parameters, *Technometrics*, 7, 423–434.

[25] Dubey, S. D. (1966). Characterization Theorems for Several Distributions and Their Applications. *Journal of Industrial Mathematics*, 16, 1–22.

[26] Eldredge, G. G. (1957). Analysis of Corrosion Pitting by Extreme Value Statistics and Its Application to Oil Well Tubing Caliper Surveys. *Corrosion*, 13, 51–76.

[27] Engelhardt, M. and Bain, L. J. (1977). Simplified Statistical Procedures for the Weibull or Extreme-Value Distribution. *Technometrics*, 19, 323–331.

[28] Fuller, W. E. (1914), Flood Flows. *Transactions of the American Society of Civil Engineering*, 77, 564.

[29] Gnedenko, B. (1943). Sur La Distribution Limite du Terme Maximum Dúne Série Aléatoire. *Annals of Mathematics*, 44, 423–453.

[30] Gumbel, E. J. (1937a). La Duée Extrême de la Vie Humaine. *Actualités Scientifiques et Industrielles*, Paris: Hermann et Cie.

[31] Gumbel, E. J. (1937b). Les Intervalles Extrêmes Entre Les Èmissions Radioactives. *Journal de Physique et le Radium*, 8, 446–452.

[32] Gumbel, E. J. (1941). The Return Period of Flood Flows. *Annals of Mathematical Statistics*, 12, 163–190.

[33] Gumbel, E. J. (1944). On the Plotting of Flood Discharges. *Transactions of the American Geophysical Union*, 25, 699–719.

[34] Gumbel, E. J. (1945). Floods Estimated by Probability Methods. *Engineering News Record*, 134, 97–101.

[35] Gumbel, E. J. (1949). The Statistical Forecast of Floods. *Bulletin No. 15*, 1–21, Ohio Water Resources Board, Columbus, Ohio.

[36] Hall, W. J. and Simons, G. (1969). On Characterizations of the Gamma Distribution. *Sankhyā The Indian Journal of Statistics*, Series A, 31, Part 4, 385–390.

[37] Harter, H. L. (1971). Some Optimization Problems in Parameter Estimation, *In Optimizing Methods in Statistics*, (J. S. Rustagi, Ed.), Academic Press, New York.

[38] Harter, H. L. and Moore, A. H. (1968). Maximum Likelihood Estimation, from Doubly Censored Samples, of the Parameters of the First Asymptotic Distribution of Extreme Values. *Journal of the American Statistical Association*, 63, 889–901.

[39] Henery, R. J. (1984). An Extreme-Value Model for Predicting the Results of Horse Races. *Applied Statistics*, 33, No. 2, 125–133.

[40] Jenkinson, A. F. (1955). The Frequency Distribution of the Annal Maximum (or Minimum) Values of Meteorological Elements. *Quarterly Journal of the Royal Meteorological Society*, 81, 158–171.

[41] Johnson, R. A. and Haskell, J. H. (1983). Sampling Properties of Estimators of a Weibull Distribution of Use in Lumber Industry, *The Canadian Journal of Statistics*, 11(2), 155–169.

[42] Kao, J. H. K. (1958). Computer Methods for Estimating Weibull Parameters in Reliability Studies, *Transactions of IRE — Reliability and Quality Control*, 13, 15–22.

[43] Kao, J. H. K. (1959). A Graphical Estimation of Mixed Weibull Parameters in Life-Testing Electron Tubes, *Technometrics*, 1, 389–407.

[44] Keating, J. P., Glaser, R. E., and Ketchum, N. S. (1990). Testing Hypothesis about the Shape Parameter of a Gamma Distribution. *Technometrics*, Vol. 32, No.1, 67–82.

[45] Khatri, C. G. and Rao, C. R. (1968). Some Characterizations of the Gamma Distribution, *Sankhyā The Indian Journal of Statistics*, Series A, 30, 157–166.

[46] Leadbetter, M. R., Lindgren, G., and Rootzén, H. (1983). Extremes and Related Properties of Random Sequences and Processes, Springer-Verlag, New York.

[47] Liberman, G. J. and Ross, S. M. (1971). Confidence Intervals for Independent Exponential Series Systems. *Journal of the American Statistical Association*, 66, 837–840.

[48] Lieblein, J. and Zelen, M. (1956). Statistical Investigation of the Fatigue Life of Deep-Grove Ball Bearings, *Journal of Research*, National Bureau of Standards, 47, 273–316.

[49] Lockhart, R. A. and Stephens, M. A. (1994). Estimation and Tests of Fit for the Three Parameter Weibull Distribution, *Journal of the Royal Statistical Society* (Series B), 56(3), 491–500.

[50] Lukacs, E. (1955). A Characterization of the Gamma Distribution. *Annals of Mathematical Statistics*, Vol. 26, 319–324.

[51] Matis, J. H., Rubink, W. L., and Makela, M. (1992). Use of the Gamma Distribution for Predicting Arrival Times of Invading Insect Populations. *Environmental Entomology*, Vol. 21, No. 3, 436–440.

[52] McDonald, J. B. and Jensen, B. C. (1979). An Analysis of Estimators of Alternative Measures of Income Inequality Associated with the Gamma Distribution Function. *Journal of the American Statistical Association*, 74, 856–860.

[53] Michael, J. R. (1983). The Stabilized Probability Plot, *Biometrika*, 17, 11–17.

[54] Öztürk, A. and Korukoğlu, S. (1988). A New Test for the Extreme Value Distribution. *Communications in Statistics, Simulation, and Computations*, 17, No. 4, 1375–1393.

[55] Pearson, K. (1933). On a Method of Determining Whether a Sample of Size n Supposed to have been Drawn from a Parent Population Having a Known Probability Integral has been Drawn at Random. *Biometrika*, 25, 379–410.

[56] Press, H. (1950). The Application of the Statistical Theory of Extreme Value to Gust-Load Problems, National Advisory Committee, Aeronautics, Technical Report 991.

[57] Proschan, F. (1963). Theoretical Explanation of Observed Decreasing Failure Rate, *Technometrics*, 5, 375–383.

[58] Pyle, J. A. (1985). Assessment Models (Chapter 12), Atmospheric Ozone (p. 653), NASA, U.S.A.

[59] Roy, D. and Mukherjee, S. P. (1986). A Note on Characterizations of the Weibull Distribution, *Sankhyā: The Indian Journal of Statistics*, Series A, 48, 250–253.

[60] Sakai, A. K. and Burris, T. A. (1985). Growth in Male and Female Aspen Clones: A Twenty-Five Year Longitudinal Study. *Ecology*, 66, 1921–1927.

[61] Schűpbach, M. and Hűsler, J. (1983). Simple Estimators for the Parameters of the Extreme-Value Distribution Based on Censored Data. *Technometrics*, 25, No. 2, 189–192.

[62] Seshadri, V. and Csörgö, M. (1969). Tests for the Exponential Distribution Using Kolmogorov-type Statistics. *Journal of the Royal Statistical Society*, Series B, 31, 499–509.

[63] Sethuraman, J. (1965). On a Characterization of the Three Limiting Types of the Extreme. *Shankhyā: The Indian Journal of Statistics*, Series A, 27, 357–364.

[64] Shapiro, S. S. and Brain, C. W. (1987). W-Test for the Weibull Distribution. *Communications in Statistics, Computation & Simulation*, Vol. 16, No. 1, 209–219.

[65] Shenton, L. R. and Bowman, K. O. (1969). Maximum Likelihood Estimator Moments for the 2-Parameter Gamma Distribution. *Shankhyā*, Series B, Vol. 31, 379–396.

[66] Shenton, L. R. and Bowman, K. O. (1972). Further Remarks on Maximum Likelihood Estimators for the Gamma Distribution. *Technometrics*, Vol. 14, 725–733.

[67] Singh, V. P. and Singh, K. (1985). Derivation of the Gamma Distribution by Using the Principle of Maximum Entropy (POME). *Water Resources Bulletin*, Vol. 21, 941–952.

[68] Smith, R. L. and Weissman, I. (1985). Maximum Likelihood Estimation of the Lower Tail of a Probability Distribution, *Journal of the Royal Statistical Society* (Series B), 47, 285–298.

[69] Stein, C. (1964). Inadmissibility of the Usual Estimators for the Variance of a Normal Distribution with Unknown Mean, *Annals of the Institute of Statistical Mathematics*, 16, 155–160.

[70] Stein, C. (1981). Estimation of the Mean of a Multivariate Normal Distribution, *Annals of Statistics*, 9, 1135–1151.

[71] Tawn, J. A. (1992). Estimating Probabilities of Extreme Sea Levels. *Applied Statistics*, 41, No. 1, 77–93.

[72] Thom, H. C. S. (1954). Frequency of Maximum Wind Speeds. *Proceedings of the American Society of Civil Engineers*, 80, 104–114.

[73] Thoman, D. R., Bain, L. J., and Antle, C. E. (1969). Inferences on the Parameter of the Weibull Distribution, *Technometrics*, 11, 445–460.

[74] Tuller, S. E. and Brett, A. C. (1984). The Characteristics of Wind Velocity that Favor the Fitting of a Weibull Distribution in Wind Speed Analysis, *Journal of Climate and Applied Meteorology*, 23, 124–134.

[75] Varian, H. R. (1975). *A Bayesian Approach to Real Estate Assessment*, Studies in Bayesian Econometrics and Statistics, North-Holland, Amstardem, 195–208.

[76] Wilks, D. S. (1989). Rainfall Intensity, the Weibull Distribution, and Estimation of Daily Surface Runoff, *Journal of Applied Meteorology*, 28, 52–58.

[77] Zanakis, S. H. and Kyparisis, J. (1986). A Review of Maximum Likelihood Estimation Methods for the Three-Parameter Weibull Distribution, *Journal of Statistical Computation and Simulation*, 25, 53–73.

==

Special Acknowledgement

The materials (data sets, partial tables, equations or other mathematical expressions) taken from:

(i) Dahiya and Gurland (1972); Engelhardt and Bain (1977); Harter and Moore (1968); Keating, Glaser and Ketchum (1990); Proschan (1963); Schűpbach and Hűsler (1983) - have been reprinted with permission from the relevant journals. Copyright by the American Statistical Association. All rights reserved.

(ii) Lockhart and Stephens (1994) - have been reprinted with permission from the Royal Statistical Society, London, UK.

(iii) Coles (1989) - have been reprinted with permission from the Oxford University Press, Oxford, UK.

==

Selected Statistical Tables

1. The Standard Normal Distribution Table

The table entry corresponding to z denotes $P(0 \leq Z \leq z)$ for a standard normal random variable Z (i.e., the shaded region = area under the $N(0,1)$ curve between 0 and z).

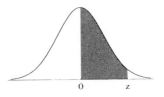

z	.00	.01	.02	.03	.04	.05	.06	.07	.08	.09
0.0	.0000	.0040	.0080	.0120	.0160	.0199	.0239	.0279	.0319	.0359
0.1	.0398	.0438	.0478	.0517	.0557	.0596	.0636	.0675	.0714	.0753
0.2	.0793	.0832	.0871	.0910	.0948	.0987	.1026	.1064	.1103	.1141
0.3	.1179	.1217	.1255	.1293	.1331	.1368	.1406	.1443	.1480	.1517
0.4	.1554	.1591	.1628	.1664	.1700	.1736	.1772	.1808	.1844	.1879
0.5	.1915	.1950	.1985	.2019	.2054	.2088	.2123	.2157	.2190	.2224
0.6	.2257	.2291	.2324	.2357	.2389	.2422	.2454	.2486	.2517	.2549
0.7	.2580	.2611	.2642	.2673	.2703	.2734	.2764	.2793	.2823	.2852
0.8	.2881	.2910	.2939	.2967	.2995	.3023	.3051	.3078	.3106	.3133
0.9	.3159	.3186	.3212	.3238	.3264	.3289	.3315	.3340	.3365	.3389
1.0	.3413	.3438	.3461	.3485	.3508	.3531	.3554	.3577	.3599	.3621
1.1	.3643	.3665	.3686	.3708	.3729	.3749	.3770	.3790	.3810	.3830
1.2	.3849	.3869	.3888	.3907	.3925	.3943	.3962	.3980	.3997	.4015
1.3	.4032	.4049	.4066	.4082	.4099	.4115	.4131	.4147	.4162	.4177
1.4	.4192	.4207	.4222	.4236	.4251	.4265	.4279	.4292	.4306	.4319
1.5	.4332	.4345	.4357	.4370	.4382	.4394	.4406	.4418	.4429	.4441
1.6	.4452	.4463	.4474	.4484	.4495	.4505	.4515	.4525	.4535	.4545
1.7	.4554	.4564	.4573	.4582	.4591	.4599	.4608	.4616	.4625	.4633
1.8	.4641	.4649	.4656	.4664	.4671	.4678	.4686	.4693	.4699	.4706
1.9	.4713	.4719	.4726	.4732	.4738	.4744	.4750	.4756	.4761	.4767
2.0	.4772	.4778	.4783	.4788	.4793	.4798	.4803	.4808	.4812	.4817
2.1	.4821	.4826	.4830	.4834	.4838	.4842	.4846	.4850	.4854	.4857
2.2	.4861	.4864	.4868	.4871	.4875	.4878	.4881	.4884	.4887	.4890
2.3	.4893	.4896	.4898	.4901	.4904	.4906	.4909	.4911	.4913	.4916
2.4	.4918	.4920	.4922	.4924	.4927	.4929	.4931	.4932	.4934	.4936
2.5	.4938	.4940	.4941	.4943	.4945	.4946	.4948	.4949	.4951	.4952
2.6	.4953	.4955	.4956	.4957	.4959	.4960	.4961	.4962	.4963	.4964
2.7	.4965	.4966	.4967	.4968	.4969	.4970	.4971	.4972	.4973	.4974
2.8	.4974	.4975	.4976	.4977	.4977	.4978	.4979	.4979	.4980	.4981
2.9	.4981	.4982	.4982	.4983	.4984	.4984	.4985	.4985	.4986	.4986
3.0	.4986	.4987	.4987	.4988	.4988	.4989	.4989	.4989	.4990	.4990
3.1	.4990	.4991	.4991	.4991	.4992	.4992	.4992	.4992	.4993	.4993
3.2	.4993	.4993	.4994	.4994	.4994	.4994	.4994	.4995	.4995	.4995

2. (Student's) t-Distribution Table

The table entries (i.e., $t_{\nu,\alpha}$ values) are the right tail α probability cut-off points for the t_ν-distribution, where $\nu = df$ (i.e., area under the t_ν-pdf on the right of $t_{\nu,\alpha}$ is α).

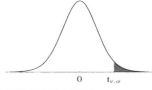

df	0.150	0.100	0.075	0.050	0.025	0.010	0.005
1	1.9626	3.0777	4.1653	6.3137	12.7062	31.8210	63.6559
2	1.3862	1.8856	2.2819	2.9200	4.3027	6.9645	9.9250
3	1.2498	1.6377	1.9243	2.3534	3.1824	4.5407	5.8408
4	1.1896	1.5332	1.7782	2.1318	2.7765	3.7469	4.6041
5	1.1558	1.4759	1.6994	2.0150	2.5706	3.3649	4.0321
6	1.1342	1.4398	1.6502	1.9432	2.4469	3.1427	3.7074
7	1.1192	1.4149	1.6166	1.8946	2.3646	2.9979	3.4995
8	1.1081	1.3968	1.5922	1.8595	2.3060	2.8965	3.3554
9	1.0997	1.3830	1.5737	1.8331	2.2622	2.8214	3.2498
10	1.0931	1.3722	1.5592	1.8125	2.2281	2.7638	3.1693
11	1.0877	1.3634	1.5476	1.7959	2.2010	2.7181	3.1058
12	1.0832	1.3562	1.5380	1.7823	2.1788	2.6810	3.0545
13	1.0795	1.3502	1.5299	1.7709	2.1604	2.6503	3.0123
14	1.0763	1.3450	1.5231	1.7613	2.1448	2.6245	2.9768
15	1.0735	1.3406	1.5172	1.7531	2.1315	2.6025	2.9467
16	1.0711	1.3368	1.5121	1.7459	2.1199	2.5835	2.9208
17	1.0690	1.3334	1.5077	1.7396	2.1098	2.5669	2.8982
18	1.0672	1.3304	1.5037	1.7341	2.1009	2.5524	2.8784
19	1.0655	1.3277	1.5002	1.7291	2.0930	2.5395	2.8609
20	1.0640	1.3253	1.4970	1.7247	2.0860	2.5280	2.8453
21	1.0627	1.3232	1.4942	1.7207	2.0796	2.5176	2.8314
22	1.0614	1.3212	1.4916	1.7171	2.0739	2.5083	2.8188
23	1.0603	1.3195	1.4893	1.7139	2.0687	2.4999	2.8073
24	1.0593	1.3178	1.4871	1.7109	2.0639	2.4922	2.7970
25	1.0584	1.3163	1.4852	1.7081	2.0595	2.4851	2.7874
26	1.0575	1.3150	1.4834	1.7056	2.0555	2.4786	2.7787
27	1.0567	1.3137	1.4817	1.7033	2.0518	2.4727	2.7707
28	1.0560	1.3125	1.4801	1.7011	2.0484	2.4671	2.7633
29	1.0553	1.3114	1.4787	1.6991	2.0452	2.4620	2.7564
30	1.0547	1.3104	1.4774	1.6973	2.0423	2.4573	2.7500
40	1.0500	1.3031	1.4677	1.6839	2.0211	2.4233	2.7045
50	1.0473	1.2987	1.4620	1.6759	2.0086	2.4033	2.6778
60	1.0455	1.2958	1.4582	1.6706	2.0003	2.3901	2.6603
70	1.0442	1.2938	1.4555	1.6669	1.9944	2.3808	2.6479
80	1.0432	1.2922	1.4535	1.6641	1.9901	2.3739	2.6387
90	1.0424	1.2910	1.4519	1.6620	1.9867	2.3685	2.6316
100	1.0418	1.2901	1.4507	1.6602	1.9840	2.3642	2.6259
120	1.0409	1.2886	1.4488	1.6577	1.9799	2.3578	2.6174
∞	1.0364	1.2816	1.4395	1.6449	1.9600	2.3263	2.5758

3. Chi-Square Distribution Table

The table entries (i.e., $\chi^2_{\nu,\alpha}$ values) are the
right tail α probability cut-off points for the
χ^2_ν-distribution, where $\nu = df$ (i.e., area under the χ^2_ν-pdf on the right of $\chi^2_{\nu,\alpha}$ is α).

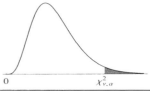

$df = \nu$	α						
	0.150	0.100	0.075	0.050	0.025	0.010	0.005
1	2.0723	2.7055	3.1701	3.8415	5.0239	6.6349	7.8794
2	3.7942	4.6052	5.1805	5.9915	7.3778	9.2103	10.5966
3	5.3170	6.2514	6.9046	7.8147	9.3484	11.3449	12.8382
4	6.7449	7.7794	8.4963	9.4877	11.1433	13.2767	14.8603
5	8.1152	9.2364	10.0083	11.0705	12.8325	15.0863	16.7496
6	9.4461	10.6446	11.4659	12.5916	14.4494	16.8119	18.5476
7	10.7479	12.0170	12.8834	14.0671	16.0128	18.4753	20.2777
8	12.0271	13.3616	14.2697	15.5073	17.5345	20.0902	21.9550
9	13.2880	14.6837	15.6309	16.9190	19.0228	21.6660	23.5894
10	14.5339	15.9872	16.9714	18.3070	20.4832	23.2093	25.1882
11	15.7671	17.2750	18.2942	19.6751	21.9201	24.7250	26.7569
12	16.9893	18.5494	19.6020	21.0261	23.3367	26.2170	28.2995
13	18.2020	19.8119	20.8966	22.3620	24.7356	27.6883	29.8195
14	19.4062	21.0641	22.1795	23.6848	26.1190	29.1412	31.3194
15	20.6030	22.3071	23.4522	24.9958	27.4884	30.5779	32.8013
16	21.7931	23.5418	24.7155	26.2962	28.8454	31.9999	34.2672
17	22.9770	24.7690	25.9705	27.5871	30.1910	33.4087	35.7185
18	24.1555	25.9894	27.2178	28.8693	31.5264	34.8053	37.1565
19	25.3289	27.2036	28.4581	30.1435	32.8523	36.1909	38.5823
20	26.4976	28.4120	29.6920	31.4104	34.1696	37.5662	39.9969
21	27.6620	29.6151	30.9200	32.6706	35.4789	38.9322	41.4011
22	28.8225	30.8133	32.1424	33.9244	36.7807	40.2894	42.7957
23	29.9792	32.0069	33.3597	35.1725	38.0756	41.6384	44.1813
24	31.1325	33.1962	34.5723	36.4150	39.3641	42.9798	45.5585
25	32.2825	34.3816	35.7803	37.6525	40.6465	44.3141	46.9279
26	33.4295	35.5632	36.9841	38.8851	41.9232	45.6417	48.2899
27	34.5736	36.7412	38.1840	40.1133	43.1945	46.9629	49.6449
28	35.7150	37.9159	39.3801	41.3371	44.4608	48.2782	50.9934
29	36.8538	39.0875	40.5727	42.5570	45.7223	49.5879	52.3356
30	37.9903	40.2560	41.7619	43.7730	46.9792	50.8922	53.6720
40	49.2439	51.8051	53.5010	55.7585	59.3417	63.6907	66.7660
50	60.3460	63.1671	65.0303	67.5048	71.4202	76.1539	79.4900
60	71.3411	74.3970	76.4113	79.0819	83.2977	88.3794	91.9517
70	82.2554	85.5270	87.6802	90.5312	95.0232	100.4252	104.2149
80	93.1058	96.5782	98.8606	101.8795	106.6286	112.3288	116.3211
90	103.9041	107.5650	109.9688	113.1453	118.1359	124.1163	128.2989
100	114.6588	118.4980	121.0166	124.3421	129.5612	135.8067	140.1695
120	136.0620	140.2326	142.9646	146.5674	152.2114	158.9502	163.6482

3. Chi-Square Distribution Table (cont.)

The table entries (i.e., $\chi^2_{\nu,\alpha}$ values) are the right tail α probability cut-off points for the χ^2_ν-distribution, where $\nu = df$ (i.e., area under the χ^2_ν-pdf on the right of $\chi^2_{\nu,\alpha}$ is α).

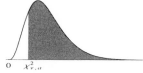

$df = \nu$	α						
	0.995	0.990	0.975	0.950	0.925	0.900	0.850
1	0.0000	0.0002	0.0010	0.0039	0.0089	0.0158	0.0358
2	0.0100	0.0201	0.0506	0.1026	0.1559	0.2107	0.3250
3	0.0717	0.1148	0.2158	0.3519	0.4720	0.5844	0.7978
4	0.2070	0.2971	0.4844	0.7107	0.8969	1.0636	1.3665
5	0.4117	0.5543	0.8312	1.1455	1.3937	1.6103	1.9938
6	0.6757	0.8721	1.2373	1.6354	1.9415	2.2041	2.6613
7	0.9893	1.2390	1.6899	2.1673	2.5277	2.8331	3.3583
8	1.3444	1.6465	2.1797	2.7326	3.1440	3.4895	4.0782
9	1.7349	2.0879	2.7004	3.3251	3.7847	4.1682	4.8165
10	2.1559	2.5582	3.2470	3.9403	4.4459	4.8652	5.5701
11	2.6032	3.0535	3.8158	4.5748	5.1243	5.5778	6.3364
12	3.0738	3.5706	4.4038	5.2260	5.8175	6.3038	7.1138
13	3.5650	4.1069	5.0088	5.8919	6.5238	7.0415	7.9008
14	4.0747	4.6604	5.6287	6.5706	7.2415	7.7895	8.6963
15	4.6009	5.2294	6.2621	7.2609	7.9695	8.5468	9.4993
16	5.1422	5.8122	6.9077	7.9617	8.7067	9.3122	10.3090
17	5.6972	6.4078	7.5642	8.6718	9.4522	10.0852	11.1249
18	6.2648	7.0149	8.2308	9.3905	10.2053	10.8649	11.9463
19	6.8440	7.6327	8.9065	10.1170	10.9653	11.6509	12.7727
20	7.4338	8.2604	9.5908	10.8508	11.7317	12.4426	13.6039
21	8.0336	8.8972	10.2829	11.5913	12.5041	13.2396	14.4393
22	8.6427	9.5425	10.9823	12.3380	13.2819	14.0415	15.2788
23	9.2604	10.1957	11.6885	13.0905	14.0648	14.8480	16.1219
24	9.8862	10.8564	12.4011	13.8484	14.8525	15.6587	16.9686
25	10.5197	11.5240	13.1197	14.6114	15.6447	16.4734	17.8184
26	11.1602	12.1982	13.8439	15.3792	16.4410	17.2919	18.6714
27	11.8076	12.8785	14.5734	16.1514	17.2414	18.1139	19.5272
28	12.4613	13.5647	15.3079	16.9279	18.0454	18.9392	20.3857
29	13.1212	14.2564	16.0471	17.7084	18.8530	19.7677	21.2468
30	13.7867	14.9535	16.7908	18.4927	19.6639	20.5992	22.1103
40	20.7065	22.1643	24.4330	26.5093	27.9258	29.0505	30.8563
50	27.9907	29.7067	32.3574	34.7643	36.3971	37.6886	30.8563
60	35.5345	37.4849	40.4817	3.1880	45.0165	46.4589	48.7587
70	43.2752	45.4417	48.7576	51.7393	53.7478	55.3289	57.8443
80	51.1719	53.5401	57.1532	60.3915	62.5676	64.2778	66.9938
90	59.1963	61.7541	65.6466	69.1260	71.4596	73.2911	76.1954
100	67.3276	70.0649	74.2219	77.9295	80.4119	82.3581	85.4406
120	83.8516	86.9233	91.5726	95.7046	98.4641	100.6236	104.0374

4. (Snedecor's) F-Distribution Table

The table entries (i.e., $F_{\nu_1,\nu_2,\alpha}$ values) are
the right tail α probability cut-off points for
the F_{ν_1,ν_2}-distribution, where ν_1 = numerator df and ν_2 = denominator df (i.e., area
under the F_{ν_1,ν_2}-pdf on the right of $F_{\nu_1,\nu_2,\alpha}$
is α).

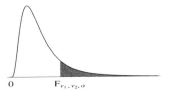

Denominator	$\alpha = 0.05$						
	Numerator $df = \nu_1$						
$df = \nu_2$	1	2	3	4	5	6	7
1	161.4476	199.5000	215.7073	224.5832	230.1619	233.9860	236.7684
2	18.5128	19.0000	19.1643	19.2468	19.2964	19.3295	19.3532
3	10.1280	9.5521	9.2766	9.1172	9.0135	8.9406	8.8867
4	7.7086	6.9443	6.5914	6.3882	6.2561	6.1631	6.0942
5	6.6079	5.7861	5.4095	5.1922	5.0503	4.9503	4.8759
6	5.9874	5.1433	4.7571	4.5337	4.3874	4.2839	4.2067
7	5.5914	4.7374	4.3468	4.1203	3.9715	3.8660	3.7870
8	5.3177	4.4590	4.0662	3.8379	3.6875	3.5806	3.5005
9	5.1174	4.2565	3.8625	3.6331	3.4817	3.3738	3.2927
10	4.9646	4.1028	3.7083	3.4780	3.3258	3.2172	3.1355
11	4.8443	3.9823	3.5874	3.3567	3.2039	3.0946	3.0123
12	4.7472	3.8853	3.4903	3.2592	3.1059	2.9961	2.9134
13	4.6672	3.8056	3.4105	3.1791	3.0254	2.9153	2.8321
14	4.6001	3.7389	3.3439	3.1122	2.9582	2.8477	2.7642
15	4.5431	3.6823	3.2874	3.0556	2.9013	2.7905	2.7066
16	4.4940	3.6337	3.2389	3.0069	2.8524	2.7413	2.6572
17	4.4513	3.5915	3.1968	2.9647	2.8100	2.6987	2.6143
18	4.4139	3.5546	3.1599	2.9277	2.7729	2.6613	2.5767
19	4.3807	3.5219	3.1274	2.8951	2.7401	2.6283	2.5435
20	4.3512	3.4928	3.0984	2.8661	2.7109	2.5990	2.5140
30	4.1709	3.3158	2.9223	2.6896	2.5336	2.4205	2.3343
40	4.0847	3.2317	2.8387	2.6060	2.4495	2.3359	2.2490
50	4.0343	3.1826	2.7900	2.5572	2.4004	2.2864	2.1992
60	4.0012	3.1504	2.7581	2.5252	2.3683	2.2541	2.1665
80	3.9604	3.1108	2.7188	2.4859	2.3287	2.2142	2.1263
100	3.9361	3.0873	2.6955	2.4626	2.3053	2.1906	2.1025
120	3.9201	3.0718	2.6802	2.4472	2.2899	2.1750	2.0868
∞	3.8415	2.9957	2.6049	2.3719	2.2141	2.0986	2.0096

4. (Snedecor's) F-Distribution Table (cont.)

The table entries (i.e., $F_{\nu_1,\nu_2,\alpha}$ values) are
the right tail α probability cut-off points for
the F_{ν_1,ν_2}-distribution, where $\nu_1 =$ numerator df and $\nu_2 =$ denominator df (i.e., area
under the F_{ν_1,ν_2}-pdf on the right of $F_{\nu_1,\nu_2,\alpha}$
is α).

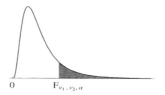

| Denominator | $\alpha = 0.05$ | | | | | | |
| | Numerator $df = \nu_1$ | | | | | | |
$df = \nu_2$	8	9	10	11	12	13	14
1	238.883	240.543	241.882	242.983	243.906	244.690	245.364
2	19.3710	19.3848	19.3959	19.4050	19.4125	19.4189	19.4244
3	8.8452	8.8123	8.7855	8.7633	8.7446	8.7287	8.7149
4	6.0410	5.9988	5.9644	5.9358	5.9117	5.8911	5.8733
5	4.8183	4.7725	4.7351	4.7040	4.6777	4.6552	4.6358
6	4.1468	4.0990	4.0600	4.0274	3.9999	3.9764	3.9559
7	3.7257	3.6767	3.6365	3.6030	3.5747	3.5503	3.5292
8	3.4381	3.3881	3.3472	3.3130	3.2839	3.2590	3.2374
9	3.2296	3.1789	3.1373	3.1025	3.0729	3.0475	3.0255
10	3.0717	3.0204	2.9782	2.9430	2.9130	2.8872	2.8647
11	2.9480	2.8962	2.8536	2.8179	2.7876	2.7614	2.7386
12	2.8486	2.7964	2.7534	2.7173	2.6866	2.6602	2.6371
13	2.7669	2.7144	2.6710	2.6347	2.6037	2.5769	2.5536
14	2.6987	2.6458	2.6022	2.5655	2.5342	2.5073	2.4837
15	2.6408	2.5876	2.5437	2.5068	2.4753	2.4481	2.4244
16	2.5911	2.5377	2.4935	2.4564	2.4247	2.3973	2.3733
17	2.5480	2.4943	2.4499	2.4126	2.3807	2.3531	2.3290
18	2.5102	2.4563	2.4117	2.3742	2.3421	2.3143	2.2900
19	2.4768	2.4227	2.3779	2.3402	2.3080	2.2800	2.2556
20	2.4471	2.3928	2.3479	2.3100	2.2776	2.2495	2.2250
30	2.2662	2.2107	2.1646	2.2829	2.0921	2.0630	2.0374
40	2.1802	2.1240	2.0772	2.2585	2.0035	1.9738	1.9476
50	2.1299	2.0734	2.0261	1.9861	1.9515	1.9214	1.8949
60	2.0970	2.0401	1.9926	1.9522	1.9174	1.8870	1.8602
80	2.0564	1.9991	1.9512	1.9105	1.8753	1.8445	1.8174
100	2.0323	1.9748	1.9267	1.8857	1.8503	1.8193	1.7919
120	2.0164	1.9588	1.9105	1.8693	1.8337	1.8026	1.7750
∞	1.9384	1.8799	1.8307	1.7886	1.7522	1.7202	1.6918

4. (Snedecor's) F-Distribution Table (cont.)

The table entries (i.e., $F_{\nu_1,\nu_2,\alpha}$ values) are the right tail α probability cut-off points for the F_{ν_1,ν_2}-distribution, where ν_1 = numerator df and ν_2 = denominator df (i.e., area under the F_{ν_1,ν_2}-pdf on the right of $F_{\nu_1,\nu_2,\alpha}$ is α).

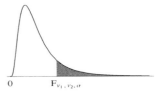

| Denominator | $\alpha = 0.05$ | | | | | | |
| | Numerator $df = \nu_1$ | | | | | | |
$df = \nu_2$	15	16	17	18	19	20	30
1	245.9499	246.4639	246.9184	247.3232	247.6861	248.0131	250.0951
2	19.4291	19.4333	19.4370	19.4402	19.4431	19.4458	19.4624
3	8.7029	8.6923	8.6829	8.6745	8.6670	8.6602	8.6166
4	5.8578	5.8441	5.8320	5.8211	5.8114	5.8025	5.7459
5	4.6188	4.6038	4.5904	4.5785	4.5678	4.5581	4.4957
6	3.9381	3.9223	3.9083	3.8957	3.8844	3.8742	3.8082
7	3.5107	3.4944	3.4799	3.4669	3.4551	3.4445	3.3758
8	3.2184	3.2016	3.1867	3.1733	3.1613	3.1503	3.0794
9	3.0061	2.9890	2.9737	2.9600	2.9477	2.9365	2.8637
10	2.8450	2.8276	2.8120	2.7980	2.7854	2.7740	2.6996
11	2.7186	2.7009	2.6851	2.6709	2.6581	2.6464	2.5705
12	2.6169	2.5989	2.5828	2.5684	2.5554	2.5436	2.4663
13	2.5331	2.5149	2.4987	2.4841	2.4709	2.4589	2.3803
14	2.4630	2.4446	2.4282	2.4134	2.4000	2.3879	2.3082
15	2.4034	2.3849	2.3680	2.3533	2.3398	2.3275	2.2468
16	2.3522	2.3335	2.3167	2.3016	2.2880	2.2756	2.1938
17	2.3077	2.2888	2.2719	2.2567	2.2429	2.2304	2.1477
18	2.2686	2.2496	2.2325	2.2172	2.2033	2.1906	2.1071
19	2.2341	2.2149	2.1977	2.1823	2.1683	2.1555	2.0712
20	2.2033	2.1840	2.1667	2.1511	2.1370	2.1242	2.0391
30	2.0148	1.9946	1.9765	1.9601	1.9452	1.9317	1.8409
40	1.9245	1.9037	1.8851	1.8682	1.8529	1.8389	1.7444
50	1.8714	1.8503	1.8313	1.8141	1.7985	1.7841	1.6872
60	1.8364	1.8151	1.7959	1.7784	1.7625	1.7480	1.6491
80	1.7932	1.7716	1.7520	1.7342	1.7180	1.7032	1.6017
100	1.7675	1.7456	1.7259	1.7079	1.6915	1.6764	1.5733
120	1.7505	1.7285	1.7085	1.6904	1.6739	1.6587	1.5543
∞	1.6664	1.6435	1.6228	1.6038	1.5865	1.5705	1.4591

4. (Snedecor's) F-Distribution Table (cont.)

The table entries (i.e., $F_{\nu_1,\nu_2,\alpha}$ values) are the right tail α probability cut-off points for the F_{ν_1,ν_2}-distribution, where ν_1 = numerator df and ν_2 = denominator df (i.e., area under the F_{ν_1,ν_2}-pdf on the right of $F_{\nu_1,\nu_2,\alpha}$ is α).

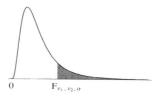

| Denominator | $\alpha = 0.05$ | | | | | | |
| $df = \nu_2$ | Numerator $df = \nu_1$ | | | | | | |
	40	50	60	80	100	120	∞
1	251.1432	251.7742	252.1957	252.7237	253.0411	253.2529	254.3144
2	19.4707	19.4757	19.4791	19.4832	19.4857	19.4874	19.4957
3	8.5944	8.5810	8.5720	8.5607	8.5539	8.5494	8.5264
4	5.7170	5.6995	5.6877	5.6730	5.6641	5.6581	5.6281
5	4.4638	4.4444	4.4314	4.4150	4.4051	4.3985	4.3650
6	3.7743	3.7537	3.7398	3.7223	3.7117	3.7047	3.6689
7	3.3404	3.3189	3.3043	3.2860	3.2749	3.2674	3.2298
8	3.0428	3.0204	3.0053	2.9862	2.9747	2.9669	2.9276
9	2.8259	2.8028	2.7872	2.7675	2.7556	2.7475	2.7067
10	2.6609	2.6371	2.6211	2.6008	2.5884	2.5801	2.5379
11	2.5309	2.5066	2.4901	2.4692	2.4566	2.4480	2.4045
12	2.4259	2.4010	2.3842	2.3628	2.3498	2.3410	2.2962
13	2.3392	2.3138	2.2966	2.2747	2.2614	2.2524	2.2064
14	2.2664	2.2405	2.2229	2.2006	2.1870	2.1778	2.1307
15	2.2043	2.1780	2.1601	2.1373	2.1234	2.1141	2.0658
16	2.1507	2.1240	2.1058	2.0826	2.0685	2.0589	2.0096
17	2.1040	2.0769	2.0584	2.0348	2.0204	2.0107	1.9604
18	2.0629	2.0354	2.0166	1.9927	1.9780	1.9681	1.9168
19	2.0264	1.9986	1.9795	1.9552	1.9403	1.9302	1.8780
20	1.9938	1.9656	1.9464	1.9217	1.9066	1.8963	1.8432
30	1.7918	1.7609	1.7396	1.7121	1.6950	1.6835	1.6223
40	1.6928	1.6600	1.6373	1.6077	1.5892	1.5766	1.5089
50	1.6337	1.5995	1.5757	1.5445	1.5249	1.5115	1.4383
60	1.5943	1.5590	1.5343	1.5019	1.4814	1.4673	1.3893
80	1.5449	1.5081	1.4821	1.4477	1.4259	1.4107	1.3247
100	1.5151	1.4772	1.4504	1.4146	1.3917	1.3757	1.2832
120	1.4952	1.4565	1.4290	1.3922	1.3685	1.3519	1.2539
∞	1.3940	1.3501	1.3180	1.2735	1.2434	1.2214	1.0000

4. (Snedecor's) F-Distribution Table (cont.)

The table entries (i.e., $F_{\nu_1,\nu_2,\alpha}$ values) are the right tail α probability cut-off points for the F_{ν_1,ν_2}-distribution, where ν_1 = numerator df and ν_2 = denominator df (i.e., area under the F_{ν_1,ν_2}-pdf on the right of $F_{\nu_1,\nu_2,\alpha}$ is α).

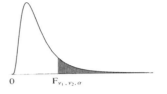

Denominator	$\alpha = 0.025$						
	Numerator $df = \nu_1$						
$df = \nu_2$	1	2	3	4	5	6	7
1	647.7890	799.5000	864.1630	899.5833	921.8479	937.1111	948.2169
2	38.5063	39.0000	39.1655	39.2484	39.2982	39.3315	39.3552
3	17.4434	16.0441	15.4392	15.1010	14.8848	14.7347	14.6244
4	12.2179	10.6491	9.9792	9.6045	9.3645	9.1973	9.0741
5	10.0070	8.4336	7.7636	7.3879	7.1464	6.9777	6.8531
6	8.8131	7.2599	6.5988	6.2272	5.9876	5.8198	5.6955
7	8.0727	6.5415	5.8898	5.5226	5.2852	5.1186	4.9949
8	7.5709	6.0595	5.4160	5.0526	4.8173	4.6517	4.5286
9	7.2093	5.7147	5.0781	4.7181	4.4844	4.3197	4.1970
10	6.9367	5.4564	4.8256	4.4683	4.2361	4.0721	3.9498
11	6.7241	5.2559	4.6300	4.2751	4.0440	3.8807	3.7586
12	6.5538	5.0959	4.4742	4.1212	3.8911	3.7283	3.6065
13	6.4143	4.9653	4.3472	3.9959	3.7667	3.6043	3.4827
14	6.2979	4.8567	4.2417	3.8919	3.6634	3.5014	3.3799
15	6.1995	4.7650	4.1528	3.8043	3.5764	3.4147	3.2934
16	6.1151	4.6867	4.0768	3.7294	3.5021	3.3406	3.2194
17	6.0420	4.6189	4.0112	3.6648	3.4379	3.2767	3.1556
18	5.9781	4.5597	3.9539	3.6083	3.3820	3.2209	3.0999
19	5.9216	4.5075	3.9034	3.5587	3.3327	3.1718	3.0509
20	5.8715	4.4613	3.8587	3.5147	3.2891	3.1283	3.0074
30	5.5675	4.1821	3.5894	3.2499	3.0265	2.8667	2.7460
40	5.4239	4.0510	3.4633	3.1261	2.9037	2.7444	2.6238
50	5.3403	3.9749	3.3902	3.0544	2.8327	2.6736	2.5530
60	5.2856	3.9253	3.3425	3.0077	2.7863	2.6274	2.5068
80	5.2184	3.8643	3.2841	2.9504	2.7295	2.6736	2.4502
100	5.2184	3.8284	3.2496	2.9166	2.6961	2.5374	2.4168
120	5.1523	3.8046	3.2269	2.8943	2.6740	2.5154	2.3948
∞	5.0239	3.6889	3.1161	2.7858	2.5665	2.4082	2.2875

4. (Snedecor's) F-Distribution Table (cont.)

The table entries (i.e., $F_{\nu_1,\nu_2,\alpha}$ values) are the right tail α probability cut-off points for the F_{ν_1,ν_2}-distribution, where ν_1 = numerator df and ν_2 = denominator df (i.e., area under the F_{ν_1,ν_2}-pdf on the right of $F_{\nu_1,\nu_2,\alpha}$ is α).

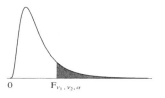

Denominator	$\alpha = 0.025$						
	Numerator $df = \nu_1$						
$df = \nu_2$	8	9	10	11	12	13	14
1	956.6562	963.2846	968.6274	973.0252	976.7079	979.8368	982.5278
2	39.3730	39.3869	39.3869	39.4071	39.4146	39.4210	39.4265
3	14.5399	14.4731	14.4189	14.3742	14.3366	14.3045	14.2768
4	8.9796	8.9047	8.8439	8.7935	8.7512	8.7150	8.6838
5	6.7572	6.6811	6.6192	6.5678	6.5245	6.4876	6.4556
6	5.5996	5.5234	5.4613	5.4098	5.3662	5.3290	5.2968
7	4.8993	4.8232	4.7611	4.7095	4.6658	4.6285	4.5961
8	4.4333	4.3572	4.2951	4.2434	4.1997	4.1622	4.1297
9	4.1020	4.0260	3.9639	3.9121	3.8682	3.8306	3.7980
10	3.8549	3.7790	3.7168	3.6649	3.6209	3.5832	3.5504
11	3.6638	3.5879	3.5257	3.4737	3.4296	3.3917	3.3588
12	3.5118	3.4358	3.3736	3.3215	3.2773	3.2393	3.2062
13	3.3880	3.3120	3.2497	3.1975	3.1532	3.1150	3.0819
14	3.2853	3.2093	3.1469	3.0946	3.0502	3.0119	2.9786
15	3.1987	3.1227	3.0602	3.0078	2.9633	2.9249	2.8915
16	3.1248	3.0488	2.9862	2.9337	2.8890	2.8506	2.8170
17	3.0610	2.9849	2.9222	2.8696	2.8249	2.7863	2.7526
18	3.0053	2.9291	2.8664	2.8137	2.7689	2.7302	2.6964
19	2.9563	2.8801	2.8172	2.7645	2.7196	2.6808	2.6469
20	2.9128	2.8365	2.7737	2.7209	2.6758	2.6369	2.6030
30	2.6513	2.5746	2.5112	2.4577	2.4120	2.3724	2.3378
40	2.5289	2.4519	2.3882	2.3343	2.2882	2.2481	2.2130
50	2.4579	2.3808	2.3168	2.2627	2.2162	2.1758	2.1404
60	2.4117	2.3344	2.2702	2.2159	2.1692	2.1286	2.0929
80	2.3549	2.2775	2.2130	2.1584	2.1115	2.0706	2.0346
100	2.3215	2.2439	2.1793	2.1245	2.0773	2.0363	2.0001
120	2.2994	2.2217	2.1570	2.1021	2.0548	2.0136	1.9773
∞	2.1918	2.1136	2.0483	1.9927	1.9447	1.9027	1.8656

4. (Snedecor's) F-Distribution Table (cont.)

The table entries (i.e., $F_{\nu_1,\nu_2,\alpha}$ values) are
the right tail α probability cut-off points for
the F_{ν_1,ν_2}-distribution, where $\nu_1 =$ numerator df and $\nu_2 =$ denominator df (i.e., area
under the F_{ν_1,ν_2}-pdf on the right of $F_{\nu_1,\nu_2,\alpha}$
is α).

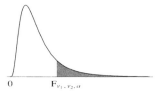

| Denominator | $\alpha = 0.025$ | | | | | | |
| | Numerator $df = \nu_1$ | | | | | | |
$df = \nu_2$	15	16	17	18	19	20	30
1	984.8668	986.9187	988.7331	990.3490	991.7973	993.1028	1001.414
2	39.4313	39.4354	39.4391	39.4424	39.4453	39.4479	39.4646
3	14.2527	14.2315	14.2127	14.1960	14.1810	14.1674	14.0805
4	8.6565	8.6326	8.6113	8.5924	8.5753	8.5599	8.4613
5	6.4277	6.4032	6.3814	6.3619	6.3444	6.3286	6.2269
6	5.2687	5.2439	5.2218	5.2021	5.1844	5.1684	5.0652
7	4.5678	4.5428	4.5206	4.5008	4.4829	4.4667	4.3624
8	4.1012	4.0761	4.0538	4.0338	4.0158	3.9995	3.8940
9	3.7694	3.7441	3.7216	3.7015	3.6833	3.6669	3.5604
10	3.5217	3.4963	3.4737	3.4534	3.4351	3.4185	3.3110
11	3.3299	3.3044	3.2816	3.2612	3.2428	3.2261	3.1176
12	3.1772	3.1515	3.1286	3.1081	3.0896	3.0728	2.9633
13	3.0527	3.0269	3.0039	2.9832	2.9646	2.9477	2.8372
14	2.9493	2.9234	2.9003	2.8795	2.8607	2.8437	2.7324
15	2.8621	2.8360	2.8128	2.7919	2.7730	2.7559	2.6437
16	2.7875	2.7614	2.7380	2.7170	2.6980	2.6808	2.5678
17	2.7230	2.6968	2.6733	2.6522	2.6331	2.6158	2.5020
18	2.6667	2.6404	2.6168	2.5956	2.5764	2.5590	2.4445
19	2.6171	2.5907	2.5670	2.5457	2.5265	2.5089	2.3937
20	2.5731	2.5465	2.5228	2.5014	2.4821	2.4645	2.3486
30	2.3072	2.2799	2.2554	2.2334	2.2134	2.1952	2.0739
40	2.1819	2.1542	2.1293	2.1068	2.0864	2.0677	1.9429
50	2.1090	2.0810	2.0558	2.0330	2.0122	1.9933	1.8659
60	2.0613	2.0330	2.0076	1.9846	1.9636	1.9445	1.8152
80	2.0026	1.9741	1.9483	1.9250	1.9037	1.8843	1.7523
100	1.9679	1.9391	1.9132	1.8897	1.8682	1.8486	1.7148
120	1.9450	1.9161	1.8900	1.8663	1.8447	1.8249	1.6899
∞	1.8326	1.8028	1.7759	1.7515	1.7291	1.7085	1.5660

4. (Snedecor's) F-Distribution Table (cont.)

The table entries (i.e., $F_{\nu_1,\nu_2,\alpha}$ values) are
the right tail α probability cut-off points for
the F_{ν_1,ν_2}-distribution, where ν_1 = numerator df and ν_2 = denominator df (i.e., area
under the F_{ν_1,ν_2}-pdf on the right of $F_{\nu_1,\nu_2,\alpha}$
is α).

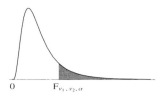

| Denominator | $\alpha = 0.025$ | | | | | | |
| | Numerator $df = \nu_1$ | | | | | | |
$df = \nu_2$	40	50	60	80	100	120	∞
1	1005.596	1008.098	1009.787	1011.911	1013.163	1014.036	1018.256
2	39.4729	39.4779	39.4812	39.4854	39.4879	39.4896	39.4979
3	14.0365	14.0099	13.9921	13.9697	13.9563	13.9473	13.9021
4	8.4111	8.3808	8.3604	8.3349	8.3195	8.3092	8.2573
5	6.1750	6.1436	6.1225	6.0960	6.0800	6.0693	6.0153
6	5.0125	4.9804	4.9589	4.9318	4.9154	4.9044	4.8491
7	4.3089	4.2763	4.2544	4.2268	4.2101	4.1989	4.1423
8	3.8398	3.8067	3.7844	3.7563	3.7393	3.7279	3.6702
9	3.5055	3.4719	3.4493	3.4207	3.4034	3.3918	3.3329
10	3.2554	3.2214	3.1984	3.1694	3.1517	3.1399	3.0798
11	3.0613	3.0268	3.0035	2.9740	2.9561	2.9441	2.8828
12	2.9063	2.8714	2.8478	2.8178	2.7996	2.7874	2.7249
13	2.7797	2.7443	2.7204	2.6900	2.6715	2.6590	2.5955
14	2.6740	2.6384	2.6142	2.5833	2.5646	2.5519	2.4872
15	2.5850	2.5488	2.5242	2.4930	2.4739	2.4611	2.3953
16	2.5085	2.4719	2.4471	2.4154	2.3961	2.3831	2.3163
17	2.4422	2.4053	2.3801	2.3481	2.3285	2.3153	2.2474
18	2.3842	2.3468	2.3214	2.2890	2.2692	2.2558	2.1869
19	2.3329	2.2952	2.2696	2.2368	2.2167	2.2032	2.1333
20	2.2873	2.2493	2.2234	2.1902	2.1699	2.1562	2.0850
30	2.0089	1.9681	1.9400	1.9039	1.8816	1.8664	1.7867
40	1.8752	1.8324	1.8028	1.7644	1.7405	1.7242	1.6371
50	1.7963	1.7520	1.7211	1.6810	1.6558	1.6386	1.5452
60	1.7440	1.6985	1.6668	1.6252	1.5990	1.5810	1.4821
80	1.6790	1.6318	1.5987	1.5549	1.5271	1.5079	1.3990
100	1.6401	1.5917	1.5575	1.5122	1.4833	1.4631	1.3470
120	1.6141	1.5649	1.5299	1.4834	1.4833	1.4327	1.3104
∞	1.4835	1.4284	1.3883	1.3329	1.2956	1.2684	1.0000

4. (Snedecor's) F-Distribution Table (cont.)

The table entries (i.e., $F_{\nu_1,\nu_2,\alpha}$ values) are the right tail α probability cut-off points for the F_{ν_1,ν_2}-distribution, where ν_1 = numerator df and ν_2 = denominator df (i.e., area under the F_{ν_1,ν_2}-pdf on the right of $F_{\nu_1,\nu_2,\alpha}$ is α).

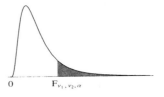

Denominator $df = \nu_2$	$\alpha = 0.01$						
	Numerator $df = \nu_1$						
	1	2	3	4	5	6	7
1	4052.185	4999.34	5403.534	5624.257	5763.955	5858.95	5928.334
2	98.5025	99.0000	99.1662	99.2494	99.2993	99.3326	99.3564
3	34.1162	30.8165	29.4567	28.7099	28.237	27.9107	27.6717
4	21.1977	18.0000	16.6944	15.9770	15.5219	5.2069	14.9758
5	16.2582	13.2739	12.0600	11.3919	10.9670	10.6723	10.4555
6	13.7450	10.9248	9.7795	9.1483	8.7459	8.4661	8.2600
7	12.2464	9.5466	8.4513	7.8466	7.4604	7.1914	6.9928
8	11.2586	8.6491	7.5910	7.0061	6.6318	6.3707	6.1776
9	10.5614	8.0215	6.9919	6.4221	6.0569	5.8018	5.6129
10	10.0443	7.5594	6.5523	5.9940	5.6363	5.3858	5.2001
11	9.6460	7.2057	6.2167	5.6683	5.3160	5.0692	4.8861
12	9.3302	6.9266	5.9525	5.4120	5.0643	4.8206	4.6395
13	9.0738	6.7010	5.7394	5.2053	4.8616	4.6204	4.4410
14	8.8616	6.5149	5.5639	5.0354	4.6950	4.4558	4.2779
15	8.6831	6.3589	5.4170	4.8932	4.5556	4.3183	4.1415
16	8.5310	6.2262	5.2922	4.7726	4.4374	4.2016	4.0259
17	8.3997	6.1121	5.1850	4.6690	4.3359	4.1015	3.9267
18	8.2854	6.0129	5.0919	4.5790	4.2479	4.0146	3.8406
19	8.1849	5.9259	5.0103	4.5003	4.1708	3.9386	3.7653
20	8.0960	5.8489	4.9382	4.4307	4.1027	3.8714	3.6987
30	7.5625	5.3903	4.5097	4.0179	3.6990	3.4735	3.3045
40	7.3141	5.1785	4.3126	3.8283	3.5138	3.2910	3.1238
50	7.1706	5.0566	4.1993	3.7195	3.4077	3.1864	3.0202
60	7.0771	4.9774	4.1259	3.6490	3.3389	3.1187	2.9530
80	6.9627	4.8807	4.0363	3.5631	3.2550	3.0361	2.8713
100	6.8953	4.8239	3.9837	3.5127	3.2059	2.9877	2.8233
120	6.8509	4.7865	3.9491	3.4795	3.1735	2.9559	2.7918
∞	6.6349	4.6052	3.7816	3.3192	3.0173	2.8020	2.6393

4. (Snedecor's) F-Distribution Table (cont.)

The table entries (i.e., $F_{\nu_1,\nu_2,\alpha}$ values) are the right tail α probability cut-off points for the F_{ν_1,ν_2}-distribution, where ν_1 = numerator df and ν_2 = denominator df (i.e., area under the F_{ν_1,ν_2}-pdf on the right of $F_{\nu_1,\nu_2,\alpha}$ is α).

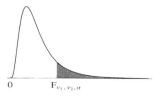

Denominator	$\alpha = 0.01$						
	Numerator $df = \nu_1$						
$df = \nu_2$	8	9	10	11	12	13	14
1	5980.954	6022.397	6055.925	6083.399	6106.682	6125.774	6143.004
2	99.3742	99.3881	99.3992	99.4083	99.4159	99.4159	99.4278
3	27.4892	27.3452	27.2287	27.1326	27.0518	26.9831	26.9238
4	14.7989	14.6591	14.5459	14.4523	14.3736	14.3065	14.2486
5	10.2893	10.1578	10.0510	9.9626	9.8883	9.8248	9.7700
6	8.1017	7.9761	7.8741	7.7896	7.7183	7.6575	7.6049
7	6.8400	6.7188	6.6201	6.5382	6.4691	6.4100	6.3590
8	6.0289	5.9106	5.8143	5.7343	5.6667	5.6089	5.5589
9	5.4671	5.3511	5.2565	5.1779	5.1114	5.0545	5.0052
10	5.0567	4.9424	4.8491	4.7715	4.7059	4.6496	4.6008
11	4.7445	4.6315	4.5393	4.4624	4.3974	4.3416	4.2932
12	4.4994	4.3875	4.2961	4.2198	4.1553	4.0999	4.0518
13	4.3021	4.1911	4.1003	4.0245	3.9603	3.9052	3.8573
14	4.1399	4.0297	3.9394	3.8640	3.8001	3.7452	3.6975
15	4.0045	3.8948	3.8049	3.7299	3.6662	3.6115	3.5639
16	3.8896	3.7804	3.6909	3.6162	3.5527	3.4981	3.4506
17	3.7910	3.6822	3.5931	3.5185	3.4552	3.4007	3.3533
18	3.7054	3.5971	3.5082	3.4338	3.3706	3.3162	3.2689
19	3.6305	3.5225	3.4338	3.3596	3.2965	3.2422	3.1949
20	3.5644	3.4567	3.3682	3.2941	3.2311	3.1769	3.1296
30	3.1726	3.0665	2.9791	2.9057	2.8431	2.7890	2.7418
40	2.9930	2.8870	2.8005	2.7274	2.6648	2.6107	2.5634
50	2.8900	2.7850	2.6981	2.6250	2.5625	2.5083	2.4609
60	2.8233	2.7185	2.6318	2.5587	2.4961	2.4419	2.3943
80	2.7420	2.6374	2.5508	2.4777	2.4151	2.3608	2.3131
100	2.6943	2.5898	2.5033	2.4302	2.3676	2.3132	2.2654
120	2.6629	2.5586	2.4721	2.3990	2.3363	2.2818	2.2339
∞	2.5113	2.4073	2.3209	2.2477	2.1847	2.1299	2.0815

4. (Snedecor's) F-Distribution Table (cont.)

The table entries (i.e., $F_{\nu_1,\nu_2,\alpha}$ values) are
the right tail α probability cut-off points for
the F_{ν_1,ν_2}-distribution, where $\nu_1 =$ numerator df and $\nu_2 =$ denominator df (i.e., area
under the F_{ν_1,ν_2}-pdf on the right of $F_{\nu_1,\nu_2,\alpha}$
is α).

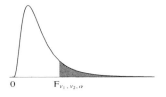

Denominator	$\alpha = 0.01$						
	Numerator $df = \nu_1$						
$df = \nu_2$	15	16	17	18	19	20	30
1	6156.974	6170.012	6181.188	6191.432	6200.746	6208.662	6260.350
2	99.4325	99.4367	99.4404	99.4436	99.4465	99.4492	99.4658
3	26.8722	26.8269	26.7867	26.7509	26.7188	26.6898	26.5045
4	14.1982	14.1539	14.1146	14.0795	14.0480	14.0196	13.8377
5	9.7222	9.6802	9.6429	9.6096	9.5797	9.5526	9.3793
6	9.7222	7.5186	7.4827	7.4507	7.4219	7.3958	7.2285
7	6.3143	6.2750	7.4827	6.2089	6.1808	6.1554	5.9920
8	5.5151	5.4766	5.4423	5.4116	5.3840	5.3591	5.1981
9	4.9621	4.9240	4.8902	4.8599	4.8327	4.8080	4.6486
10	4.5581	4.5204	4.4869	4.4569	4.4299	4.4054	4.2469
11	4.2509	4.2134	4.1801	4.1503	4.1234	4.0990	3.9411
12	4.0096	3.9724	3.9392	3.9095	3.8827	3.8584	3.7008
13	3.8154	3.7783	3.7452	3.7156	3.6888	3.6646	3.5070
14	3.6557	3.6187	3.5857	3.5561	3.5294	3.5052	3.3476
15	3.5222	3.4852	3.4523	3.4228	3.3961	3.3719	3.2141
16	3.4089	3.3720	3.3391	3.3096	3.2829	3.2587	3.1007
17	3.3117	3.2748	3.2419	3.2124	3.1857	3.1615	3.0032
18	3.2273	3.1904	3.1575	3.1280	3.1013	3.0771	2.9185
19	3.1533	3.1165	3.0836	3.0541	3.0274	3.0031	2.8442
20	3.0880	3.0512	3.0183	2.9887	2.9620	2.9377	2.7785
30	2.7002	2.6632	2.6301	2.6003	2.5732	2.5487	2.3860
40	2.5216	2.4844	2.4511	2.4210	2.3937	2.3689	2.2034
50	2.4190	2.3816	2.3481	2.3178	2.2903	2.2652	2.0976
60	2.3523	2.3148	2.2811	2.2507	2.2230	2.1978	2.0285
80	2.2709	2.2332	2.1993	2.1686	2.1408	2.1153	1.9435
100	2.2230	2.1852	2.1511	2.1203	2.0923	2.0666	1.8933
120	2.1915	2.1536	2.1194	2.0885	2.0604	2.0346	1.8933
∞	2.0385	2.0000	1.9652	1.9336	1.9048	1.8783	1.6964

4. (Snedecor's) F-Distribution Table (cont.)

The table entries (i.e., $F_{\nu_1,\nu_2,\alpha}$ values) are
the right tail α probability cut-off points for
the F_{ν_1,ν_2}-distribution, where ν_1 = numerator df and ν_2 = denominator df (i.e., area
under the F_{ν_1,ν_2}-pdf on the right of $F_{\nu_1,\nu_2,\alpha}$
is α).

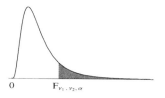

Denominator	$\alpha = 0.01$						
	Numerator $df = \nu_1$						
$df = \nu_2$	40	50	60	80	100	120	∞
1	6286.427	6302.26	6312.97	6326.474	6333.925	6339.513	6365.590
2	99.4742	99.4792	99.4825	99.4867	99.4892	99.4908	99.4992
3	26.4108	26.3542	26.3164	26.2688	26.2402	26.2211	26.1252
4	13.7454	13.6896	13.6522	13.6053	13.5770	13.5581	13.4631
5	9.2912	9.2378	9.2020	9.1570	9.1299	9.1118	9.0204
6	7.1432	7.0915	7.0567	7.0130	6.9867	6.9690	6.8800
7	5.9084	5.8577	5.8236	5.7806	5.7547	5.7373	5.6495
8	5.1156	5.0654	5.0316	4.9890	4.9633	4.9461	4.8588
9	4.5666	4.5167	4.4831	4.4407	4.4150	4.3978	4.3105
10	4.1653	4.1155	4.0819	4.0394	4.0137	3.9965	3.9090
11	3.8596	3.8097	3.7761	3.7335	3.7077	3.6904	3.6024
12	3.6192	3.5692	3.5355	3.4928	3.4668	3.4494	3.3608
13	3.4253	3.3752	3.3413	3.2984	3.2723	3.2548	3.1654
14	3.2656	3.2153	3.1813	3.1381	3.1118	3.0942	3.0040
15	3.1319	3.0814	3.0471	3.0037	2.9772	2.9595	2.8684
16	3.0182	2.9675	2.9330	2.8893	2.8627	2.8447	2.7528
17	2.9205	2.8694	2.8348	2.7908	2.7639	2.7459	2.6530
18	2.9205	2.7841	2.7493	2.7050	2.6779	2.6597	2.5660
19	2.7608	2.7093	2.6742	2.6296	2.6023	2.5839	2.4893
20	2.6947	2.6430	2.6077	2.5628	2.5353	2.5168	2.4212
30	2.2992	2.2450	2.2079	2.1601	2.1307	2.1108	2.0062
40	2.1142	2.0581	2.0194	1.9694	1.9383	1.9172	1.8047
50	2.0066	1.9490	1.9090	1.8571	1.8248	1.8026	1.6831
60	1.9360	1.8772	1.8363	1.7828	1.7493	1.7263	1.6006
80	1.8489	1.7883	1.7459	1.6901	1.6548	1.6305	1.4942
100	1.7972	1.7353	1.6918	1.6342	1.5977	1.5723	1.4272
120	1.7628	1.7000	1.6557	1.5968	1.5592	1.5330	1.3805
∞	1.5923	1.5231	1.4730	1.4041	1.3581	1.3246	1.0000

4. (Snedecor's) *F*-Distribution Table (cont.)

The table entries (i.e., $F_{\nu_1,\nu_2,\alpha}$ values) are the right tail α probability cut-off points for the F_{ν_1,ν_2}-distribution, where ν_1 = numerator df and ν_2 = denominator df (i.e., area under the F_{ν_1,ν_2}-pdf on the right of $F_{\nu_1,\nu_2,\alpha}$ is α).

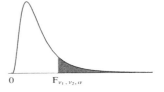

Denominator	$\alpha = 0.005$						
	Numerator $df = \nu_1$						
$df = \nu_2$	1	2	3	4	5	6	7
1	16210.72	19999.50	21614.74	22499.58	23055.80	23437.11	23714.57
2	198.5013	199.0000	199.1664	199.2497	199.2996	199.3330	199.3568
3	55.5520	49.7993	47.4672	46.1946	45.3916	44.8385	44.4341
4	31.3328	26.2843	24.2591	23.1545	22.4564	21.9746	21.6217
5	22.7848	18.3138	16.5298	15.5561	14.9396	14.5133	14.2004
6	18.6350	14.5441	12.9166	12.0275	11.4637	11.0730	10.7859
7	16.2356	12.4040	10.8824	10.0505	9.5221	9.1553	8.8854
8	14.6882	11.0424	9.5965	8.8051	8.3018	7.9520	7.6941
9	13.6136	10.1067	8.7171	7.9559	7.4712	7.1338	6.8849
10	12.8265	9.4270	8.0807	7.3428	6.8724	6.5446	6.3025
11	12.2263	8.9122	7.6004	6.8809	6.4217	6.1015	5.8648
12	11.7542	8.5096	7.2257	6.5211	6.0711	5.7570	5.5245
13	11.3735	8.1864	6.9257	6.2334	5.7909	5.4819	5.2529
14	11.0602	7.9216	6.6803	5.9984	5.5622	5.2573	5.0313
15	10.7980	7.7007	6.4760	5.8029	5.3721	5.0708	4.8472
16	10.5754	7.5138	6.3033	5.6378	5.2117	4.9134	4.6920
17	10.3841	7.3536	6.1556	5.4967	5.0746	4.7789	4.5594
18	10.2181	7.2148	6.0278	5.3746	4.9560	4.6627	4.4448
19	10.0725	7.0935	5.9161	5.2681	4.8526	4.5614	4.3448
20	9.9439	6.9865	5.8177	5.1743	4.7616	4.4721	4.2569
30	9.1797	6.3547	5.2388	4.6234	4.2276	3.9492	3.7416
40	8.8279	6.0664	4.9758	4.3738	3.9860	3.7129	3.5088
50	8.6258	5.9016	4.8259	4.2316	3.8486	3.5785	3.3765
60	8.4946	5.7950	4.7290	4.1399	3.7599	3.4918	3.2911
80	8.3346	5.6652	4.6113	4.0285	3.6524	3.3867	3.1876
100	8.2406	5.5892	4.5424	3.9634	3.5895	3.3252	3.1271
120	8.1788	5.5393	4.4972	3.9207	3.5482	3.2849	3.0874
∞	7.8829	5.3011	4.2819	3.7175	3.3522	3.0935	2.8990

4. (Snedecor's) F-Distribution Table (cont.)

The table entries (i.e., $F_{\nu_1,\nu_2,\alpha}$ values) are the right tail α probability cut-off points for the F_{ν_1,ν_2}-distribution, where ν_1 = numerator df and ν_2 = denominator df (i.e., area under the F_{ν_1,ν_2}-pdf on the right of $F_{\nu_1,\nu_2,\alpha}$ is α).

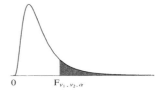

Denominator	$\alpha = 0.005$						
	Numerator $df = \nu_1$						
$df = \nu_2$	8	9	10	11	12	13	14
1	23925.41	24091.00	24224.49	24334.36	24426.37	24504.54	24571.77
2	199.3746	199.3885	199.3996	199.4087	199.4163	199.4227	199.4282
3	44.1256	43.8824	43.6858	43.5236	43.3874	43.2714	43.1716
4	21.3519	21.1390	20.9667	20.8243	20.7046	20.6027	20.5148
5	13.9609	13.7716	13.6181	13.4912	13.3844	13.2934	13.2148
6	10.5657	10.3914	10.2500	10.1329	10.0342	9.9501	9.8774
7	8.6781	8.5138	8.3803	8.2696	8.1764	8.0967	8.0278
8	7.4959	7.3385	7.2106	7.1044	7.0149	6.9383	6.8721
9	6.6933	6.5410	6.4171	6.3142	6.2273	6.1530	6.0887
10	6.1159	5.9675	5.8466	5.7462	5.6613	5.5886	5.5257
11	5.6821	5.5367	5.4182	5.3196	5.2363	5.1649	5.1030
12	5.3450	5.2021	5.0854	4.9883	4.9062	4.8358	4.7748
13	5.0760	4.9350	4.8199	4.7240	4.6429	4.5733	4.5128
14	4.8566	4.7172	4.6033	4.5084	4.4281	4.3591	4.2992
15	4.6743	4.5363	4.4235	4.3294	4.2497	4.1813	4.1218
16	4.5206	4.3838	4.2719	4.1785	4.0993	4.0313	3.9722
17	4.3893	4.2535	4.1423	4.0495	3.9708	3.9032	3.8444
18	4.2759	4.1409	4.0304	3.9381	3.8598	3.7925	3.7340
19	4.1770	4.0428	3.9328	3.8410	3.7630	3.6960	3.6377
20	4.0899	3.9564	3.8470	3.7555	3.6779	3.6111	3.5530
30	3.5800	3.4504	3.3439	3.2547	3.1787	3.1132	3.0560
40	3.3497	3.2219	3.1167	3.0284	2.9531	2.8880	2.8312
50	3.2188	3.0920	2.9875	2.8996	2.8247	2.7598	2.7031
60	3.1344	3.0082	2.9041	2.8166	2.7418	2.6771	2.6204
80	3.0320	2.9066	2.8030	2.7158	2.6412	2.5767	2.5200
100	2.9721	2.8472	2.7439	2.6569	2.5825	2.5179	2.4613
120	2.9329	2.8082	2.7051	2.6183	2.5439	2.4794	2.4228
∞	2.7465	2.6232	2.5209	2.4346	2.3604	2.2959	2.2392

4. (Snedecor's) F-Distribution Table (cont.)

The table entries (i.e., $F_{\nu_1,\nu_2,\alpha}$ values) are the right tail α probability cut-off points for the F_{ν_1,ν_2}-distribution, where ν_1 = numerator df and ν_2 = denominator df (i.e., area under the F_{ν_1,ν_2}-pdf on the right of $F_{\nu_1,\nu_2,\alpha}$ is α).

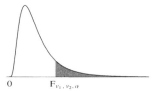

Denominator	$\alpha = 0.005$						
	Numerator $df = \nu_1$						
$df = \nu_2$	15	16	17	18	19	20	30
1	24630.21	24681.46	24726.79	24767.17	24803.35	24835.97	25043.62
2	199.4329	199.4370	199.4407	199.4440	199.4469	199.4495	199.4662
3	43.0846	43.0082	42.9406	42.8803	42.8262	42.7775	42.4658
4	20.4382	20.3709	20.3113	20.2581	20.2103	20.1672	19.8915
5	13.1463	13.0860	13.0326	12.9849	12.9421	12.9034	12.6556
6	9.8139	9.7581	9.7086	9.6644	9.6246	9.5887	9.3582
7	7.9677	7.9148	7.8678	7.8258	7.7881	7.7539	7.5344
8	6.8142	6.7632	6.7180	6.6775	6.6411	6.6082	6.3961
9	6.0324	5.9828	5.9388	5.8994	5.8639	5.8318	5.6247
10	5.4706	5.4221	5.3789	5.3402	5.3054	5.2740	5.0705
11	5.0489	5.0011	4.958	4.9205	4.8862	4.8552	4.6543
12	4.7213	4.6741	4.6321	4.5945	4.5606	4.5299	4.3309
13	4.4600	4.4132	4.3716	4.3343	4.3007	4.2703	4.0727
14	4.2468	4.2004	4.1591	4.1221	4.0888	4.0585	3.8619
15	4.0697	4.0237	3.9826	3.9458	3.9126	3.8825	3.6867
16	3.9204	3.8746	3.8338	3.7972	3.7641	3.7341	3.5388
17	3.7929	3.7472	3.7066	3.6701	3.6372	3.6073	3.4124
18	3.6827	3.6372	3.5967	3.5603	3.5274	3.4977	3.3030
19	3.5865	3.5412	3.5008	3.4645	3.4317	3.4020	3.2075
20	3.5019	3.4567	3.4164	3.3802	3.3474	3.3177	3.1234
30	3.0057	2.9610	2.9211	2.8852	2.8526	2.8230	2.6278
40	2.7810	2.7365	2.6966	2.6607	2.6281	2.5984	2.4014
50	2.6531	2.6085	2.5686	2.5326	2.4999	2.4701	2.2716
60	2.5704	2.5258	2.4859	2.4498	2.4170	2.3872	2.1874
80	2.4700	2.4254	2.3853	2.3491	2.3162	2.2862	2.0844
100	2.4112	2.3666	2.3264	2.2901	2.2571	2.2270	2.0239
120	2.3727	2.3279	2.2877	2.2514	2.2183	2.1881	1.9839
∞	2.1889	2.1438	2.1032	2.0664	2.0328	2.0020	1.7913

4. (Snedecor's) F-Distribution Table (cont.)

The table entries (i.e., $F_{\nu_1,\nu_2,\alpha}$ values) are the right tail α probability cut-off points for the F_{ν_1,ν_2}-distribution, where $\nu_1 =$ numerator df and $\nu_2 =$ denominator df (i.e., area under the F_{ν_1,ν_2}-pdf on the right of $F_{\nu_1,\nu_2,\alpha}$ is α).

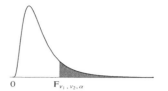

Denominator	$\alpha = 0.005$						
	Numerator $df = \nu_1$						
$df = \nu_2$	40	50	60	80	100	120	∞
1	25148.15	25211.09	25253.13	25305.79	25337.45	25358.57	25463.18
2	199.4745	199.4795	199.4829	199.4870	199.4895	199.4912	199.4994
3	42.3082	42.2130	42.1494	42.0696	42.0215	41.9894	41.8302
4	19.7517	19.6673	19.6107	19.5397	19.4969	19.4683	19.3264
5	12.5297	12.4535	12.4024	12.3382	12.2995	12.2737	12.1451
6	9.2408	9.1696	9.1219	9.0619	9.0256	9.0014	8.8808
7	7.4224	7.3544	7.3087	7.2512	7.2165	7.1933	7.0774
8	6.2875	6.2215	6.1771	6.1212	6.0875	6.0649	5.9519
9	5.5185	5.4539	5.4104	5.3555	5.3223	5.3001	5.1889
10	4.9659	4.9021	4.8591	4.8049	4.7721	4.7501	4.6398
11	4.5508	4.4876	4.4450	4.3911	4.3585	4.3366	4.2268
12	4.2281	4.1653	4.1229	4.0692	4.0367	4.01494	3.9052
13	3.9704	3.9078	3.8655	3.8120	3.7795	3.7576	3.6478
14	3.7560	3.6975	3.6552	3.6017	3.5692	3.5473	3.4372
15	3.5850	3.5225	3.4802	3.4266	3.3940	3.3721	3.2615
16	3.4372	3.3747	3.3324	3.2787	3.2460	3.2240	3.1128
17	3.3107	3.2482	3.2058	3.1519	3.1191	3.0970	2.9853
18	3.2013	3.1387	3.0962	3.0421	3.0092	2.9870	2.8745
19	3.1057	3.0430	3.0003	2.9461	2.9130	2.8907	2.7775
20	3.0215	2.9586	2.9158	2.8614	2.8282	2.8058	2.6918
30	2.5240	2.4594	2.4151	2.3583	2.3234	2.2997	2.1775
40	2.2958	2.2295	2.1838	2.1248	2.0884	2.0635	1.9334
50	2.1644	2.0967	2.0498	1.9890	1.9512	1.9253	1.7880
60	2.0788	2.0100	1.9621	1.8998	1.8608	1.8341	1.6904
80	1.9739	1.9033	1.8539	1.7892	1.7484	1.7202	1.5654
100	1.9119	1.8400	1.7896	1.7230	1.6808	1.6516	1.4875
120	1.8709	1.7981	1.7468	1.6789	1.6356	1.6055	1.4335
∞	1.6715	1.5923	1.5351	1.4567	1.4046	1.3668	1.0528

Index

Printed and bound by CPI Group (UK) Ltd, Croydon, CR0 4YY

24/10/2024

01778277-0011